Reductions by the Alumino- and Borohydrides in Organic Synthesis

Reductions by the Alumino- and Borohydrides in Organic Synthesis

Second Edition

Jacqueline Seyden-Penne

WILEY-VCH

NEW YORK / CHICHESTER / WEINHEIM / BRISBANE / SINGAPORE / TORONTO

Jacqueline Seyden-Penne
Le Vallat de Vermenoux
84220 Goult
France

English Language Editor
Dennis P. Curran
Department of Chemistry
University of Pittsburgh
Parkman Avenue & University Drive
Pittsburgh, PA 15260

This book is printed on acid-free paper. ♾

Published simultaneously in Canada.

Library of Congress Cataloging in Publication Data:

Seyden-Penne, J.
[Réductions par les alumino- et borohydrures en synthèses organique. English]
Reductions by the alumino- and borohydrides in organic synthesis / Jacqueline Seyden-Penne. — 2nd ed.
p. cm.
Includes bibliographical references and index.
ISBN 0-471-19036-5 (cloth : alk. paper)
1. Reduction (Chemistry) 2. Hydrides. 3. Organic compounds—Synthesis. I. Title.
QD63.R4S4913 1997
547′.23—dc21 96-49776

Printed in the United States of America

10 9 8 7 6 5 4 3 2 1

Contents

Preface

Alumino- and borohydrides and, to a lesser extent, boranes form a part of the chemist's classic arsenal of reducing agents employed in organic synthesis. A number of these compounds are commercially available, but the study of their properties, the introduction of improved reagents, and the development of new reaction conditions continue to be important areas of research. Selectivity is imperative in modern organic synthesis, especially when multifunctional molecules are involved. The reagents chosen at each stage of a chemical transformation must not affect other functional groups in the molecule. Moreover, functional groups can influence a reaction process by altering regioselectivity or stereoselectivity.

In this book, we compare the synthetic potential of the most important commercial hydrides and their readily available derivatives. All these hydrides are easy to use, and the book is organized so that the reader can match the appropriate reagent to a given transformation. The book emphasizes:

- Compatibility between the reduction of the target group and the other functional groups present in the molecule;
- The possibilities for partial reduction;
- The regio- and stereoselectivity of reductions that are altered or controlled by other neighboring groups;
- Asymmetric reductions. These reactions have rapidly developed since the First Edition. In addition to chiral hydrides, other strategies for asymmetric reduction include the use of reagents such as chiral chloroboranes or hydrogenation in the presence of catalysts bearing chiral ligands [S3].

This second edition has been broadly updated, but it is no longer exhaustive. As in the previous edition, the examples are selected in order to cover problems that are frequently encountered in synthesis.

The present book is organized in the following fashion:

- Chapter 1 introduces the most useful reagents and indicates their stability and solubility characteristics and their main applications;
- Chapters 2–5 present the reduction of the main functional groups by these reagents, with reference to features of selectivity (chimio-, regio-, stereo-, and enantioselectivity) and compatibility;
- At the end of the book, synoptic tables indicate how to obtain the main functional groups by hydride reduction.

I am particularly grateful to Mr. Fenouil (Lavoisier-Tec-Doc), who allowed me to publish this Second Edition with a free hand, and to the staff of the library of the University of Aix-Marseille-St-Jérôme, who allowed me to work there as often as I wanted. I am also grateful to the members of the Orsay laboratory, who supplied all the documents that I needed, namely, Robert Bloch, Yves Langlois, and above all Tekla Strzalko. My husband, Bob, handled the production aspects of the work, typing the manuscript and drawing the figures on the computer. I also thank Suzanne Curran and Valerie Wadyko for correcting the files according to the proposals of Dennis Curran, who revised my text and my English. Again, I greatly appreciated the improvements he brought to this book.

JACQUELINE SEYDEN-PENNE

Goult, France

Foreword

Although it may be difficult to imagine now, it was not that long ago that the basic reduction of one organic functional group to another was a demanding proposition. Choices of reagents were very limited, and reaction conditions were harsh. Enter the alumino- and borohydrides. Lithium aluminohydride and sodium borohydride were introduced by Schlesinger and Brown in 1953. Lithium aluminohydride was useful because it reduced so many things, while the milder sodium borohydride effected certain kinds of selective reductions in organic molecules. Soon the complexity of molecules grew, and along with this complexity came the need for more reducing agents with different properties and selectivities. So a few new alumino- and borohydrides were introduced. But the spiral did not stop there. The complexity of molecules grew rapidly, reductions became more and more demanding, and even better and more selective reducing agents were introduced in response to this demand. The response to the need for chiral reducing agents has recently sent this spiral to new heights.

So it would appear that synthetic organic chemists should be happy, because for a given kind a reduction—even a very demanding one—there is probably already an alumino- or borohydride reducing agent and a set of reaction conditions that is up to the task. But there is still unhappiness because finding the right combination from the maze of catalogs, papers, and experimental procedures can itself be a daunting task.

From out of this maze springs this book. Professor Jacqueline Seyden-Penne is an acknowledged expert in the area. The book is a major update of the First Edition, which was published in 1991 by VCH Publishers (a translation from the popular first French edition). It includes the important developments that have occurred in the intervening half-dozen years (notably in the area of asymmetric reductions). Professor Seyden-Penne first describes the features of more than two dozen of the most powerful and commonly used alumino- and borohydrides, and then goes on to detail in individual chapters their reactions with important classes of organic molecules. There is a strong emphasis on selectivity at every level (chemo-, regio-, diastereo-, and enantioselection), and experimental practicality is also directly addressed. Synoptic tables present much information at a glance, and extensive references (about 1000) lead the reader back to the original papers and experimental procedures.

The book is in effect a road atlas that allows the organic chemist to maneuver rapidly through the maze of information on reductions of organic compounds by alumino- and borohydrides to locate the desired goal. For anyone trying to navigate in this area, this road atlas is indispensable.

DENNIS P. CURRAN

Pittsburgh, PA

Abbreviations

Ac	acetyl
AcOEt	ethyl acetate
Ar	aryl
BOC	*t*-butyloxycarbonyl
Bz	benzoyl
DMA	dimethylacetamide
DME	dimethoxyethane
DMF	dimethylformamide
DMSO	dimethylsulfoxide
Et	ethyl
Et_2O	diethyl ether
HMPA	hexamethylphosphorotriamide
i-Pr	isopropyl
Me	methyl
MeCN	acetonitrile
MEM	methoxymethyl
Ph	phenyl
s-Bu	*sec*-butyl
Sia	*iso*-amyl
TBDMS	*t*-butyldimethylsilyl

t-Bu	*tert*-butyl
THF	tetrahydrofuran
THP	tetrahydropyranyl
Tol	*p*-methylphenyl

Reductions by the Alumino- and Borohydrides in Organic Synthesis

Chapter 1

Description and Characteristics of the Main Reagents

This chapter lists and describes the characteristics of the main reagents. Cross references are made to the corresponding sections of the other chapters for more complete details.

1.1 LITHIUM AND SODIUM ALUMINOHYDRIDES: $LiAlH_4$ (LAH), $NaAlH_4$ (SAH)

Lithium aluminohydride ($LiAlH_4$, LAH) is soluble in ethers. In diethylether and dioxane it forms tight ion pairs, but in THF and in DME it forms loose ion pairs [AD1, WS1]. LAH is used either in solution, as a suspension, or in a solid–liquid phase transfer medium (benzene, 15-crown-5) [DC1, GL4]. It is also used adsorbed onto silica gel [KH2, KH3]; however, its reducing power is so diminished under the latter conditions that it can selectively reduce ketoesters to hydroxyesters or amide esters into amide alcohols [KS5].

LAH reacts violently with water and must be handled away from moisture. Decomposition of an excess of LAH can be carried out either by careful treatment with water-saturated diethylether or by addition of ethyl acetate, which is reduced to ethanol, before treatment with water. Crude reaction mixtures can be treated either in acidic or basic media, by complexation with tartaric acid, or even by the addition of a stoichiometric quantity of water to form LiOH and $Al(OH)_3$, which precipitate and are coated by solid $MgSO_4$ and Na_2SO_4, through which they are filtered [H3]. If the reaction leads to aminoalcohols, which are good ligands for aluminum, it is sometimes difficult to recover the product of the reduction, but treatment with $(HOCH_2CH_2)_3N$ before the addition of water allows isolation of the product in good yield [PJ1].

LAH shows very high reducing power and consequently does not appear to be very selective, even when the conditions of medium and temperature are varied. Alcohols and phenols react with LAH in controlled amounts to produce alkoxyaluminum hydrides, whose reducing power can be modulated (see the following). Reaction with secondary amines forms aminoaluminohydrides. Some of these have been characterized by X-ray crystallography [HS5]. With tertiary amines, complexes can be formed. For example, N-methylpyrrolidine gives an air-stable complex [FS1] whose reducing properties are similar to those of LAH. The use of this complex does not require special procedures for exclusion of moisture and air and after reduction, workup is done by addition of water. Treatment of LAH with pyridine produces a special reagent, lithium tetrakis N-dihydropyridinoaluminohydride [LL1]. There is a review devoted to the rearrangements of various carbon skeletons observed during reduction by LAH [C2].

Sodium aluminohydride ($NaAlH_4$, SAH) in THF is somewhat less reactive than LAH toward carboxylic acids, anhydrides, epoxides, amides, and nitro compounds [CB5], and it can be used for selective reductions. However, it is as sensitive to moisture as LAH; so similar precautions must be taken.

1.2 LITHIUM AND SODIUM ALKOXY- AND AMINOALUMINOHYDRIDES

The reaction of stoichiometric quantities of alcohols with LAH leads to the formation of alkoxyaluminohydrides. The problem most often encountered in this reaction is disproportionation according to the following equilibria [HM3]:

$$\begin{aligned}
LiAlH_4 + ROH &\rightleftharpoons Li(RO)AlH_3 + H_2 \\
Li(RO)AlH_3 + ROH &\rightleftharpoons Li(RO)_2AlH_2 + H_2 \\
Li(RO)_2AlH_2 + ROH &\rightleftharpoons Li(RO)AlH_3 + H_2 \\
Li(RO)AlH_3 + ROH &\rightleftharpoons ROLi + (RO)_3Al + H_2
\end{aligned}$$

Because of this disproportionation, some solutions of alkoxyaluminohydrides contain essentially the alcoholates and LAH, and thus they present the same characteristics as LAH itself. This is especially the case when R = Et or *i*-Pr [WS1].

The following reagents are nevertheless stable:

- $Li(MeO)_3AlH$ is a dimer in THF [BK5, M1, M3]: Its interest resides in the 1,2 attack of α-enones (Section 3.2.9).
- $Li(t\text{-}BuO)_3AlH$ (LTBA) is a monomer in THF, and its reductive properties have been well studied [BK5, M1, M3, W3]. Its principal applications are the reduction of acid chlorides and imidazolides to aldehydes at low temperature. Because of its bulkiness, a high stereoselectivity during the reduction of carbonyl compounds often makes the reaction more selective than with LAH. At low temperature, aldehydes can be reduced in the presence of ketones, and only slightly hindered ketones can even be reduced in the presence of more hindered ones (Section 3.2.1). Likewise, LTBA attacks saturated ketones more rapidly than α-enones (Section 3.2.9). LTBA leaves ethers, acetals, epoxides, chlorides

and bromides, and nitro derivatives intact. Aliphatic esters are reduced only slowly; in contrast, phenyl esters are converted into aldehydes (Section 3.2.5). $Na(t\text{-}BuO)_3AlH$ can be prepared in a similar way. Sparingly soluble in THF, it may be used in DME–THF mixtures and is recommended for reductions of acid chlorides to aldehydes [CB6].

- $Li(t\text{-}BuEt_2O)_3AlH$ is a bulky reagent that has been used in stereoselective reductions of prochiral ketones [BD2], and it reduces aldehydes selectively in the presence of ketones [K4].
- $Li(EtO)_3AlH$ (LTEA) and $Li(EtO)_2AlH_2$ can be produced in situ and have some interesting properties, but because they rapidly undergo disproportionation, they must be used very soon after their formation to reduce sufficiently reactive substrates. They reduce nitriles into imines, which can then be hydrolyzed to aldehydes (Section 4.3), and they also convert tertiary amides into aldehydes (Section 3.2.8).
- Reducing agents having special properties are obtained by the reaction of alkoxyaluminohydrides with CuBr [CA1, SS1]. These reduce the double and triple bonds of α,β-unsaturated carbonyl compounds (Sections 3.2.9, 4.2, 4.4) and allow one to obtain N-acyldihydro-1,4-pyridines (Section 3.3.3.3).

Various sodium aminoaluminohydrides have been proposed for selective reduction of esters and aromatic nitriles to the corresponding aldehydes [CK3, CK5, CJ1, YA2]. Chiral alkoxy- and aminoaluminohydrides have been used in asymmetric reductions of ketones and imines, and these will be described in the corresponding chapters (Sections 3.2.3 and 3.3.1).

1.3 SODIUM BIS(METHOXYETHOXY)ALUMINOHYDRIDE: $Na(OCH_2CH_2OCH_3)_2AlH_2$ (Red-Al)

An interesting feature of sodium bis(methoxyethoxy)aluminohydride is its solubility in aromatic hydrocarbons [M1, MC1, W3]. It is also soluble in ethers. Most frequently, reductions are carried out in a benzene or toluene solution to which are added various cosolvents. The reaction of Red-Al with water is less violent than that of LAH, which facilitates workup. As with LAH, hydrolysis can be carried out in acidic or basic media or with a minimal amount of water. In the last case, the addition of a small amount of acid to neutralize the NaOH that forms is recommended.

The features of Red-Al are the following: It easily reduces halogenated derivatives even if acetylenic (Section 2.1); tertiary amides lead to aldehydes (Section 3.2.8); and propargylic alcohols and amines are reduced to corresponding allylic alcohols and amines (Section 4.1). Epoxides remain intact unless they carry an alcohol functional group at the α position: The reduction is then regioselective (Section 2.3). Aromatic nitriles are reduced, but aliphatic nitriles are not affected (Section 4.3).

In the presence of CuBr in THF, Red-Al gives rise to an interesting reagent [SS1] that is especially good for selective reduction of the carbon–carbon double and triple bonds of unsaturated ketones, esters, or nitriles (Sections 3.2.9, 4.2, 4.4), leaving the functional group unchanged.

1.4 DIISOBUTYL ALUMINUM HYDRIDE: *i*-Bu_2AlH (DIBAH)

This reagent [BK5, W1, W3, YG1] is both soluble and stable in toluene or hexane. It is also soluble in ethers (diethylether, THF, DME, glymes), but these solutions are stable only at low temperature. It is a particularly strong Lewis acid. At high temperature, DIBAH hydroaluminates carbon–carbon double and triple bonds [HH1]. The usual workup after reduction consists of addition of methanol then water to the solution, followed by separation of the aluminum salts that have precipitated. Alternatively, the mixture can be treated with dilute aqueous HCl followed by extraction, or else addition of tartaric acid in ethanol followed by addition of $NaSO_4$ and celite and then filtration [BL2].

This reagent presents the following characteristics: It allows carbon–halogen bonds to remain unperturbed (Section 2.1). It can cleave aromatic ethers (ArOMe) to give phenols (Section 2.4) and acetals to give ethers (Section 2.4). Nitriles are reduced to imines, hydrolysis of which gives aldehydes (Sections 4.3, 4.4). Esters are generally reduced selectively to aldehydes at low temperature; however, if they are α,β-unsaturated, allylic alcohols are produced (Sections 3.2.5, 3.2.9). The reduction of acid esters to lactones can be easily performed [SO2]. Lactones are reduced to lactols (Section 3.2.5) and imides to α'-hydroxyamides (Section 3.2.8). DIBAH is the reagent of choice for selectively reducing the carbonyl of α,β-unsaturated aldehydes and ketones (Sections 3.2.9, 4.2) in toluene at low temperature. By way of contrast, in the presence of HMPA, sometimes with addition of a catalytic amount of MeCu, DIBAH reduces α,β-ethylenic ketones and esters to saturated ketones and esters (Section 3.2.9) and α,β-acetylenic ketones and esters to α,β-ethylenic derivatives (Section 4.2).

Because of the Lewis acid properties of DIBAH, the reduction of functionalized carbonyl compounds often shows an interesting stereoselectivity (Section 3.2.4).

DIBAH forms ate complexes by action of *n*-BuLi in hexane [KA1]. In THF–hexane, these ate complexes selectively reduce esters to alcohols, tertiary amides to aldehydes (at 0°C), and α-enones to allyl alcohols (at −78°C). Primary and secondary amides as well as nitriles are unaffected at low temperatures. Primary halides are only reduced at room temperature; so these reagents perform selective reductions according to the reaction conditions (Sections 2.1, 3.2.5, 3.2.9). The uses of DIBAH-*i*-Bu_3Al ate complexes have also been described [PP2].

1.5 ALUMINUM HYDRIDE (AlH_3), AMINOHYDRIDES, AND ALUMINUM CHLOROHYDRIDES (AlH_2Cl, $AlHCl_2$)

The reagents AlH_3, $AlHCl_2$, and AlH_2Cl are obtained by reaction of a limited quantity of $AlCl_3$ with a solution of LAH in diethylether. AlH_3 can also be prepared by the action of H_2SO_4 on LAH in THF [BY1], but the so-formed reagent slowly cleaves THF at room temperature [CB7]. This drawback has been overcome by generation of $AlH_3 \cdot Et_3N$. A solution of this reagent in THF is stable for at least 1 month [CB7]. These reagents are just as sensitive as LAH toward water and must be decomposed under the same conditions as LAH. The ready generation of a dimethylethylamine-AlH_3 or N-methylpyrrolidine-AlH_3 complex, which can be used in toluene-THF and whose reducing properties are similar to those of AlH_3 in THF, has been described [MP2].

These reagents are strong Lewis acids that cleave THF and acetals (Section 2.4). Nevertheless, they leave bromo- and chloroderivatives intact (Section 2.1). The regioselectivity of the opening of epoxides is opposite to that observed for LAH in THF (Section 2.3). Diarylcarbinols can be reduced to hydrocarbons (Section 2.4), and α,β-unsaturated carbonyl compounds to allylic alcohols (Section 3.2.9). The reduction of amides to amines is easier than with LAH (Section 3.2.8), especially in the case of α,β-ethylenic amides or of β-lactams. These reagents do not reduce NO_2 groups.

Aluminum bis-(N-methylpiperazino)hydride, obtained by combining 2 equivalents of N-methylpiperazine and a solution of AlH_3 in THF, is especially recommended for the reduction of esters or acids to aldehydes (Sections 3.2.5, 3.2.6) [MM3].

1.6 SODIUM AND POTASSIUM BOROHYDRIDES: $NaBH_4$, KBH_4

The sodium and potassium borohydrides [BK5, PS1, W3, W4] are soluble in water, alcohols, glymes, and DMF. They are not very soluble in diethylether and are slightly soluble in cold THF, but are more soluble under heating. Basic aqueous solutions are relatively stable, but solutions in methanol or ethanol are rapidly decomposed to borates, which in turn reduce only very reactive substrates. Solutions in *i*-PrOH or glymes are more stable and are often used. If the substrates or products of the reaction are fragile in an alkaline medium, the solutions can be buffered by $B(OH)_3$ [DS1]. These reagents are useful in phase transfer systems (liquid–liquid or solid–liquid) [BK8, ML1], on solid supports in the presence of THF or diethylether [BI1], on resins [NS1], in micelles [FR2, NS4], or in microemulsions [FR2, JW1]. An increase in the degree of reducing power of $NaBH_4$ in hot THF by addition of methanol after reflux has been noted [SO1].

The most frequent workup treatment after reduction is the addition of an acid. When the alkoxyboranes or aminoboranes are formed, the decomposition of these intermediates may require heating in a strong acid medium or even treatment by H_2O_2 in an alkaline medium [PS1, H3]—a problem that often arises with reducing reagents derived from boron.

Sodium and potassium borohydrides are above all used for reducing aldehydes and ketones (Sections 3.2.1, 3.2.2); α,β-ethylenic ketones are converted to mixtures [W3]. In alcoholic media or THF, they leave epoxides, esters and lactones, acids, amides, and most nitro compounds unreacted, but they reduce halides (Section 2.1), anhydrides (Section 3.2.6), quarternary pyridinium salts (Section 3.3), double bonds conjugated to two electron-withdrawing groups (Sections 3.2.9, 4.4), and CUPd and C—Hg bonds (Section 5.3). However, in the presence of hot methanol in THF, $NaBH_4$ reduces esters to alcohols [SO1], and in refluxing pyridine some tertiary amides are reduced [KI1].

Compounds able to undergo solvolysis to sufficiently stable cations are reduced via these carbocations by $NaBH_4$ in alcoholic media sometimes in the presence of acid. Diarylketones (Section 3.2) or the di- or triarylcarbinols are reduced to hydrocarbons (Section 2.4), imines and the iminium salts are reduced to amines (Sections 3.3.1, 3.3.2), and imides to α'-hydroxyamides (Section 3.2.8).

In the presence of organic acids, sodium and potassium borohydrides form acyloxyborohydrides that show some remarkable characteristics [GN1]. Their reaction path depends on the quantity of acid present, which leads to either monoacyloxy- ($NaRCOOBH_3$) or trisacyloxyborohydrides [$Na(RCOO)_3BH$]. The reduction can be performed in the presence of a cosolvent (dioxane, THF, ethanol) or in pure organic acid (AcOH, CF_3COOH most frequently). Acyloxyborohydrides are easily decomposed by water. Aldehydes and ketones react more slowly with these reagents than with the borohydrides in alcoholic media [GN1]. Given an acidic medium, these reagents reduce di- and triarylketones and alcohols to hydrocarbons (Sections 2.4, 3.2.1), acetals to ethers (Section 2.4), and nitriles to amines (Section 4.3). Their most interesting application consists of the reduction of C=N double bonds to amines. Imines, oximes, enamines, iminium salts, and numerous nitrogen heterocyclic compounds are reduced (Sections 3.3.1–3.3.4). These are the reagents of choice for effecting reductive aminations (Section 3.3.1) or the reductions of tosylhydrazones to hydrocarbons (Section 3.3.4). Depending on the substrate, $NaBH_4$ may be used, but it is preferable to substitute $NaCNBH_3$ while operating under the same conditions [GN1].

Under the action of Lewis acids such as BF_3, $AlCl_3$, I_2, and Me_3SiCl, the borohydrides are converted into boranes, which then become the reducing agents (see the following).

1.7 LITHIUM BOROHYDRIDE: $LiBH_4$

$LiBH_4$ is soluble in alcohols and ethers [BK5, PS1, W3]. In an diethylether or THF medium, the Li^+ cation is a stronger Lewis acid than Na^+, which gives to this reagent an increased reducing power. Epoxides, esters, and lactones may then be reduced (Sections 2.3, 3.2.5), while amides and nitriles remain intact unless one adds hot DME or methanol. Under these conditions, tertiary amides give alcohols (Section 3.2.8) and nitriles give amines (Section 4.3).

$LiBH_4$ can also be activated by adding $(MeO)_3B$ or Et_3B in diethylether. With this reagent, esters are rapidly reduced, tertiary amides and nitriles are also reduced, but sulfone, sulfoxide, and NO_2 groups remain intact [BN3, YP2].

1.8 TETRABUTYLAMMONIUM BOROHYDRIDE: n-Bu_4NBH_4

This reagent is soluble in alcohols, ethers, CH_2Cl_2, and toluene [PS1, RG1]. In hot CH_2Cl_2, it decomposes slowly to borane. It is usable on solid supports [BI1].

n-Bu_4NBH_4 is a very mild reducing agent. The reactivity order in CH_2Cl_2 is as follows: RCOCl > RCHO > RCOR′ >> RCOOR′, esters being reduced only under reflux. This reagent reduces aldehydes selectively in the presence of ketones (Section 3.2.1). In organic acid media, tetrabutylammonium acyloxyborohydrides are formed. Under reflux in C_6H_6, these reagents also reduce aldehydes selectively without affecting the ketones (Section 3.2.1) [GN1]. Borohydrides supported on exchange resin [GB5, GW3, YK5, YP3] exhibit a similar, although weaker, reducing power to the standard reagents.

1.9 CALCIUM BOROHYDRIDE

Calcium borohydride is generated in methanol or ethanol from $CaCl_2$ and $NaBH_4$ [BR3]. It reduces esters to alcohols, leaving acid salts intact, thus allowing the formation of lactones from hemiesters [LR1] (Section 3.2.5). It has also been used in stereoselective reduction of α,β-epoxyketones [TF2] (Section 3.2.4).

1.10 ZINC BOROHYDRIDE: $Zn(BH_4)_2$

Zinc borohydride [BK5, KH1, ON1, R3, W3], which exists in the dimeric form **1.1**, (on page 11) is obtained by adding $ZnCl_2$ in diethylether to a solution of $LiBH_4$ in this solvent. It has also been prepared from $NaBH_4$ and $ZnCl_2$ in THF or DME, but under these conditions the reagent is a mixture of several components [SB3]. It has also been used on silica gel [R3]. Its complex with polypyrazine is stabilized and can be used as a reagent [TL1]. This relatively strong Lewis acid reduces α,β-ethylenic ketones to allylic alcohols (Section 3.2.9). It also reduces esters and azides in DME [R3, RS1] as well as acids into alcohols in THF [NM3] or in DME in the presence of $(CF_3CO)_2O$ [R3]. As a good chelating agent, it can be used in some very stereoselective reductions of ketones bearing heteroatoms at the α or β position, especially α- and β-ketoesters, ketoamides, or even epoxyketones (Section 3.2.4). Ester, amide, nitrile, and nitro groups and halogens are not usually affected; however, the reduction of tertiary halides can be carried out [KH1].

A complex $Zn(BH_4)_2{\cdot}1.5$ DMF has been described [HJ1]. This shows a greater selectivity than $Zn(BH_4)_2$ in diethylether and does not react with the α-enones. In MeCN, this complex allows the reduction of aldehydes in the presence of ketones, the reduction of some sterically unhindered ketones in the presence of other less accessible ketones, or even the reduction of aliphatic ketones in the presence of aromatic ones (Section 3.2.1).

1.11 SODIUM AND TETRABUTYLAMMONIUM CYANOBOROHYDRIDES: $NaCNBH_3$, n-Bu_4NCNBH_3

The Na and tetrabutylammonium cyanoborohydrides [BK5, HN1, L1, PS1, W3] are soluble in water, alcohols, organic acids, THF, and polar aprotic solvents. They are insoluble in diethylether and hydrocarbons and may be used under phase transfer conditions [HM1]. One feature of the cyanoborohydrides is their stability in acid media at about pH 3. It is thus necessary to treat the crude reaction mixture with a strong acid to decompose the intermediates formed. The use of resin-supported cyanoborohydride has also been described [HN3].

These reagents are interesting because aldehydes and ketones are affected in acidic media only, which permits the reduction of carbon–halogen bonds (Section 2.1) without affecting carbonyl groups, esters, or nitriles.

In organic acid media, $NaCNBH_3$ is converted to acyloxycyanoborohydrides whose reactivity is comparable to that of $NaBH_4$ in CF_3COOH, especially concerning the reduction of imines to amines, tosylhydrazones to saturated hydrocarbons, oximes to hydroxylamines, or reductive amination. Depending on the substrate, $NaBH_4$ or $NaCNBH_3$ is recommended (Sections 3.3.1, 3.3.4) [GN1].

1.12 ZINC CYANOBOROHYDRIDE

Zinc cyanoborohydride [KO1, LD1] is formed by reaction of $ZnCl_2$ in diethylether with a solution of $NaCNBH_3$ in this solvent [KO1] or by the reaction of ZnI_2 with $NaCNBH_3$ in CH_2Cl_2 [LD1].

In ether media (diethylether or THF), the nature of the reagent is ill defined. It reduces aldehydes, ketones, and acid chlorides, but leaves esters, anhydrides, and amides unchanged. In methanol, the reduction of enamines and imines to amines may be effected in the same way as the reduction of tosylhydrazones to hydrocarbons (Section 3.3.4).

The reagent formed by reaction of ZnI_2 with $NaCNBH_3$ in CH_2Cl_2 allows the reduction of aromatic aldehydes and ketones as well as benzylic, allylic, and tertiary alcohols to hydrocarbons, probably by a radical process [LD1] (Section 2.4). Some comparable reductions are carried out in ether media starting from tertiary, benzylic, or allylic halides (Section 2.1).

1.13 CUPROUS BIS(DIPHENYLPHOSPHINE) BOROHYDRIDE AND CYANOBOROHYDRIDE

These cuprous borohydrides [DF1, FH1, FH2, HM2, SP1, W4] are isolated complexes of the structure **1.2** (on page 11), which transfer only a single hydride. They can be supported on ion-exchange resins [SP1].

In neutral media, they leave carbonyl derivatives intact but reduce tosylhydrazones to the corresponding hydrocarbons under reflux of $CHCl_3$ (Section 3.3.4). This reduction is compatible with α-enone, epoxide, or lactone groups present in the molecule [GL3]. In cold acetone, these reagents reduce acid chlorides to aldehydes [FH1] (Section 3.2.7). In the presence of Lewis acids or gaseous HCl in CH_2Cl_2, they reduce aldehydes and ketones. The selective reduction of aldehydes in the presence of ketones can also be realized (Section 3.2.1). These reagents also reduce aromatic azides to amines (Section 5.2).

1.14 POTASSIUM TRIISOPROPOXYBOROHYDRIDE: $K(i\text{-}PrO)_3BH$

This borohydride [BC3], obtained in THF by adding 3 moles of *i*-PrOH to a solution of KBH_4, essentially reduces aldehydes, ketones, and halogenated derivatives. Its principal use is for the reduction of the haloboranes RR′BCl or RR′BBr to boranes RR′BH (Section 5.7). This process allows sequential hydroborations, first by a halogenoborane, which is then reduced to a hydrogenoborane that can undergo a new hydroboration, giving access to mixed trialkylboranes. This reagent also transfers KH similarly to hindered trialkylboranes, thereby forming KR_3BH.

1.15 LITHIUM AMINOBOROHYDRIDES

Lithium aminoborohydrides are obtained by the reaction of *n*-BuLi with amine–boranes [FF2, FH5, NT2]. They can be generated in situ as THF solutions or as solids when formed in diethylether or hexane (*n*-BuLi must then be used in substoichiometric amounts). They are stable under dry air and are slowly decomposed by water [NT2] or methanol so that workup of the reactions mixtures can be carried out with 3M HCl. They reduce alkyl halides (Section 2.1), epoxides (Section 2.3), aldehydes, and ketones (Section 3.2.1) (in the latter case with an interesting stereoselectivity [HF1]), and esters to primary alcohols (Section 3.2.5). α,β-Unsaturated aldehydes, ketones, and esters are reduced to allyl alcohols (Section 3.2.9) [FF2, FS2]. Depending on the bulkiness of the amines associated with the reagent and to the substrate, tertiary amides give amines or alcohols (Section 3.2.8) [FF1, FF2]. Amines are also formed from imines (Section 3.3.1) [FB1] and from azides (Section 5.2) [AF1]. However, carboxylic acids remain untouched.

1.16 LITHIUM TRIETHYLBOROHYDRIDE: $LiEt_3BH$ (SUPERHYDRIDE)

$LiEt_3BH$ [BK5, BK6, BN4, KB3, KB5, W3] is soluble in ethers (diethylether, THF, glymes) and hydrocarbons. Rapidly decomposed by water or alcohols, it must be handled away from moisture. The workup of the crude reaction mixture consists of hydrolysis, sometimes in the presence of acid, followed by the action of alkaline H_2O_2 to oxidize Et_3B (a byproduct of the reduction) to ethanol and boric acid, both of which are soluble in water.

Although it is much more reactive than $LiBH_4$, the triethylborohydride shows an analogous reactivity spectrum. It reacts particularly well with primary and secondary alkyl halides and tosylates, even when hindered, with an inversion of configuration (Section 2.1), and with epoxides at the least sterically hindered site (Section 2.3). It reduces ammonium salts to tertiary amines. The reduction of cyclic or functionalized ketones and imines by $LiEt_3BH$ in THF can be very stereoselective (Sections 3.2.2, 3.3.1), but in general Li(*s*-Bu_3)BH is preferable. Tertiary amides are reduced first to aldehydes then to alcohols (Section 3.2.8), and nitriles are reduced to imines, which are hydrolyzed to give aldehydes (Section 4.3). The use of KEt_3BH for chemoselective reduction of carboxylic acid esters has been suggested [YY1].

1.17 LITHIUM AND POTASSIUM TRI(*s*-BUTYL) BOROHYDRIDES (Li AND K SELECTRIDES): Li OR K(s-Bu)$_3$BH

The Li and K Selectrides [BK5, W3] are soluble in ether media (diethylether, THF, glymes). The treatment after reduction is identical to that employed for $LiEt_3BH$.

The principal interest of these reagents resides in their bulkiness. The reductions of slightly hindered cyclic ketones and imines occurs on the equatorial face (Sections 3.2.2, 3.3.1), and aliphatic carbonyl compounds are reduced with a high stereoselectivity (Section 3.2.2). The Li and K Selectrides selectively reduce the carbon–carbon double bond of α-enones and α,β-ethylenic esters unless the β position is disubstituted (Section 3.2.9); in the latter case, the carbonyl of the α-enones is reduced.

Li and K trisiamyl borohydrides, which are even bulkier, are sometimes used [KB8].

1.18 LITHIUM ALKYLBOROHYDRIDES

These can be easily prepared by reaction of di- or trialkylboranes with lithium aminoborohydrides [HA1]. The properties of two types of reagents have been explored: Li(n-Bu)BH_3 [KM2] and the boratabicyclononane Li 9-BBN-H **1.3** (on page 11) [BM1, KB1]. No special features have been pointed out in relation to other reducing agents.

The treatment of the crude reaction mixture after reduction by Li 9-BBN-H requires the action of H_2O_2 in an alkaline medium to convert the intermediate borane to water-soluble or volatile compounds.

Chiral Li alkylborohydrides have been used in asymmetric reductions (Section 3.2.3) [BJ1, BR4].

1.19 BORANE: BH_3

Rarely used in its gaseous dimeric form (B_2H_6), borane is generally employed as a solvate with THF or Me_2S. $BH_3{\cdot}THF$ is employed in ether media. $BH_3{\cdot}Me_2S$ is soluble in ethers, hydrocarbons, and CH_2Cl_2. Borane can also be generated in situ by reaction of $NaBH_4$ with iodine [BB7], HCl, $MeSO_3H$, or sulfuric acid [AM2] or trimethylsilyl chloride [DA2]. Under such conditions, there is no need to use dry solvents.

Borane reduces carboxylic acids in the cold without attacking esters or nitriles, and it reduces halogenated derivatives (Section 3.2.6). Enantioenriched amino acids can be transformed into amino alcohols without epimerization [AM2, DA2, JJ2]. Borane easily reduces amides in refluxing THF (Section 3.2.8). Esters can also be reduced at higher temperatures (Section 3.2.5). An important limitation is competing hydroboration of carbon–carbon double and triple bonds [BK7, HH1, L2], although this can be avoided when reducing acids at 0°C [BP5].

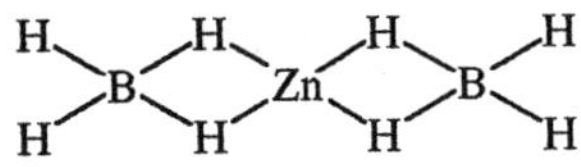

1.1

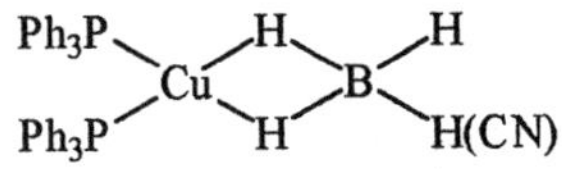

1.2

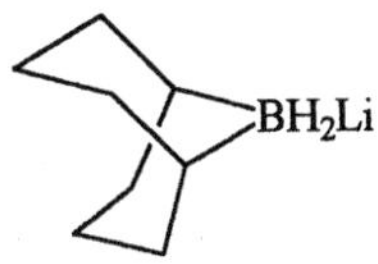

1.3

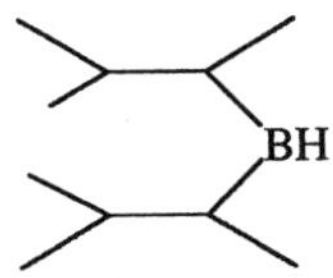

1.4

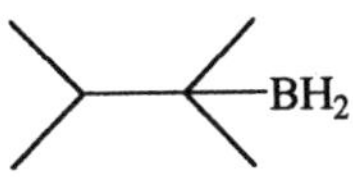

1.5

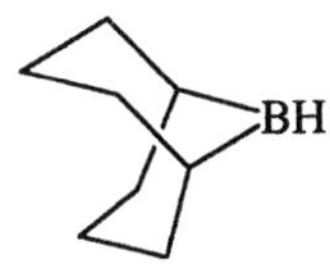

1.6

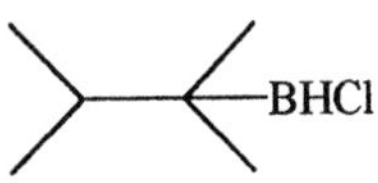

1.7

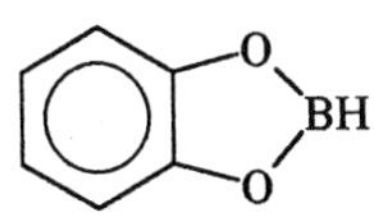

1.8

1.20 AMINE–BORANES: $R_3N{\bullet}BH_3$

These complexes are more stable than the borane complexes with diethylether or Me_2S. They are soluble in water and alcohols and stable in the presence of acetic acid. Their decomposition requires the action of a strong acid or decomplexation by an amino alcohol.

With respect to reactivity, the amine–boranes lie somewhere between $BH_3{\cdot}THF$ and $NaBH_4$. They reduce aldehydes and ketones without affecting ester, ether, SPh, and NO_2 groups (Section 3.2.1). The reduction of ketones can be accelerated by the addition of Lewis acids or when carried out in acetic acid [PS1]. On alumina or silica supporrts, amine–boranes can selectively reduce aldehydes without affecting keto groups (Section 3.2.1) [BS1]. Chiral amino acids can be reduced to amino alcohols without epimerization [PS1].

$Ph_2NH{\cdot}BH_3$ is a recommended reagent because its stability and reactivity are superior to those of amine–boranes formed from aliphatic amines [CU1]. Pyridine–borane reacts slowly with carbonyl compounds and has been suggested for carrying out reductive aminations (Section 3.3.1) [PR2]; however, in the presence of AcOH, it reduces aldehydes, leaving ketones untouched [CW1].

Some amino alcohols react with borane to generate oxazaborolidines, which have been mainly used in asymmetric reduction of ketones (Section 3.2.3) and imines (Section 3.3.1) [NN1, S3]. In addition, they can also perform some chemoselective reductions [IW1].

1.21 SUBSTITUTED BORANES

Substituted boranes are obtained by hydroboration of relatively hindered olefins such as trimethylethylene, tetramethylethylene, and 1,5-cyclooctadiene, which, by action of BH_3, lead, respectively, to diisoamylborane, Sia_2BH **1.4** (on page 11), thexylborane, $ThexBH_2$ **1.5** (on page 11), and 9-BBN **1.6** (on page 11). These reagents are used in THF. Thexylchloroborane is obtained by reaction of $ClBH_2{\cdot}SMe_2$ with tetramethylethylene. $ThexBHCl{\cdot}SMe_2$ **1.7** (on page 11) in solution in CH_2Cl_2 or in THF, where it is less stable, is also recommended, as is $Cl_2BH{\cdot}Me_2S$ [SB3]. The crude reaction mixture is hydrolyzed in a hot acid medium.

The reactions of these reagents reflect their sterically hindered and Lewis acidic characters. This is why the reduction of relatively hindered acyclic ketones by Sia_2BH **1.4** shows the opposite stereoselectivity to that observed with the alumino- or borohydrides (Section 3.2.2) [HW1]; the reduction of hindered cyclanones by $ThexBHCl{\cdot}SMe_2$ leads to the least stable alcohol [BN5]. α,β-Ethylenic aldehydes and ketones are reduced by 9-BBN or $ThexBHCl{\cdot}SMe_2$ to allylic alcohols, with a better selectivity than that observed with $BH_3{\cdot}SMe_2$ or $ThexBH_2$ (Section 3.2.9). Acids are selectively reduced to aldehydes by $ThexBHCl{\cdot}SMe_2$ (Section 3.2.6) [BC5]. Tertiary amides are reduced by 9-BBN to alcohols and by Sia_2BH and $ThexBH_2$ to aldehydes (Section 3.2.8), while BH_3 transforms these tertiary amides to amines and ThexBHCl reacts with them slowly. $Cl_2BH{\cdot}SMe_2$ is recommended for selective reduction of azides (Section 5.2) [SB3].

Catecholborane **1.8** (on page 11) is a mild reducing agent that is not sensitive to moisture [KB7]. It can be used without solvent or in $CHCl_3$, and it reduces aldehydes, ketones, hydrazones, and acetals. It also reduces acids if used in excess at room temperature. Esters are reduced in refluxing THF, and alkenes are hydroborated in similar conditions.

1.22 ALUMINO- AND BOROHYDRIDES IN THE PRESENCE OF TRANSITION METAL SALTS

Solutions or suspensions of LAH in diethylether or THF in the presence of iron salts, $CoCl_2$, $TiCl_3$, or $NiCl_2$ [AL1, GO2] are used as reducing agents. Similarly, Li or $NaBH_4$ in methanol, THF, or DMF may be used in the presence of salts or complexes containing nickel, cobalt, tin, copper, palladium, or lanthanides [AL1, CY2, DG1, GO2, PV1, YC2, YL5]. The structures of these reagents are often not well known. However, it is thought that Ni_2B is formed from $NaBH_4$ and $NiCl_2$ in MeOH. Titanium salts and complexes are also proposed as addends [B4, B5, BH5, BS6, DK3, LS4, RB3, RC2].

Each reagent shows some particular characteristics, but a certain number of transformations merit emphasis. These include:

- The reduction of alkenes with LAH–$FeCl_2$, $CoCl_2$, $TiCl_3$ or $NiCl_2$, or $NaBH_4$–$CoCl_2$, all of which do not modify aromatic derivatives (Section 3.1);
- The reduction of the aromatic moieties with $NaBH_4$–$RhCl_3$ in ethanol;
- The reduction of aromatic nitrogen-containing heterocycles with $NaBH_4$–$NiCl_2$ in methanol, which does not perturb aromatic carbon-containing rings (Section 3.3.3);
- The reduction of aromatic or alicyclic halogenated derivatives with $NaBH_4$–$NiCl_2$ in DMF either in the presence of Ph_3P or with LAH in the presence of various transition metal salts (Section 2.1);
- The reduction of nitriles and nitro derivatives to amines with $NaBH_4$–$CoCl_2$ in methanol (Sections 4.3, 5.1);
- The reduction of oximes and nitro derivatives to amines with $NaBH_4$ in the presence of nickel or copper salts (Sections 3.3.4, 5.1);
- The reduction of arylketones to hydrocarbons with $NaBH_4$–$PdCl_2$ in methanol (Section 3.2.1);
- The reduction of allylic acetates to saturated hydrocarbons with $NaBH_4$–$NiCl_2$ (Section 2.2);
- The reduction of azides to amines with $NaBH_4$–$Ni(OAc)_2$ (Section 5.2);
- The reduction of α-enones to allylic alcohols with $NaBH_4$–$CeCl_3$ in methanol or with $(i\text{-PrO})_2TiBH_4$, generated from $(i\text{-PrO})_2TiCl_2$ and benzyltriethylammonium borohydride in a 1:2 ratio, in CH_2Cl_2 (Section 3.2.9) [RB3].

Chapter 2

Cleavage of the Carbon–Heteroatom Single Bond

2.1 HALIDES: $\geqslant$C—X

LAH in THF reduces chlorides, bromides, and iodides to hydrocarbons, whatever their degree of substitution (primary, secondary, tertiary, aromatic, vinyl, and cyclopropyl). The order of reactivity for a given hydrocarbon residue is iodide > bromide > chloride. For a given halogen, the order of reactivity is: $ArCH_2X \approx allylX > RCH_2X > R_2CHX > R_3CX$.

Aliphatic and alicyclic iodides and bromides are reduced at room temperature, while the aromatic, vinyl, and cyclopropyl bromides as well as the chlorides can be reduced only under reflux. For example, the selective reductions shown in Figure 2.1 can be performed [P1].

In the presence of $CeCl_3$ in cold DME or under reflux in THF, LAH reduces all the halides [GO2].

The mechanism of these reductions is a bimolecular nucleophilic substitution for the reaction of LAH with most primary and secondary halides [BK5, PC1]. A single-electron transfer (SET) has been proposed in the reduction of sterically hindered primary iodides [AD3, AG1, AW1], although some doubts have been cast [PC1] on this mechanism with bromocyclopropanes [HW2] and aromatic or vinyl halides [C1], especially in the presence of $CeCl_3$ [GO2]. In this case, some rearrangements may be observed. SET does take place in the reduction of geminal dihalides by LAH [AD4] as well as in the reduction of bromocyclopropanes in the strict absence of molecular oxygen [PN1]. In the presence of oxygen, the C—Br bond of 2,2-diphenyl-1-bromocyclopropanecarboxylic acid is left unchanged [PN1].

The alkoxyaluminohydrides reduce aliphatic and alicyclic iodides and bromides but not the corresponding chlorides. An exception is Red-Al in benzene, which reduces all halides, as well as the cyclopropyl and aromatic derivatives. The reduc-

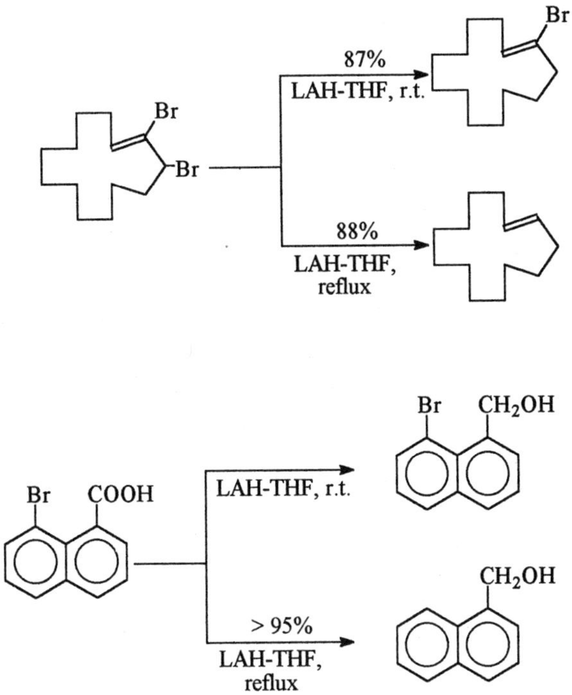

Figure 2.1

tion of 1-bromoalkynes by the latter reagent cleanly gives the debrominated alkyne, unlike other aluminohydrides that give a mixture. Difluoroalkenes, meanwhile, are converted to the monofluoro derivatives [M1] (Figure 2.2).

LAH or AlH_3–amine complexes reduce alkyl bromides and iodides as well as benzyl chloride and bromide [FS1]. They leave other chlorides unchanged, thus allowing the selective reduction of **2.1** to the corresponding chloroalcohol [MP2].

On the other hand, the selective reduction of an α,β-ethylenic-α-chloroester **2.2** to an α-chloroallylalcohol **2.3** [DW1] comes from the inability of DIBAH to react with halogenated derivatives in cold toluene [YG1] (Figure 2.2).

DIBAH–*n*-BuLi ate complex in hexane–THF reduces primary bromides and chlorides at room temperature to hydrocarbons. Secondary halides react more slowly, while tertiary and aryl halides remain unchanged [KA1].

The reduction of the fluorides requires the electrophilic assistance of a Lewis acid in breaking the C—F bond: AlH_3 in Et_2O and LAH–$CeCl_3$ in DME are adequate reducing agents [GO2, PI]. AlH_3, however, leaves the aliphatic C—Br and C—Cl bonds intact [BK5, PC1] (Figure 2.2).

The alkaline borohydrides are less reactive toward halides. $NaBH_4$ in DME or DMSO or in the presence of polyethylene glycols [SF2] reduces only primary or secondary bromides upon heating; with chlorides the reaction is even slower [PS1]. $NaBH_4$–$Ni(OAc)_2$ in MeOH has shown some efficiency [FM1]. The dibromo-

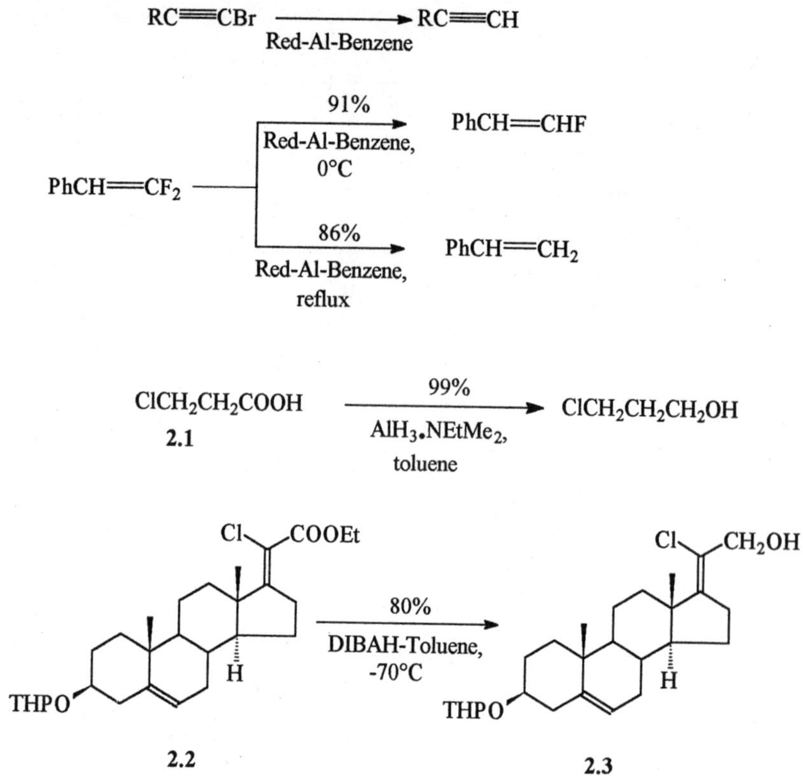

Figure 2.2

cyclopropanes can be selectively reduced to monobromocyclopropanes by heating with $NaBH_4$ in DMF [PS1]. The reduction of aromatic halides under these conditions requires UV irradiation, and these reductions undoubtedly take place via a radical pathway. Reductions of primary, secondary, and aryl bromides and iodides by $NaBH_4$ in hot toluene in the presence of benzo-15-crown-5 and a polymer-bound tin halide catalyst have been described [BW1]. Glycosyl bromides are reduced by titanocene borohydride [CS2]. Aryl bromides and iodides are also dehalogenated by $NaBH_4$–$CuCl_2$ in MeOH [NH1], while chlorides and fluorides remain unaffected. Aryl bromides are inert in the presence of $NaBH_4$–$ZrCl_4$ in THF [IS1]. $LiBH_4$ leaves halogens intact in the selective reduction of **2.4** [BK5] (Figure 2.3). Lithium aminoborohydrides reduce aliphatic iodides as well as benzyl bromide at room temperature [FF2].

$LiEt_3BH$ in THF is the reagent of choice for the reduction of primary and secondary halides; the latter reduction takes place by an S_N2 mechanism with inversion of configuration and without rearrangement as shown in Figure 2.3 [BK1, BK5]. The neopentyl or norbornyl skeletons, which easily undergo rearrangement, thereby remain unchanged (Figure 2.3). Similarly, hexen-5-yl iodide **2.5**, capable of cyclization via a radical pathway, is transformed into a linear olefin, without mod-

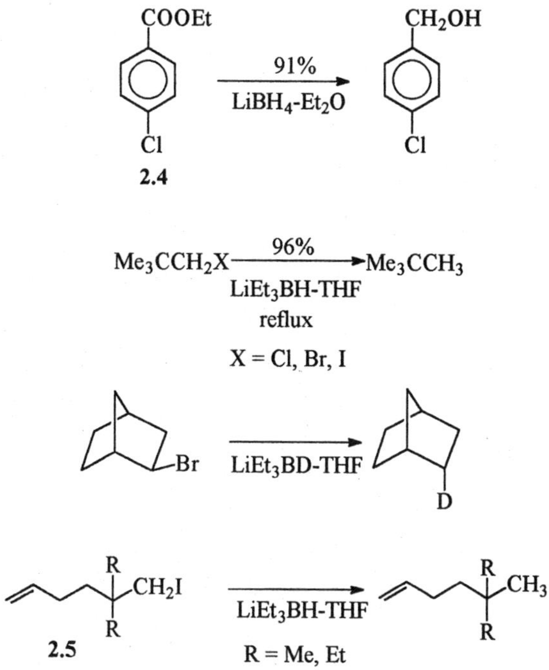

Figure 2.3

ification of the carbon skeleton [AG1] (Figure 2.3). When the same reduction is performed with LAH in THF, it gives 81% cyclic product [BK5].

It is beneficial to use 2 equivalents of $LiEt_3BH$ per mole to reduce a given halide because the byproduct of the reduction is BEt_3. This forms a complex with $LiEt_3BH$ $[Et_3BH{\cdot}BEt_3]Li^+$, which is much less reactive. Aromatic and tertiary halides remain intact under these conditions [BK1].

The selective reduction of the primary halides can also be accomplished with $NaCNBH_3$ in HMPA or DMSO or even by $NaBH_4$ in warm DMSO. Epoxides, nitriles, amides, ketones, and esters are not affected under these conditions [HK1, L1], as illustrated in Figure 2.4. n-Bu_4NCNBH_3 or resin-supported cyanoborohydride is even more selective, since each reduces only the primary iodides and bromides, leaving the chlorides unchanged [HN3].

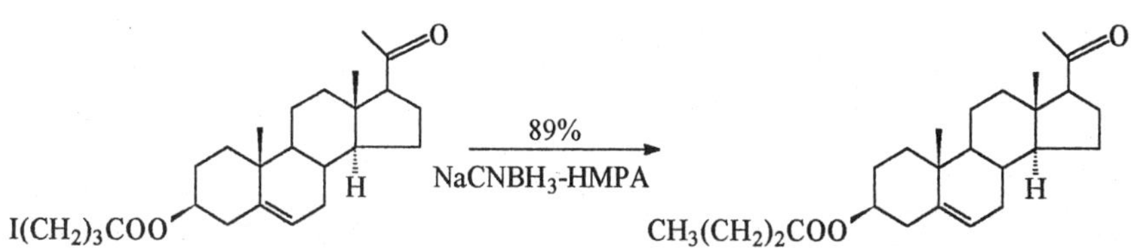

Figure 2.4

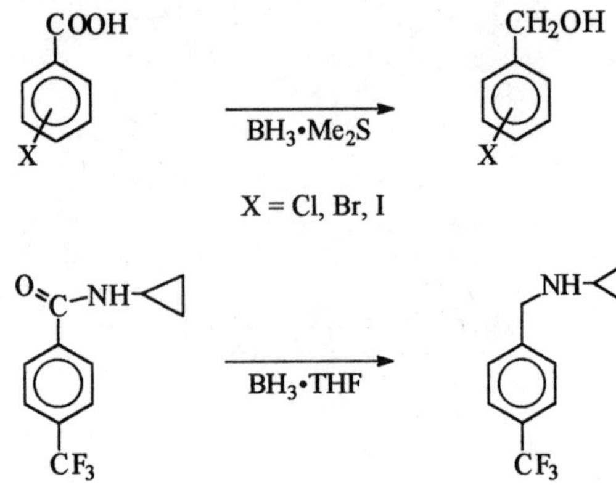

Figure 2.5

Borane leaves the halides intact in an ether medium, allowing the selective reductions shown in Figure 2.5 [P1].

In the presence of transition metal complexes such as $(Ph_3P)_4Ni$, halogenated aromatic derivatives are reduced by $NaBH_4$ in DMF [W4] or $NaBH_4–PdCl_2$ in MeOH [GO2]. The DDT and 2,4 D classes of pesticides are also dechlorinated by $NaBH_4–Ni_2B$ in alcohol [GO2] or by $NaBH_2(OCH_2CH_2OMe)_2–NiCl_2$ in THF under reflux [TP1]. Titanium complexes also catalyze the dechlorination of polychlorinated aryl halides by $NaBH_4$ in DMF via a nonradical process [LS5], leading to dimethylamino-substituted byproducts along with hydrocarbons. In DMA or in ethers, a radical-based reaction takes place, leading only to dechlorinated products [LS4, LS5].

In protic media (alcohol or aqueous diglyme), tertiary halides undergo solvolysis and lead to the corresponding carbocations, which are reduced by $NaBH_4$. If the carbocations are able to undergo rearrangement much faster than reduction, a rearranged alkane product is obtained [BB1]. Borane in CF_3COOH shows a similar behavior [MM1] (Figure 2.6).

Borohydrides associated with a Lewis acid such as $Zn(BH_4)_2$, $NaCNBH_3–ZnI_2$, or $NaCNBH_3–SnCl_2$ can also induce, in ether, the cleavage of the C—X bond of halides, which lead to sufficiently stable carbocations. An analogous mechanism can be proposed to explain the reaction of LAH with secondary allylic chlorides such as **2.6** in ether, a process that takes place with rearrangement [HN2]. In contrast, primary derivatives such as **2.7** are reduced without rearrangement [HN2] (Figure 2.6). Similarly, propargylic chlorides are converted to allenes (Figure 2.6).

$Zn(BH_4)_2$ in Et_2O reduces tertiary and benzylic halides at the corresponding carbon sites, but the allylic derivatives give polymers [KH1]. However, $NaCNBH_3–ZnI_2$ in Et_2O or $NaCNBH_3$ in the presence of $SnCl_2$ selectively reduces tertiary, benzylic, and allylic halides without affecting primary or secondary halides, esters, and amides [KK1, KK6]. The ate complex formed by the reaction of *n*-BuLi with 9-BBN in hexane has an identical behavior: tertiary, allylic, and benzylic halides are reduced, while primary and secondary halides remain intact [TY1].

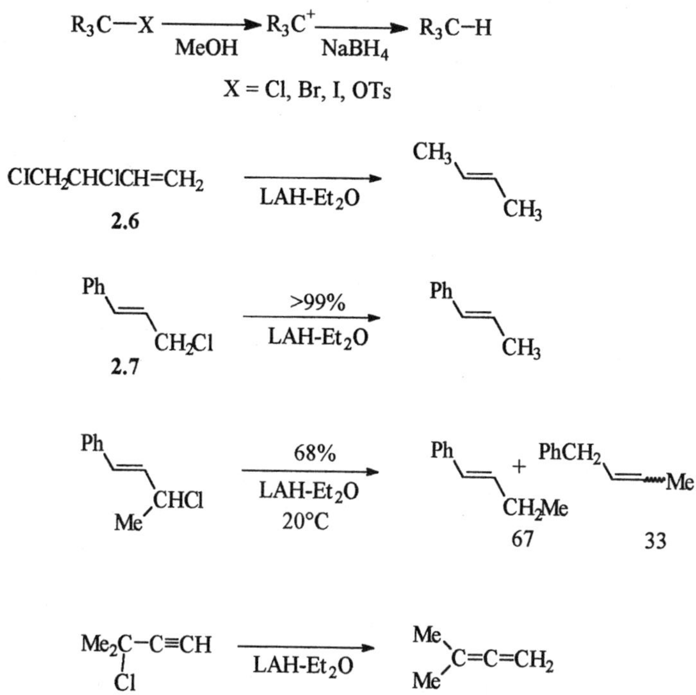

Figure 2.6

2.2 SULFONATES AND ESTERS: $\geqslant$C—OSO$_2$R; $\geqslant$C—OCOR

LAH in Et_2O reduces sulfonates, requiring the electrophilic assistance of the Li^+ cation in the cleavage of the C—O bond. This is why it is possible to reduce at will the C—Br bond or C—OTs of the bifunctional compound **2.8** by changing the solvent [K1] (Figure 2.7). In DME, where the Li^+ cation is well solvated, electrophilic assistance does not take place.

$LiBH_4$, $LiEt_3BH$, or DIBAH in THF also reduce primary and secondary sulfonates to hydrocarbons [BK5, BN3, KB1, YG1] even if they bear benzyloxy substituents [YS2]. However, if the substrate is too sterically hindered such as **2.9** (Figure 2.8), the attack of the reducing reagent takes place on the sulfur, and the corresponding alcohol is formed [GL5, SH4, WS2]. This phenomenon is not observed with $LiEt_3BH$, as shown in Figure 2.8 [KB1]. In the case of **2.10** [HS3], which is a hindered mesylate, the reaction with $LiEt_3BH$ did not produce the alkane but rather the corresponding alcohol. The authors therefore recommend the use of isopropanesulfonate **2.11**, which, when treated with $LiEt_3BH$, is not attacked at the sulfonate site (Figure 2.8).

$NaBH_4$ in hot DMSO can also reduce primary sulfonates [HH2, PS1], and this

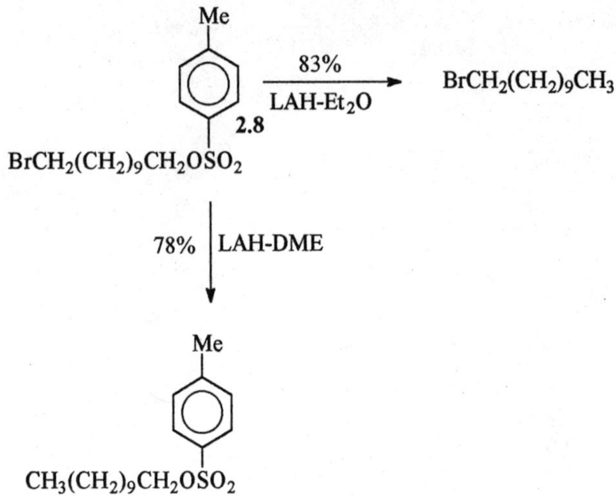

Figure 2.7

method has been applied to various sugar derivatives [KS2, WW1]. Primary allylic tosylates such as **2.12** are reduced to the corresponding olefins by LAH [HN2], but secondary tosylates do not react at all (Figure 2.9).

Acetates, whether primary or secondary, allylic, propargylic, or benzylic, are also reduced by $NaBH_4$–$NiCl_2$ in MeOH to hydrocarbons [HP2, I2], but the double bond, in general, is not preserved (Figure 2.9).

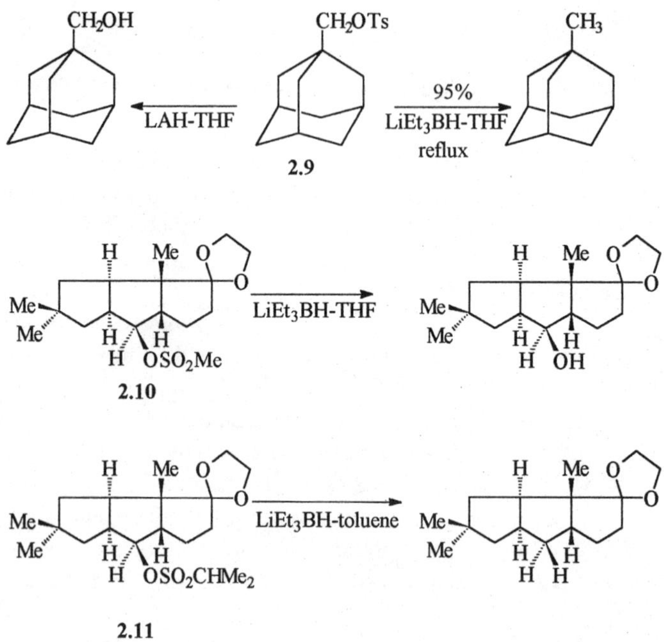

Figure 2.8

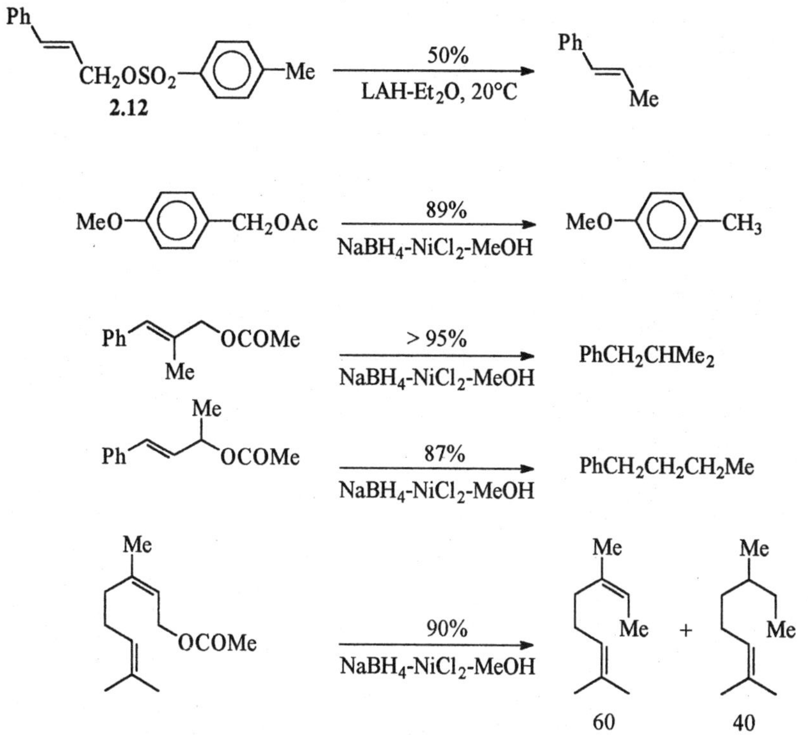

Figure 2.9

The problems of solvolysis and possible rearrangements with sulfonates are similar to those of halides. For example, the reduction of tricyclic tosylate **2.13**, whose structure is such that its double bond can participate in the reaction, leads to the formation of a cyclopropane via an intermediate carbocation [KN1] (Figure 2.10).

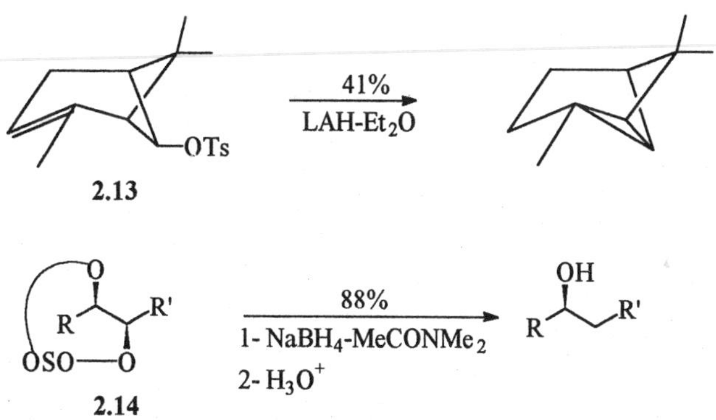

R = H, *n*-alkyl, Ph R' = COO *i*-Pr

Figure 2.10

Cyclic sulfates of 1,2-diols **2.14** are transformed into monoalcohols by $NaCNBH_3$ in refluxing THF at pH 4–5 followed by hydrolysis, or regioselectively to β-hydroxyesters by $NaBH_4$ in DMA when R = COO*i*-Pr [GS3] (Figure 2.10).

2.3 EPOXIDES: 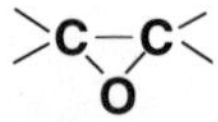

The cleavage of the C—O bond of epoxides requires the electrophilic assistance of a reagent, which can either be a Lewis acid (Li^+) or behave as such (AlH_3, DIBAH). Reduction of epoxides by SAH and by borohydrides is slow [BK5, CB5] unless one adds strong Lewis acid. For example, $NaCNBH_3$ in the presence of $BF_3{\cdot}Et_2O$ [HT1] or $LiBH_4$ in the presence of BEt_3 [YO1] or of methoxyborane [BN3] is used. Therefore, alkali borohydrides may reduce carbonyl compounds, leaving epoxides unchanged. Lithium aminoborohydrides are, however, efficient in reducing epoxides [FF2]. Their reduction with BH_3 is also assisted by the presence of $BF_3{\cdot}Et_2O$ [L2, PS1], but is more difficult when it is carried out with bulky substituted boranes (Sia_2BH, 9-BBN, or ThexBHCl) [BK5, BN5, G4]. Red-Al is not very efficient either, except with the epoxides carry an alcohol functional group at the α position [FK1, M1]. $Zn(BH_4)_2$ on silica gel or on $AlPO_4$ also reduces epoxides [CC11, R3].

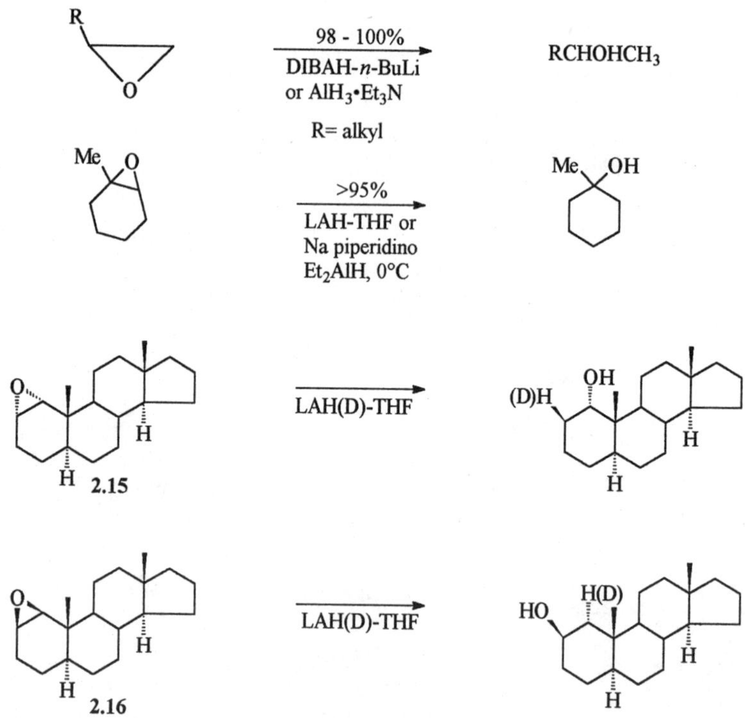

Figure 2.11

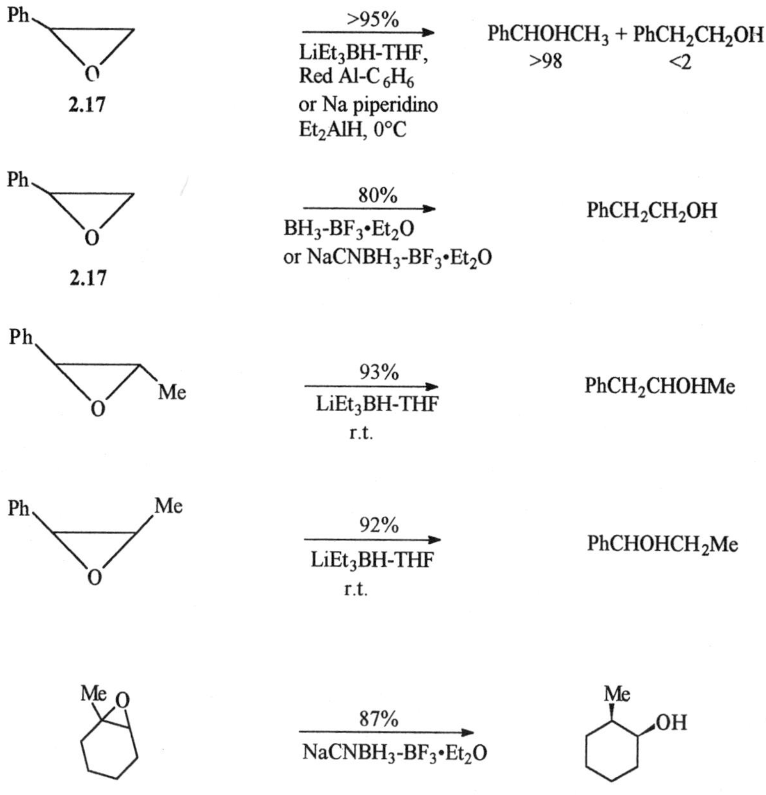

Figure 2.12

The regioselectivity of the opening of disymmetrical epoxides depends essentially on the strength of the Lewis acid–base interaction between the partners. If this interaction is rather weak, then the reduction takes place at the least substituted epoxide's carbon. Such is the case with LAH or $LiEt_3BH$ in THF or Li 9-BBN-H [G4] or with the complexes LAH-N-methylpyrrolidine [FS1], $AlH_3 \cdot Et_3N$ [CB7] and Na piperidinoEt_2AlH [YA2] (Figure 2.11). The mechanism of the reaction is S_N2 assisted by the Lewis acid; its stereoselectivity is therefore a *trans*-diaxial opening (Furst–Plattner rule) [G4], as shown in Figure 2.11 by reduction of steroidal epoxides **2.15** and **2.16** [BK5, BK6, BM1, BN4, RP1, W1].

With a stronger Lewis acid, the regioselectivity is reversed, and the reduction takes place at the most substituted epoxide carbon. The hydride attacks preferentially the carbon that is better able to stabilize a carbocation. This is the case when one uses BH_3, even in the presence of 2-aminoethanol, $NaCNBH_3$ in the presence of BF_3, AlH_3 in Et_2O, DIBAH in THF, toluene, or hexane, $LiBH_4$–BEt_3 or $Zn(BH_4)_2$ on SiO_2 [E2, G4, HT1, IW1, L2, M1, PS1, R3, YG1, YO1, W1] (Figure 2.12). The regioselective reduction of styrene oxide **2.17** to 2-phenylethanol can be performed

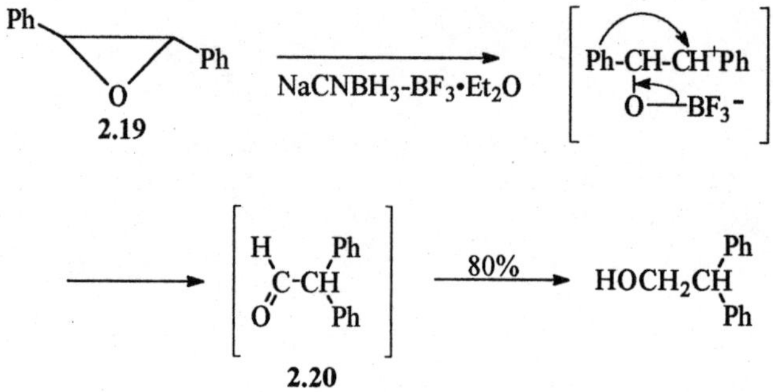

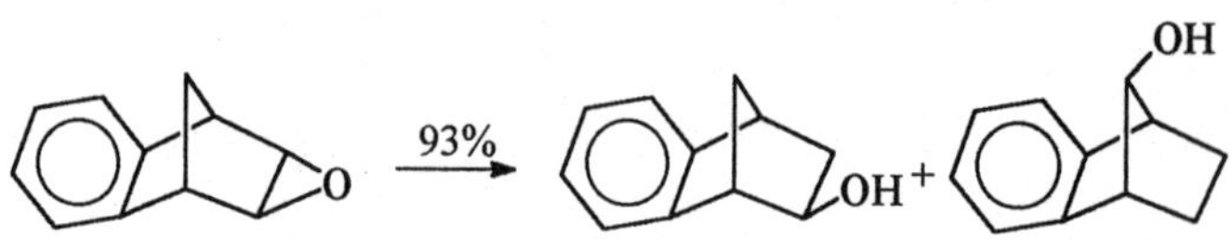

2.21

LAH	15	85
$LiEt_3BH$	93	<0.1

Figure 2.13

with BH_3 or $NaCNBH_3$ in the presence of $BF_3{\cdot}Et_2O$ or by $Zn(BH_4)_2$ on SiO_2. The other reagents give mixtures of primary and secondary alcohols. Such is also the case in the reduction of *cis*-2-methylstyrene oxide by LAH. Unexpectedly, reaction of this epoxide with $LiEt_3BH$ gives 1-phenyl-propanol [BN4] (Figure 2.12). The reduction of the epoxide of 1-methylcyclohexene under these conditions leads to *cis*-2-methylcyclohexanol **2.18** [HT1] (Figure 2.12).

In certain cases, whenever the Lewis acidity of the reagent is high enough and whenever the structure of the molecule is favorable, the reaction involves the formation of a carbocation, which can undergo migration leading from epoxide **2.19** to an aldehyde **2.20** that is later reduced [HT1] (Figure 2.13). The carbocation can rearrange in a different way so that the alcohol obtained has a modified carbon skeleton. Such is the case in the reduction of **2.21.** However, the use of $LiEt_3BH$ minimizes these rearrangements [G4] (Figure 2.13).

Epoxides undergo decomposition under the influence of the acyloxyboranes in organic acids [MM1]. Being bulky, LTBA in THF leaves the epoxide unattacked in the cold and leads to the selective reduction shown in Figure 2.14 [M1]. The

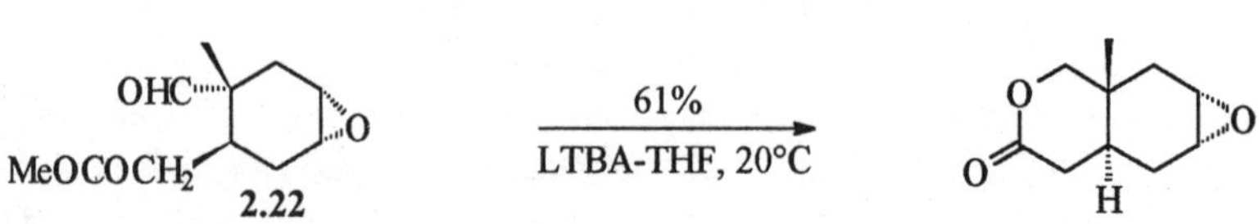

Figure 2.14

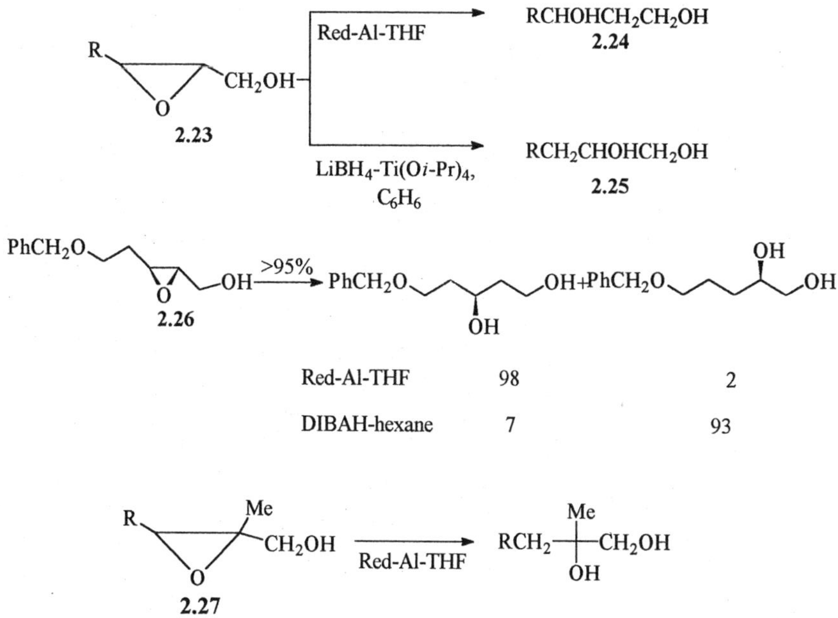

Figure 2.15

primary alcohol that is formed from the aldehyde **2.22** undergoes lactonization, but the epoxide and the ester are not reduced.

The presence of a functional group in the vicinity of the epoxide can lead to interesting results. Such is the case for the epoxy-2,3 alcohols **2.23,** which can be obtained in a nonracemic form by asymmetric epoxidation of the corresponding allylic alcohols [KS3]. The action of LAH in THF or better yet of Red-Al in the same solvent [MM2, V1] or preferably in DME [GS4] selectively leads to the 1,3-diols **2.24**, while DIBAH [FK1] or $LiBH_4$–$(i\text{-PrO})_4Ti$ in C_6H_6 [DL1] gives access to the 1,2-diols **2.25** (Figure 2.15). The hydride attack is stereospecific, and in the nonracemic chiral molecule **2.26**, the reaction proceeds with inversion [FK1] (Figure 2.15). If the alcohol residue is transformed into a methyl ether, Red-Al does not promote any reduction [FK1].

A limitation of the Red-Al method is steric hindrance. If the carbon atom bearing the primary alcohol is disubstituted such as in **2.27**, the other regioisomer is formed [V1] (Figure 2.15).

Vinylic epoxides such as **2.28** can be reduced by attack on the epoxide carbon atoms according to the usual rules, or they can undergo conjugate reduction, as shown in Figure 2.16 [LK1]. LAH attacks the epoxide at the least substituted carbon, and DIBAH in THF mainly attacks the epoxide at the most substituted one, whereas DIBAH in hexane gives only the conjugate reduction. Acetylenic epoxides are reduced by LAH into homopropargylic alcohols [HD1].

Epoxytosylates **2.29** can be reduced by DIBAH (3 equiv.) at −40°C in CH_2Cl_2 to

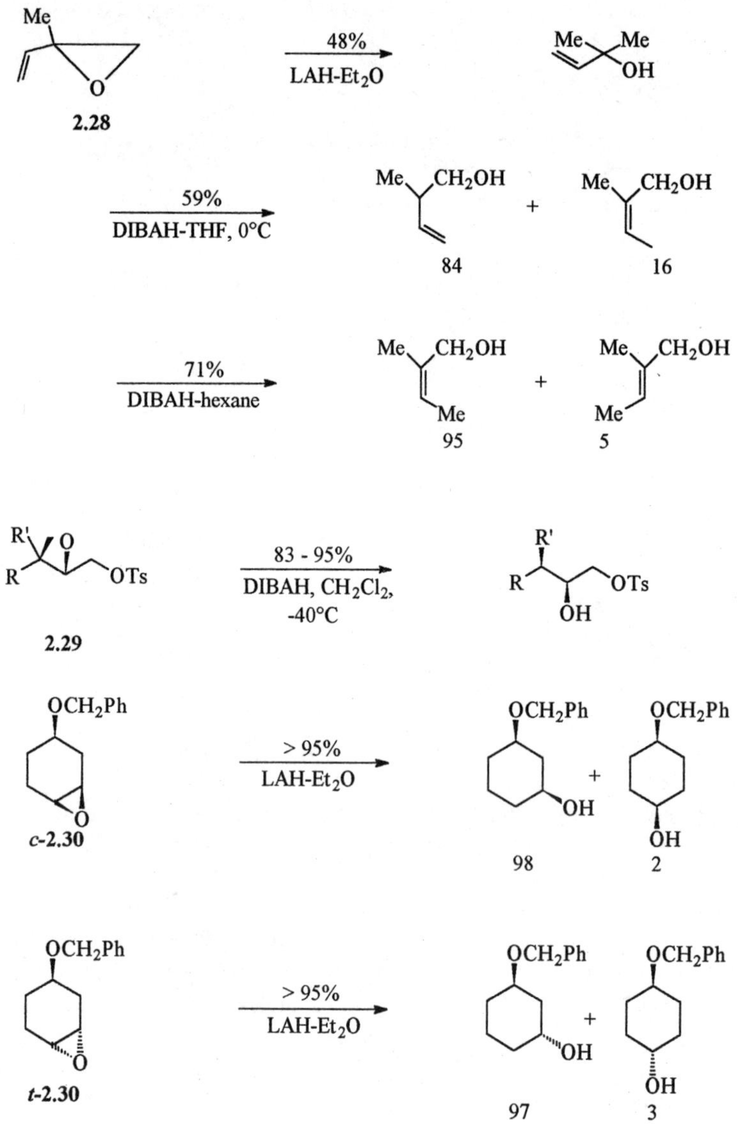

Figure 2.16

2-hydroxytosylates [CJ2]. This reaction is stereospecific. If the reaction is run in hexane, overreduction of the tosylate to a methyl group takes place (Figure 2.16).

The reduction of *cis*- or *trans*-4-benzyloxycyclohexeneoxide *c*-**2.30** and *t*-**2.30** with LAH in ether or pentane at room temperature is regioselective towards *cis*- *or trans*-3-benzyloxycyclohexanol (Figure 2.16). However, in the presence of 12-crown-4, which hinders chelation, the *cis* derivative is preferentially transformed into *cis*-4-benzyloxycyclohexanol [CC5, CC8] (Figure 2.16). Similar but less selective reductions are observed with five-membered analogs [CC8].

2-Methylglycidic acid is regio- and stereoselectively reduced to 2-deutero-3-hydroxybutanoic acid by $NaBD_4$-DO^- in D_2O [MV3], while in the presence of LiBr the regioselectivity is lower. F-Alkyl-α,β-epoxyesters are also reduced to diols by $NaBH_4$ in alcoholic media [LP1].

Oxetanes are also reduced by aluminohydrides [SP2]. When 2-substituted by an aryl group, the Lewis acidity of the reagent and the electronic character of the aryl substituent determined the relative amounts of primary and secondary alcohols so formed [BL5, SS8].

2.4 ALCOHOLS, ETHERS, AND ACETALS: $\geqslant$C—OR; 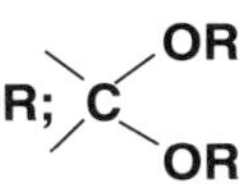

2.4.1 Alcohols

Alcohols are generally converted to alcoholates by the alumino- and borohydrides. The cleavage of the C—O bond can take place upon warming with Red-Al [M1], or it can occur under solvolytic conditions starting with appropriate alcohols such as benzylic or allylic alcohols that give stable carbocations. The carbocations thus formed are then reduced to hydrocarbons. Therefore, the diaryl- and triarylcarbinols are reduced by $NaBH_4$ in CF_3COOH [GN1] or $(CF_3COO)_2BH$ in THF-CF_3COOH [MM1] (Figure 2.17).

When using suitable experimental conditions, electron-donor-substituted primary benzyl alcohols can also be reduced to substituted toluenes [NB2]. However, $NaBH_4$–CF_3SO_3H in Et_2O is superior to $NaBH_4$ in CF_3COOH in reducing 2-aryladamantanols to the corresponding hydrocarbons [OW1]. Under the same conditions, adamantylmethanol leads to homoadamantane. Similarly, other carbocyclic substituted methanols give ring-expanded cycloalkanes [OW2].

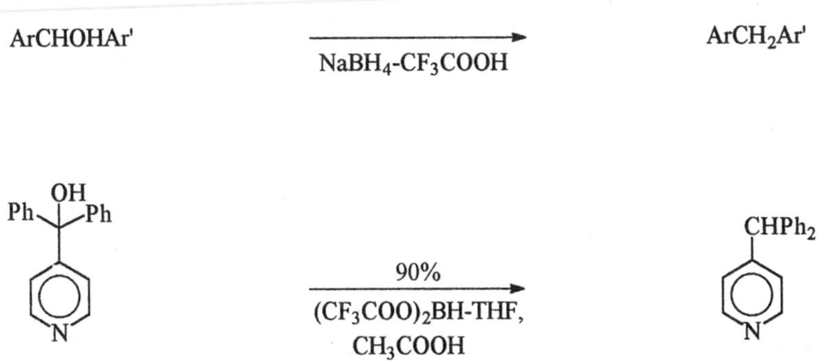

Figure 2.17

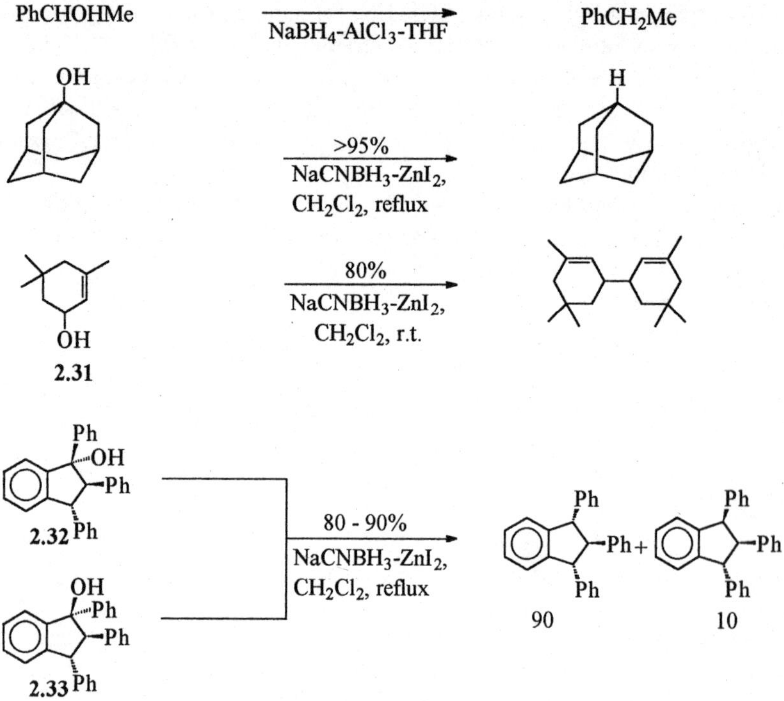

Figure 2.18

$NaBH_4$–$AlCl_3$ in THF or AlH_3 in Et_2O also reduce diarylcarbinols or the arylalkylcarbinols to hydrocarbons [E2, M1, OS2] (Figure 2.18). In the presence of $NaCNBH_3$–ZnI_2, allylic, benzylic, and even tertiary alcohols are reduced to hydrocarbons via the corresponding carbocations [LD1] (Figure 2.18). In some cases, the reaction probably takes place by a radical process. The dimerization observed from **2.31** and the stereoconvergence of the reduction of indan-1-ols **2.32** and **2.33** [AB1] have been interpreted in this way (Figure 2.18). Ferrocenyl alcohols suffer reductive deoxygenation with $NaCNBH_3$–$TiCl_4$ [B6].

Allyl alcohols can also be transformed into olefins by $NaCNBH_3$–$BF_3 \cdot Et_2O$, but some isomerizations can occur [SV1]. The reduction of primary alcohols as well as allyl and benzyl alcohols into hydrocarbons by $NaBH_4$ can be carried out via alkoxyphosphonium salts **2.34** generated in situ [HS6] (Figure 2.19).

The cobalt complexes derived from tertiary propargylic alcohols **2.35** are reduced by $NaBH_4$ in CF_3COOH to hydrocarbons via the corresponding carbocations, which, after decomplexation, yield substituted acetylenes. These compounds are much less easily prepared otherwise [N3] (Figure 2.19). The reduction can also be carried out by $BH_3 \cdot Me_2S$ in CF_3COOH [PL1]. Such methodology also applies to iron complexes [DS4].

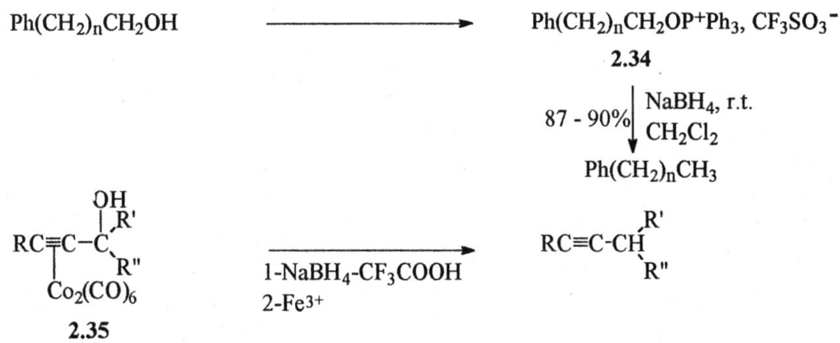

Figure 2.19

2.4.2 Ethers

Aliphatic ethers are generally inert in the presence of alumino- and borohydrides, although oxabicyclic [3.2.1] compounds are reduced by DIBAH [LC1]. Aromatic ethers can be cleaved to give phenols using DIBAH [MH2, W1], while benzylic ethers are cleaved by Red-Al in xylene under reflux [M1]. An example is given in Figure 2.20: The hindered ketone group of **2.36** is not reduced in these conditions. $LiEt_3BH$ or Li (s-Bu)$_3$BH in excess also cleaves arylmethylethers [MZ1]. The reaction is run in glyme or THF at 67°C. Selective demethylation of **2.37** and **2.38** can be performed (Figure 2.20). C—Cl or C—Br bonds are left intact.

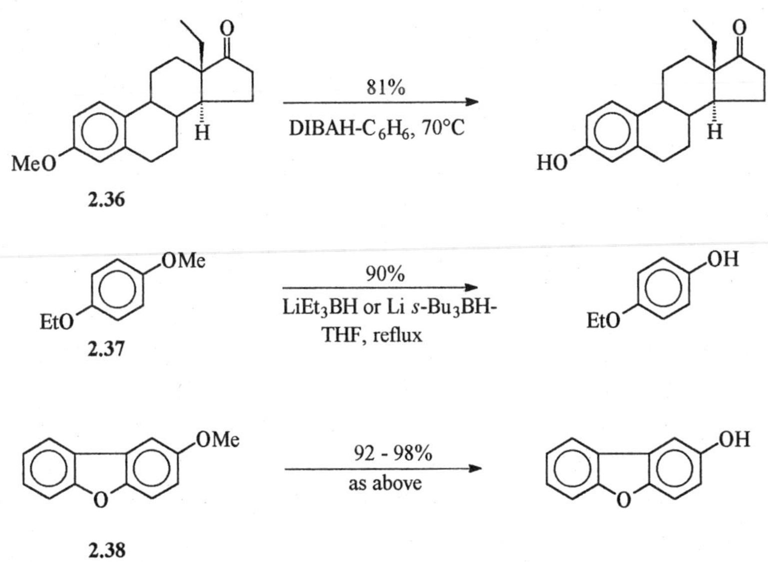

Figure 2.20

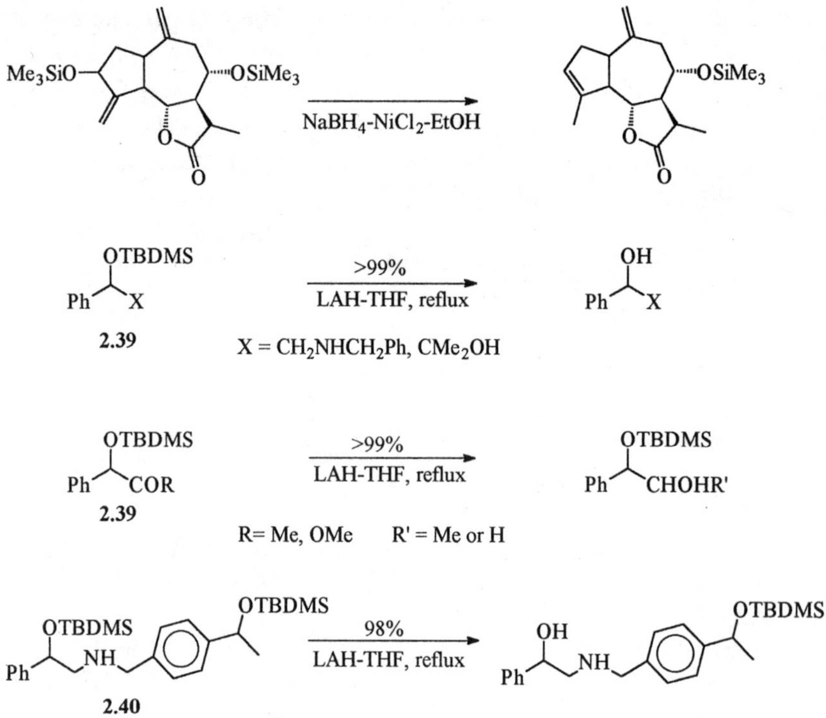

Figure 2.21

In the presence of $Pd(PPh_3)_4$, allyl ethers are reduced by LAH or $LiEt_3BH$ in THF (Section 5.3). Methyl or trimethylsilyl allyl ethers are reduced to the corresponding unsaturated hydrocarbons by $NaBH_4$–$NiCl_2$ in EtOH [GO2]. Lactones and epoxides remain intact under these conditions, as do ethers of saturated alcohols (Figure 2.21).

Trimethylsilyl ethers ($ROSiMe_3$) are cleaved by aluminohydrides and borohydrides [G3]. Aluminohydrides and DIBAH also reduce some unsymmetrical methylated silyl ethers [CG1, CG2, F1]. However, more hindered silyl ethers such as $ROSiMe_2t$-Bu or $ROSiPh_2t$-Bu seem to resist these reducing reagents, particularly if the reactions are carried out at low temperature [G3]. If a functional group bearing a relatively acidic proton (amine, alcohol) or precursor of an alkoxyaluminoborohydride (ketone, ester) is adjacent to the $OSiMe_2t$-Bu group, intramolecular cleavage occurs, so that LAH in THF can transform TBDMS ethers **2.39** into alcohols [VB1] (Figure 2.21). The selective cleavage of **2.40** has been easily performed (Figure 2.21).

2.4.3 Acetals and Orthoesters

Acetals associate with DIBAH at low temperatures and are cleaved into ethers only at room temperature or by heating, depending on the substrate [TA1, W1]. An

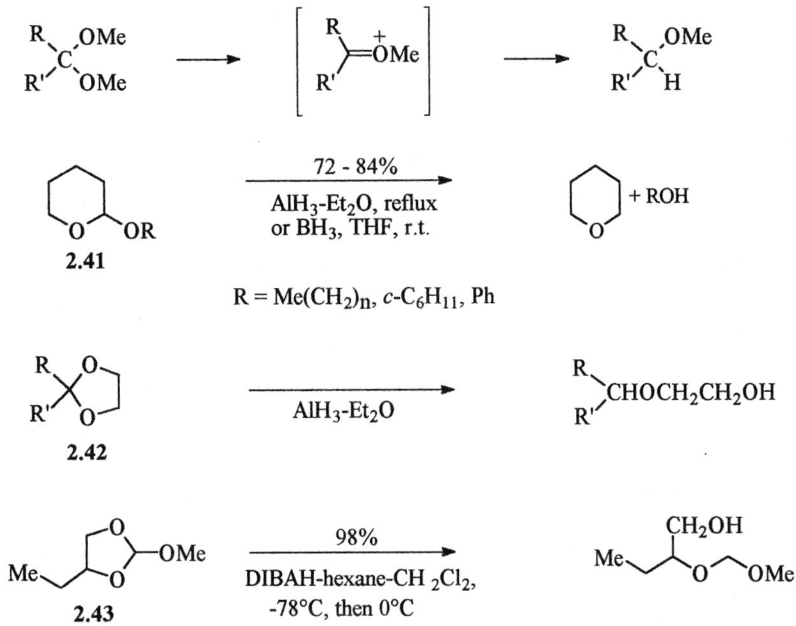

Figure 2.22

excess of reagent must be used. Similarly, orthoesters are cleaved by DIBAH at 0°C [TN1]. The reducing agents AlH_3 [E2, EB1, MK3], $BH_3{\cdot}THF$ [H3, L2, PS1], $BH_3{\cdot}Me_2S$–$BF_3{\cdot}Et_2O$ [SK6], $AlBr_2H$, $AlCl_2H$ [MF1], and $ClBH_2{\cdot}Me_2S$ [BB3] have sufficient Lewis acid character to convert acetals to ethers. The mechanism of these reductions involves the formation of an oxonium ion, which is then reduced (Figure 2.22). On the other hand, LAH and $LiEt_3BH$ leave acetals untouched, except in the presence of $TiCl_4$ [MA1, NG2] or other Lewis acids that eventually induce the formation of tricoordinated aluminohydrides.

An application of these reactions is the regeneration of alcohols from the tetrahydropyranyl (THP) ethers **2.41** used as protecting groups. Reaction of THP ethers with AlH_3 (most frequently formed in situ starting from LAH in ether solution by adding $AlCl_3$ or $BF_3{\cdot}Et_2O$) provides the expected alcohols. If the product alcohols are likely to form carbocations, these are in turn reduced to hydrocarbons (tertiary, benzylic, allylic alcohols) (Figure 2.22). Improvements in this reaction involve the use of $BH_3{\cdot}THF$ [CB8] or $NaCNBH_3$–$BF_3{\cdot}Et_2O$ [SS5]. Under the AlH_3 conditions, dioxolanes **2.42** lead to glycol monoethers (Figure 2.22).

The cleavage of orthoesters **2.43** is regioselective when they are dissymmetrical: Primary alcohols are formed rather than secondary ones [TN1] (Figure 2.22).

Other reducing reagents leave acetals intact [BK5, M1] except in acid media, where the oxonium ions are likely to be generated. Oxonium ions can be formed and reduced with $NaCNBH_3$ in MeOH–HCl, $NaCNBH_3$–$BF_3{\cdot}Et_2O$ [HJ3, NG2, SV2], $NaBH_4$–CF_3COOH in THF, $NaCNBH_3$–CF_3COOH in DMF, or $NaCNBH_3$–

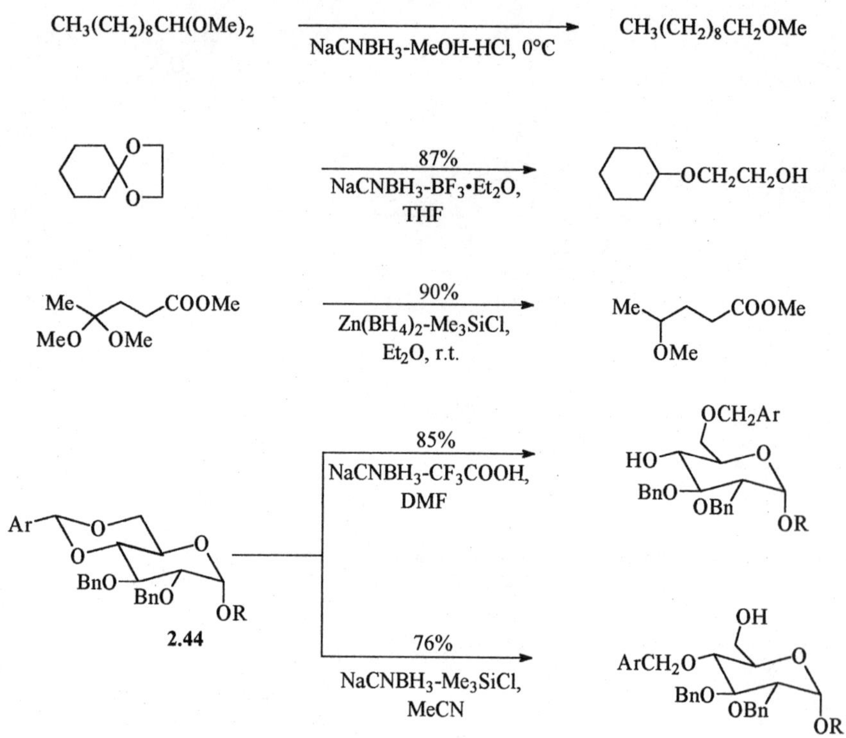

Figure 2.23

Me_3SiCl in MeCN [GN1, JS1, MK3]. The last three reagents cleave only the dioxolane derivatives of aromatic ketones, whereas the first ones are more reactive and have a broader scope of application (Figure 2.23).

At ambient temperature, $Zn(BH_4)_2$ in Et_2O, in the presence of trimethylsilyl chloride, transforms acetals and ketals into ethers. Under these conditions, esters remain intact, but the double bonds are reduced [KU1] (Figure 2.23). According to the experimental conditions, the regioselectivity of the cleavage can vary [JS1], as shown by the reduction of sugar derivative **2.44** (Figure 2.23).

$NaCNBH_3$–$TiCl_4$ can also be used for acetal reduction. Esters are left unchanged under these conditions. For instance, a benzylidene acetal formed from tartaric acid **2.45** can be transformed into a chiral monoether [AS1] (Figure 2.24). Other examples are described in the literature [MA1]. The carbon–oxygen bond of hemithioketals **2.46** is hydrogenolyzed by AlH_3 in Et_2O, while the CUS bond remains unchanged [E2] (Figure 2.24). Dithianes, in contrast, are not affected by aluminohydrides, borohydrides, or boranes [NG2].

The treatment of the acetal derivatives of chiral diols **2.47** with DIBAH, or better with $AlHBr_2$ or $AlHCl_2$ (obtained by combining 3 equivalents of AlX_3 with LAH in Et_2O), leads to an enantioselective cleavage to an ether–alcohol **2.48**, which can then be oxidized to a nonracemic alcohol with Swern reagent [MF1, MI1] (Figure

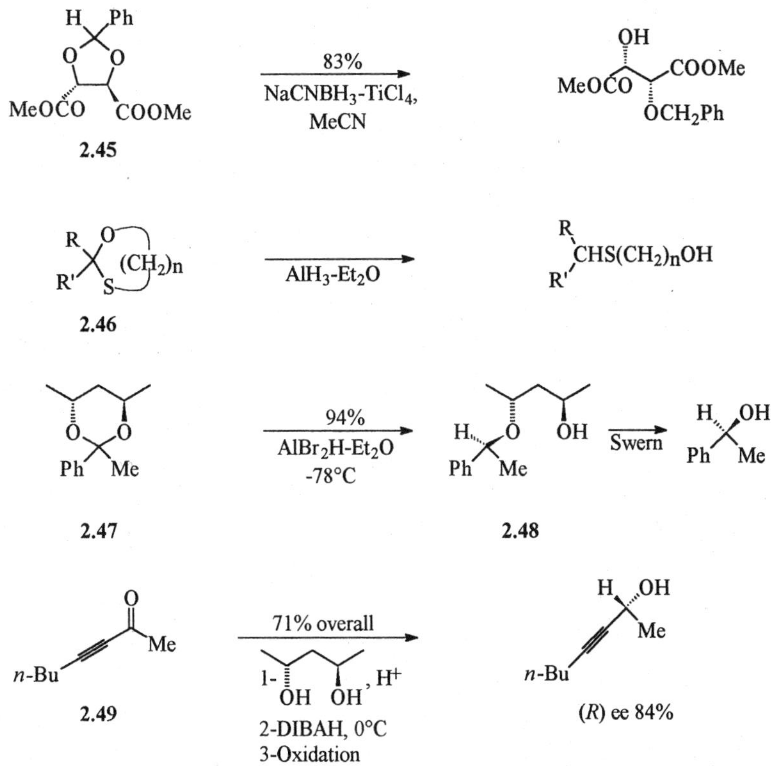

Figure 2.24

2.24). The method is applicable to acetals of α,β-acetylenic ketones such as **2.49** [IM1] (Figure 2.24). The other enantiomers can be obtained by using $Et_3SiH–TiCl_4$ on the same acetals [IM1, IM2, MI1]. This methodology provides a reduction of ketones to chiral alcohols that is complementary to the reduction effected by the chiral alumino- and borohydrides (Section 3.2.3).

The stereoselectivity of the reduction of bicyclic 1,3-dioxolanes **2.50** has been studied [IM3, KU2, KU3]. DIBAH in excess leads predominantly to *trans*-substituted isomers, while $Et_3SiH–TiCl_4$ or $Zn(BH_4)_2–TiCl_4$ in CH_2Cl_2 gives the reverse stereoselectivity (Figure 2.25). Other reagents such as $AlHBr_2$ or $AlHCl_2$ are less stereoselective [KU4].

A solvent effect has been observed in the DIBAH reduction of the bicyclic 1,3-dioxane **2.51** [IM3] (Figure 2.25). In $CHCl_3$ or CH_2Cl_2, the *trans* isomer **2.53** is predominantly formed, while in THF, the *cis* isomer **2.52** is the major product. With $AlHBr_2$ in Et_2O, the stereoselectivity is even higher [IM3] (Figure 2.25). These results have been interpreted in terms of conformational effects, emphasizing the importance of the size of the acetal ring (six-membered vs. five-membered).

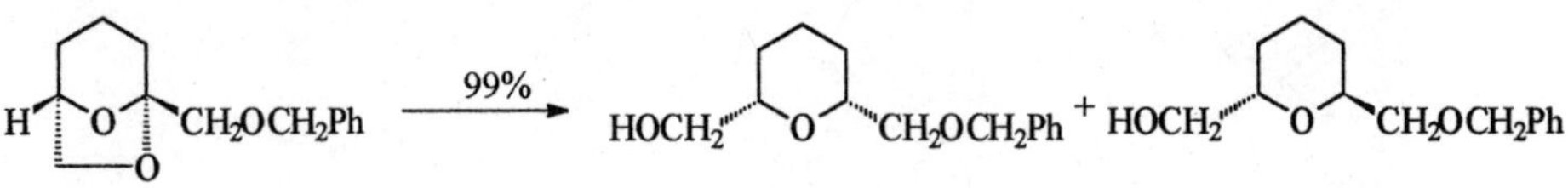

2.50

DIBAH-CH_2Cl_2, 0°C	4	96
Et_3SiH-$TiCl_4$ or $Zn(BH_4)_2$-$TiCl_4$ CH_2Cl_2, -78°C	99	1

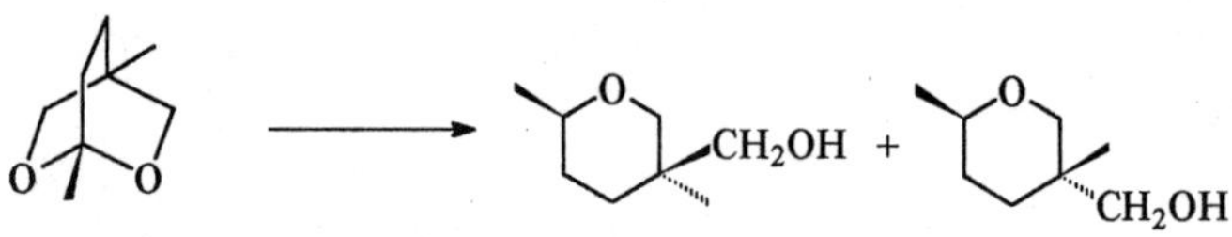

2.51 2.52 2.53

	2.52	2.53
DIBAH-$CHCl_3$	8	92
DIBAH-hexane	44	56
DIBAH-THF	82	18
$AlHBr_2$-Et_2O	97	3

Figure 2.25

2.4.4 Ozonides

Ozonides are reduced to alcohols by $LiAlH_4$ [MS3] and $NaBH_4$ in alcohols [CH4, H3]. However, it appears that the best reagent is $BH_3{\cdot}Me_2S$ in CH_2Cl_2 at room temperature, a method that is compatible with carboxylic esters [FG1] (Figure 2.26). Ozonolysis of olefins followed by reduction can be performed sequentially in a one-flask operation.

2.5 AMMONIUM SALTS: N^+R_3,X^-

LAH in THF reduces ammonium salts to amines. However, the best reagent for this reaction is $LiEt_3BH$ in THF at 25°C. Methylammonium salts are selectively demethylated [BK5, CP1, NM1, PS1] (Figure 2.27). This method has found numerous applications in synthesizing natural products [NM1].

The reduction of ammonium salts, formed by reacting Me_2SO_4 with benzylic Mannich bases **2.54**, by $NaCNBH_3$ in HMPA at 70°C, leads to methylated aromatic

$$CH_2{=}CH{-}(CH_2)_8COOMe \xrightarrow[\text{1- } O_3 \text{; 2- } BH_3{\bullet}Me_2S\text{-}CH_2Cl_2]{78 - 96\%} HOCH_2(CH_2)_8COOMe$$

Figure 2.26

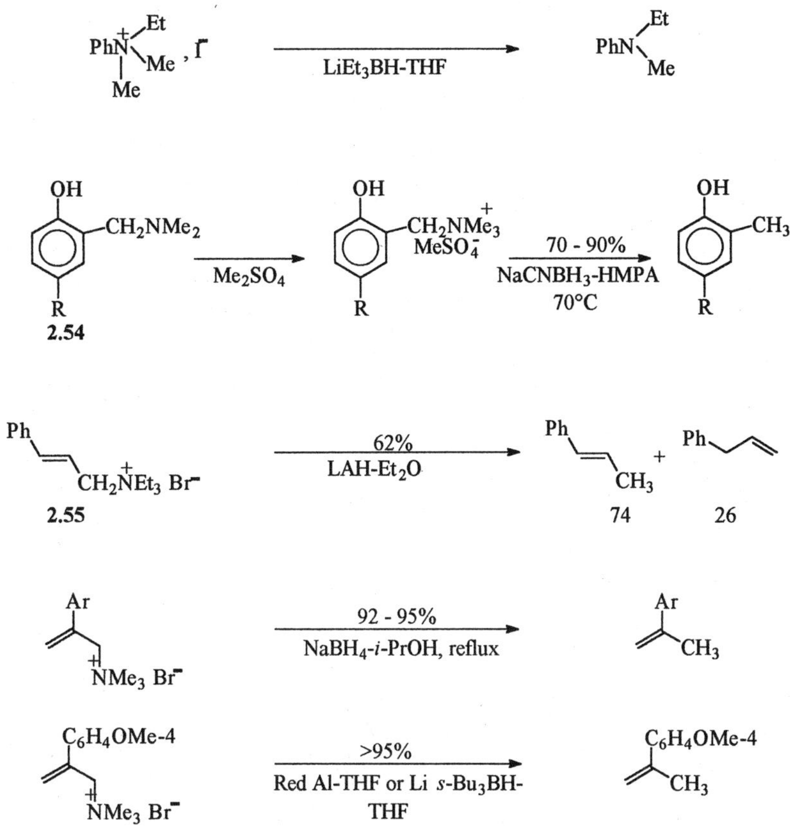

Figure 2.27

derivatives [YI1] (Figure 2.27). This method preserves the R group in the following cases: Cl, COOEt, CH_2CN, and NO_2. Reduction of allylic derivatives such as **2.55** with LAH in an ether medium can lead to mixtures of regioisomers [HN2] (Figure 2.27). If milder reducing agents such as $NaBH_4$, Red-Al, or Li(*s*-Bu_3)BH are used, the reaction can be regioselective [GL6] (Figure 2.27). $NaCNBH_3$ in *i*-PrOH under reflux does not react at all with ammonium salts.

2.6 PHOSPHORUS DERIVATIVES: $\geqslant$C—P

Single P—C bond cleavage has been described. For example, phosphonium salts **2.56** are reduced to phosphines by LAH in THF under reflux; the cleaved bond corresponds to reduction of the most stable carbanion [H2] (Figure 2.28). The P—C bonds of allylic phosphonates **2.57** or phosphonium salts can also be cleaved by

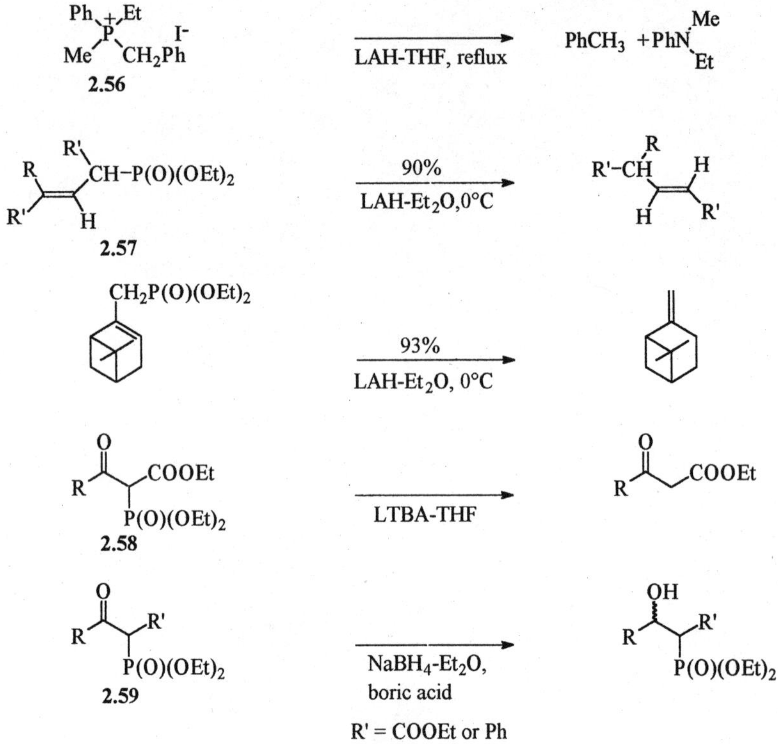

Figure 2.28

LAH in Et_2O. This reduction occurs with allylic transposition and leads to a *trans*-olefin [HJ2, HN2, KN2] (Figure 2.28).

LTBA can also cleave selectively the P—C bond of acyl phosphonates **2.58**, while preserving other functional groups [DS1] (Figure 2.28). Nevertheless, the reduction of similar compounds such as **2.59** by $NaBH_4$ in EtOH buffered by boric acid does preserve the P—C bond and leads to diastereomeric α-phosphorylated alcohols [BS5, DS1] (Figure 2.28). After enolization by NaH, the P—C bond cleavage of acylphosphonates $(EtO)_2P(O)CH_2COR$ can be realized with LAH [HS7].

Chapter 3

Reduction of Double Bonds

3.1 NONCONJUGATED CARBON–CARBON DOUBLE BONDS: $>C=C<$

Boranes add to carbon–carbon double bonds even if they are not activated by an electron-withdrawing group. These hydroboration reactions lie outside the scope of this book; nevertheless, it is important to recognize that most boranes cannot be used when it is necessary to preserve the C=C bond in a molecule (unless it is particularly hindered). However, $(CF_3COO)_2BH{\cdot}THF$ leaves the double bonds of styrene, 1-decene, and $Ph_2C{=}CH_2$ intact [MM1].

The other hydrides and borohydrides do not affect the isolated C=C bonds except in the presence of transition metals [CY2, GO2, M1, W4]. As shown in Figure 3.1, only the least hindered double bond of **3.1** is reduced (Figure 3.1). Unsubstituted and substituted styrenes can be reduced by $LiEt_3BH$ in hot THF or by $NaBH_4$–$BiCl_3$ [RP2, RP4], but the Selectrides, $Li(s\text{-}Bu)_3BH$ and $K(s\text{-}Bu)_3BH$ leave them intact. The double bonds of certain allene alcohols **3.2** and **3.3** can also be reduced [M1] (Figure 3.1). However, mixtures of alcohols and dienes are formed.

The reduction of the double bonds conjugated to electron-withdrawing groups is examined later (Section 3.2.9).

3.2 CARBON–OXYGEN DOUBLE BONDS: $>C=O$

3.2.1 Aldehydes and Ketones

Aldehydes and ketones are generally reduced to primary and secondary alcohols by all the reagents studied with the following exceptions:

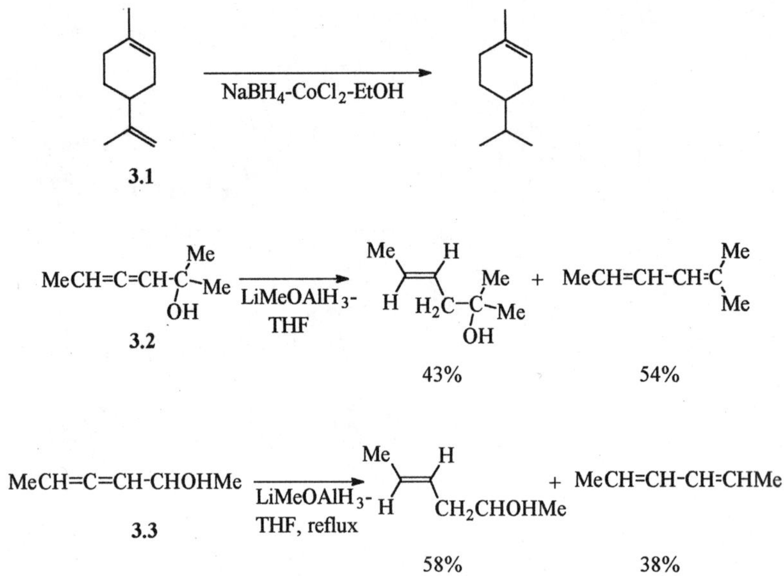

Figure 3.1

- Reduction with sodium and ammonium cyanoborohydrides in neutral or basic protic media [L1] allows the reduction of halides while leaving the carbonyl groups intact (Section 2.1);
- Reduction with $(Ph_3P)_2CuBH_4$ in neutral media allows the selective reduction of acid chlorides [FH1] (Section 3.2.7). The reduction of aldehydes by this complex, however, takes place in an acid medium or in the presence of Lewis acids.

Although single-electron transfer is proposed in the reduction of aromatic ketones by AlH_3, BH_3, and LAH-pyridine [AG2], the reductions of aldehydes and ketones by alumino- and borohydrides and boranes occur mostly by nucleophilic attack of hydride on the carbonyl carbon. This process has been the subject of numerous theoretical [ES1, HW1, N2, W2] and mechanistic [CB1, N5, W2, W4] studies.

In certain cases, the reduction can take place without electrophilic catalysis (n-Bu_4NBH_4 or phase-transfer conditions), but most frequently it requires the coordination of the carbonyl group by a Lewis acid before nucleophilic attack [S2]. The Lewis acid may be the cation associated with the reagent, an added acid, or even the boron or aluminum atom of tricoordinate reagents (AlH_3, DIBAH, boranes). The importance of this phenomenon has been shown by the introduction of coordinating macrocyclic molecules into solutions of LAH and $LiBH_4$. This considerably retards the reduction of carbonyl compounds in an ether medium [DC1, HP1]. Electrophilic catalysis is more important when the lowest unoccupied molecular orbital (LUMO)

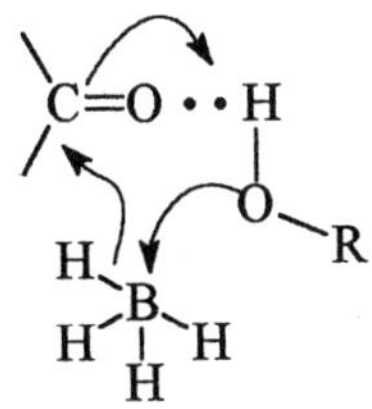

Figure 3.2

of the carbonyl compound is relatively high lying [LS2]; electrophilic assistance by the Li cation lowers the LUMO level. The observed sequence of the relative binding strength to the Li cation is the following [LS2]: cyclohexanone > 4-MeC_6H_4CHO > PhCHO > 4-ClC_6H_4CHO.

In protic media, it is the solvent that plays the role of the acid catalyst and provides electrophilic assistance by hydrogen bonding [ES1, PS1, W2]. In alcoholic media, it has been shown that the transition state for the reduction by borohydrides involves an alcohol molecule that is converted to the corresponding borate (Figure 3.2).

Reductions by LAH in ether solvents have transition states that are reactantlike [HW1, N2], whereas for reductions involving borohydrides, the transition states occur later along the reaction coordinate [CB1, ES1, W2, YH1]. With reagents having tetracoordinated aluminum or boron, the formation of the C—H bond is the rate-determining step. Chloral (Cl_3CCHO), whose lowest-lying vacant orbital is low in energy, is indeed reduced more rapidly than pivalaldehyde (Me_3CCHO), whose LUMO lies higher in energy [BK5]. A review of these mechanistic considerations has been published [W6].

In contrast, with reducing agents whose central atom is tricoordinated and thus display a strong Lewis acidity (boranes, AlH_3, DIBAH), the coordination of the reactants with the carbonyl oxygen is the dominant factor controlling rate [D3]. Pivalaldehyde, whose carbonyl oxygen is more basic, reacts more rapidly with BH_3·THF than does chloral [BK5]. This difference in behavior has some important implications with regard to the stereoselectivity of these reductions by these two types of reagents (Section 3.2.2).

Aldehydes and ketones may be reduced to alcohols by LAH and SAH in an ether medium [BK5, CB5], by LAH on a solid support [KH2, W4], and by alkoxy- and aminoaluminohydrides [M1, YA2], AlH_3 in Et_2O or AlH_3·Et_3N [CB7], Red-Al in C_6H_6 [M1], borohydrides in the solid state [TK2], under PTC conditions [BI1, BK8, FR2, GB5, IL2, ML1, YP3], in alcoholic media or ethers or glymes [BK5, R3]), by aminoborohydrides [FF2, FH5], by boranes and acyloxyboranes [BK5, GN1, IW1, KB7, MM1, PS1], and by trialkylborohydrides [BK5], or ate complexes [BM1, KA1]. In the presence of metal alkoxides, the rate of reduction of ketones by catecholborane or BH_3·THF is enhanced [LD2].

The reduction by alkaline cyanoborohydrides takes place at pH values less than 4 [L1]. The reduction of ketones by $Zn(CNBH_3)_2$ is efficient only in Et_2O [KK5] and

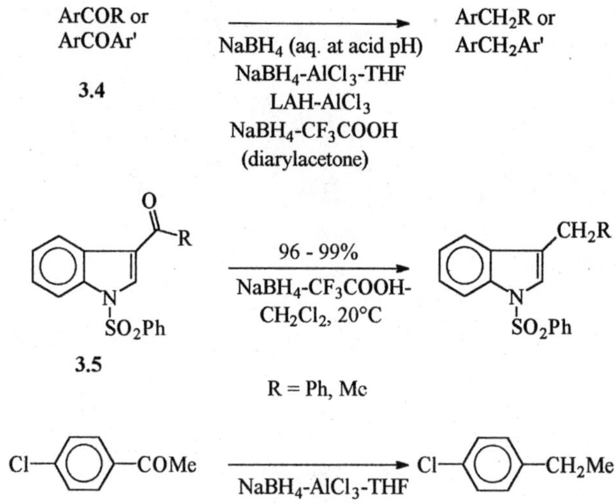

3.6

Figure 3.3

is relatively slow when $NaBH_4$ is used in organic acid media [GN1]. These reductions are in general sensitive to steric hindrance around the carbonyl, so that the experimental conditions must be appropriate.

In acid media or in the presence of Lewis acids, diaryl ketones and alkylaryl ketones **3.4–3.6** are reduced by LAH or $NaBH_4$ to the corresponding hydrocarbons [DN2, E2, GK1, GN1, OS2] (Figure 3.3). However, LAH–$AlCl_3$ gives poorer yields [GN1, KG1, KL1]. $NaBH_4$–$ZrCl_4$ in THF reduces PhCHO and $PhCOCH_3$ to the corresponding alcohols [IS1]. Flavanones are reduced by $NaCNBH_3$ in CF_3COOH either to flavanes or 1,3-diarylpropanes, depending on their substituents [LB1]. Aromatic aldehydes and ketones substituted by electron-donating groups are also reduced to hydrocarbons by $BH_3{\cdot}THF$ [L2] and $NaCNBH_3$–ZnI_2 in CH_2Cl_2 [LD1]. $NaBH_4$ in MeOH in the presence of $PdCl_2$ gives analogous results [GO2] as does ion-exchange resin borohydride–$Ni(OAc)_2$ in MeOH [BK9]. Acylferrocenes are reduced to the corresponding hydrocarbons by AlH_3, $NaCNBH_3$–$TiCl_4$ in THF, or $NaBH_4$–CF_3COOH [B6, B7, RE2].

Arylalkyl ketones (ArCOR) do not react with $Me_3N{\cdot}BH_3$ alone, but benzyl bromides ($ArCHBrCH_3$) are formed when the reaction is run in the presence of Br_2 [LG1]. Moreover, t-$BuNH_2{\cdot}BH_3$–$AlCl_3$ in CH_2Cl_2 also reduces arylalkyl ketones to the corresponding hydrocarbons ($ArCH_2R$). This transformation is compatible with ester groups, as well as chloro, bromo, nitro, and phenylthio substituents in the aryl ring; however, carboxylic acids are reduced to primary alcohols [LT1].

It is possible to reduce aldehydes and ketones selectively in the presence of isolated double bonds, halides, sulfonates, phosphonates, acetals, esters, amides, nitriles, acids, and NO_2 groups by using $NaBH_4$ or n-Bu_4NBH_4 in various media [DA3, JB2, PS1, TY3, WG2]. The examples given in Figure 3.4 illustrate this

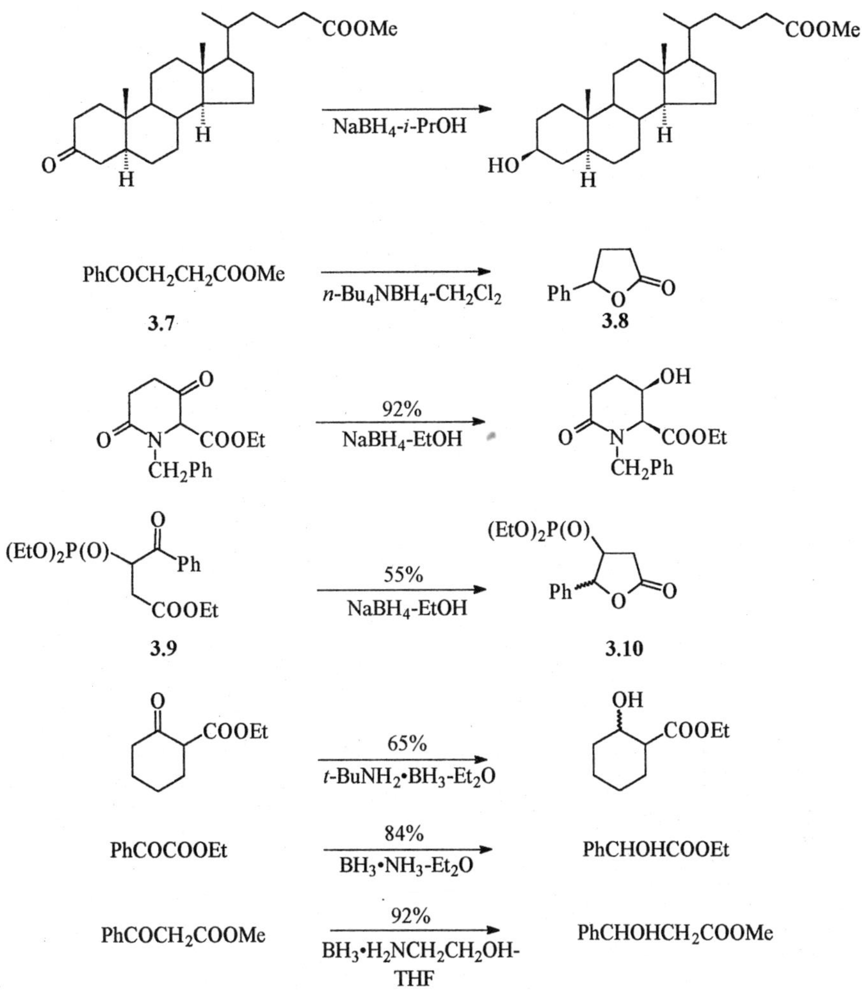

Figure 3.4

compatibility. The in situ formation of lactones **3.8** and **3.10** results from the selective reduction of the ketone functionality of **3.7** and **3.9**. Likewise, the amine- or amino alcohol–boranes effect the reduction of ketones in the presence of esters [A1, IW1] (Figure 3.4). On the other hand, $LiBH_4$ also reduces the ester group of **3.9** [JB2].

$NaBH_4$ on alumina appears to be a very mild reducing agent. Under these conditions, it is possible to avoid the hydrolysis of esters, which sometimes occurs due to the basicity of $NaBH_4$ in aqueous–alcoholic solutions. This is particularly interesting for the case of enol acetates such as **3.11**, which are fragile in protic media [W4] (Figure 3.5). The selective reduction of the chlorinated ketolactone **3.12** is another illustration of this useful selectivity [WV1] (Figure 3.5). Borane–eth-

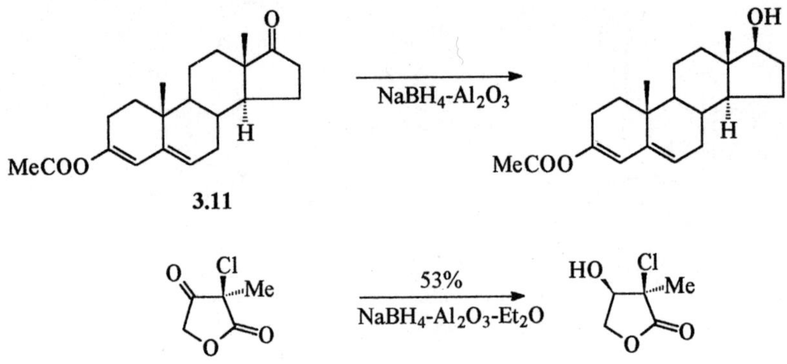

3.11

3.12

Figure 3.5

anolamine also leaves C—Cl and C—Br bonds untouched so that 2-haloketones can be transformed into halohydrins [IW1].

LAH on silica gel reduces ketoesters to hydroxyesters in Et_2O [KH3]. The reduction of epoxy ketones to epoxy alcohols is easily accomplished by action of $Zn(BH_4)_2$ in Et_2O or $NaBH_4$ in MeOH, sometimes in the presence of $CeCl_3$. The stereoselectivity of the reaction is usually high [BB6, BC2, CP3, NT1] (Section 3.2.4).

The selective reduction of aldehydes in the presence of ketones has been observed using the following systems:

- $NaBH_4$ in cold *i*-PrOH [BK5], or in EtOH–CH_2Cl_2 at −78°C [WR2];
- *n*-Bu_4NBH_4 in CH_2Cl_2 [RG1, SP1] or exchange resin borohydride in MeOH [GB5, YP3];
- *n*-Bu_4NCNBH_3 in an aqueous 0.1 N HCl solution [W3];
- $(Ph_3P)_2CuBH_4$ in acid medium [FH1];
- $Na(AcO)_3BH$ or better *n*-$Bu_4N(AcO)_3BH$ in C_6H_6 under reflux [GF1, GN1, NG1], as shown in Figure 3.6 (if the molecule bears an alcoholic functional group at the α or β position to the ketone, this is also reduced [SM2]);
- $Na(OAr)_3BH$ in THF [YK1];
- $Li(OCEt_3)_3AlH$ or LBTA in THF [K4, M1];
- $BH_3{\cdot}Me_2S$ or BH_3–LiCl [HC1, YC1];
- $Zn(BH_4)_2$ in THF at −10°C [R3] (Figure 3.6);
- Amine–boranes in Et_2O at 0°C, *t*-$BuNH_2{\cdot}BH_3$ [A1] being the most effective; pyridine–borane on Al_2O_3 [BS1] or in the presence of AcOH [CW1];
- $NaBH_4$–$SnCl_2$ in THF, which reduces aromatic aldehydes without affecting ketones [OH1].

A noteworthy result is that ketones can be reduced without affecting aldehyde groups by using $NaBH_4$–$CeCl_3$ in aqueous MeOH or EtOH at −15°C [GL1, GL2].

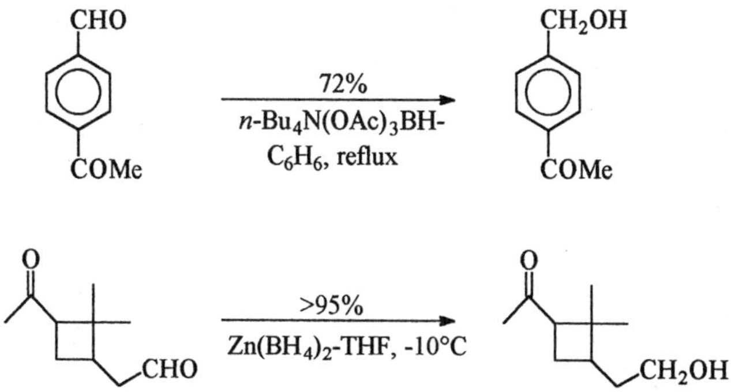

Figure 3.6

This is due to a rapid and selective transformation of aldehydes under these conditions to ketals or hemiketals, which are not reduced (Figure 3.7). The same type of situation permits the relatively rapid formation of ketals of unhindered ketones in the presence of $HC(OEt)_3$, whereby the selective reduction of the most hindered ketone group of **3.13** is possible [GL1] (Figure 3.7).

Because of the sensitivity of the reduction of some ketones to steric hindrance, it is possible with a judicious choice of reducing reagents to reduce selectively the least hindered carbonyls in a di- or triketone. The most effective reducing reagents in this regard are the complex $Zn(BH_4)_2 \cdot 1.5$ DMF in MeCN [HJ1], amine–boranes in Et_2O [A1], LTBA in THF [M1] or $K(s\text{-}Bu)_3BH$, as shown in Figure 3.8. The examples **3.14–3.17** are chosen from steroid series, in which the ketone at the 3-position is selectively reduced in accord with the stereochemical rules to be discussed later (Section 3.2.2) [GO1, TK1, WB1]. In the case of progesterone **3.17**, it is necessary to use the very bulky K Sia_3BH to observe the selective reduction of

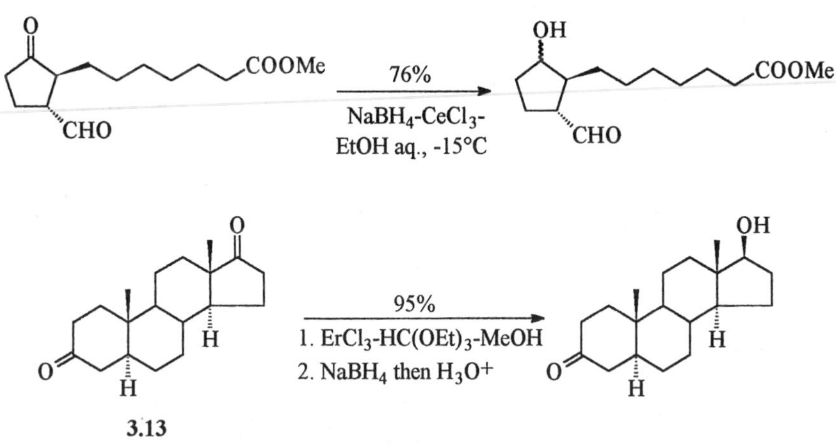

3.13

Figure 3.7

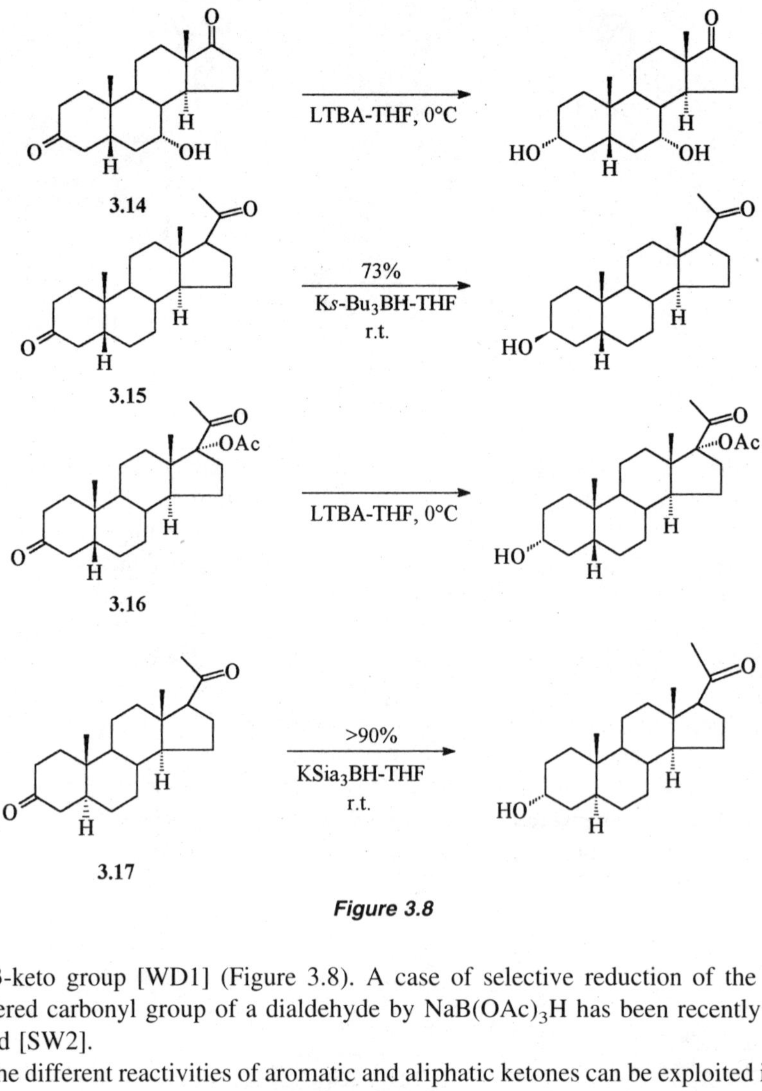

Figure 3.8

the 3-keto group [WD1] (Figure 3.8). A case of selective reduction of the most hindered carbonyl group of a dialdehyde by $NaB(OAc)_3H$ has been recently published [SW2].

The different reactivities of aromatic and aliphatic ketones can be exploited in the same way by carrying out the selective reduction of the latter. Reduction with $Zn(BH_4)_2{\cdot}1.5$ DMF in MeCN or $Zn(CNBH_3)_2$ in Et_2O in the presence of trace amounts of water is a good choice [HJ1, KK5], as is $NaBH_4$ in *i*-PrOH or $LiBH_4$ in diglyme [PS1] (Figure 3.9). Titanocene borohydride also reduces aliphatic ketones faster than aromatic ones [BS6].

If very bulky Lewis acids such as methylaluminum *bis*-(2,6-di-*t*-butyl-4-methylphenoxide) (MAD) are added, the reverse reactivity is observed. Selective complexation of the most accessible carbonyl group takes place, so that this group is no longer accessible for reduction. Under these conditions, DIBAH or $AlHBr_2$ reduces PhCO-*t*-Bu in the presence of PhCOMe or camphor [MA2]. However, discrimination between an aldehyde and a ketone is unsuccessful under such conditions.

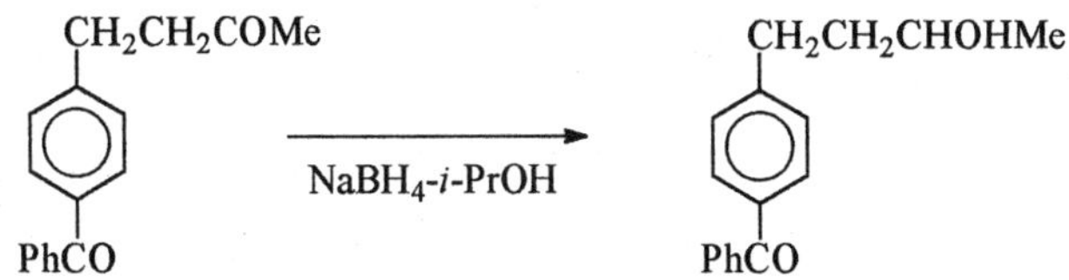

Figure 3.9

The competition between ketone and α-enone will be examined later (Section 3.2.9).

3.2.2 Stereoselectivity of the Reduction of Aldehydes and Ketones

The stereoselectivity of the reductions of aldehydes and ketones has been the object of in-depth mechanistic and theoretical studies [BR4, CB1, CH5, CL4, CL5, ES1, HW1, M5, N2, N5, NN1, W2, WH1, WT1]. According to the Lewis acid strength of the reducing agent, two models can interpret the observed results:

- When the reductions are carried out with reagents whose central atom (Al or B) is tetracoordinated, the Felkin–Anh model is usually invoked [CP2, HW1, M5, MW1, N2, NN1, WH1, WH2, WT1]. This aids in comparison of the interactions involved in the nucleophilic attack of the hydride on the C=O bond;
- When the reductions are carried out by means of reagents whose central atom is tricoordinated, Houk's model is often used. This model takes into account the predominant interactions during the coordination of the carbonyl oxygen with the Lewis acid before any hydride transfer [HP3, HW1, M5, NN1].

When the reduction substrate can follow either of the two pathways, it is possible to obtain selectively one isomer or the other by carrying out the reduction either with an alumino- or borohydride, or with DIBAH or a borane. The first author to exploit this dichotomy was M. Midland [MK1]. The principal types of interactions to be taken into consideration are those of stereoelectronic origins, steric, torsional, and orbital interactions, and also the position of the transition state (early or late) along the reaction coordinate (see the following).

***The Felkin–Anh Model* [CP2, G5, M5, N2, N5, NN1, WH1, WH2, WP2, WT1]** The attack of a hydride H^- on a prochiral carbonyl group can be accomplished either on the *Re* or *Si* face of the carbonyl, leading to a pair of diastereomers, as shown in the models in Figure 3.10. In these models L represents the most bulky group, P the most polar, and S the smallest group. Initially Cram [CA2] proposed model **3.18** to interpret the formation of the major isomer. This is called "Cram" in Figure 3.10, and the other isomer is labeled "anti-Cram."

The model to which most authors actually refer is a modification of the 1952 Cram scheme. This transition-state model, proposed by Felkin and Cherest and

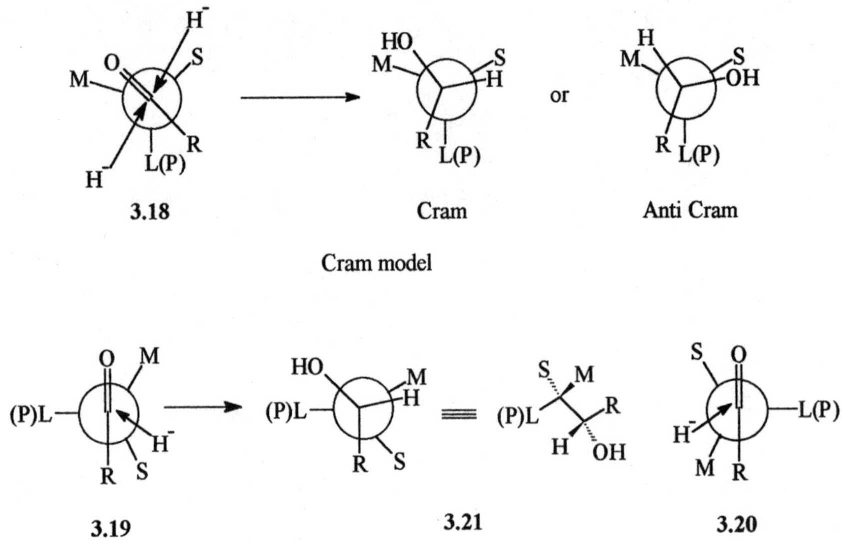

Felkin-Anh model

Figure 3.10

supported by calculations of Nguyen Trong Anh and Eisenstein, considers that the transition state most resembles the ketone and hydride reagents. The attack of the hydride takes place *anti* to the most bulky (L) or polar (P) group. In agreement with the proposals of Dunitz and Burgi, this does not take place perpendicular to the plane of the carbonyl group, but with an attack angle of about 109°. To minimize steric and torsional interactions, the attack preferentially involves the ketone conformer **3.19** and not **3.20**, as indicated in the Newman projections shown in Figure 3.10. The favored attack on **3.19** thus leads predominantly to the "Cram" stereoisomer **3.21** (Figure 3.10).

In the absence of other steric constraints, stereoisomer **3.21** is favored when the reductions are performed on acyclic prochiral ketones. Several examples are given in Figure 3.11 [SK1]. The stereoselectivity of the reduction is improved as the reducing reagent becomes more bulky. In the reduction of **3.22** and **3.23**, the phenyl or unsaturated substituent plays the role of the bulky (L) or polar (P) groups; the methyl group being M and hydrogen being S. Some other examples are given by Cherest and Prudent [CP2]. However, if the carbon skeleton of the ketone to be reduced is substituted in such a manner that conformer **3.19** is very sterically hindered, the reduction takes place on conformer **3.20**, and the stereoselectivity is reversed. Examples of this are given in steroid series (Figure 3.11). When R is an unsaturated group such as in **3.24**, the interaction between this substituent and the axial methyl at position 18 is not sufficient to disfavor the participation of conformer **3.19** during the reduction; by using Li(s-Bu)$_3$BH, which is bulky, one obtains preferentially the (S)-22-alcohol [TO1]. However, when R is a branched saturated chain as in **3.25**, a steric interaction between this chain and the 18-methyl group

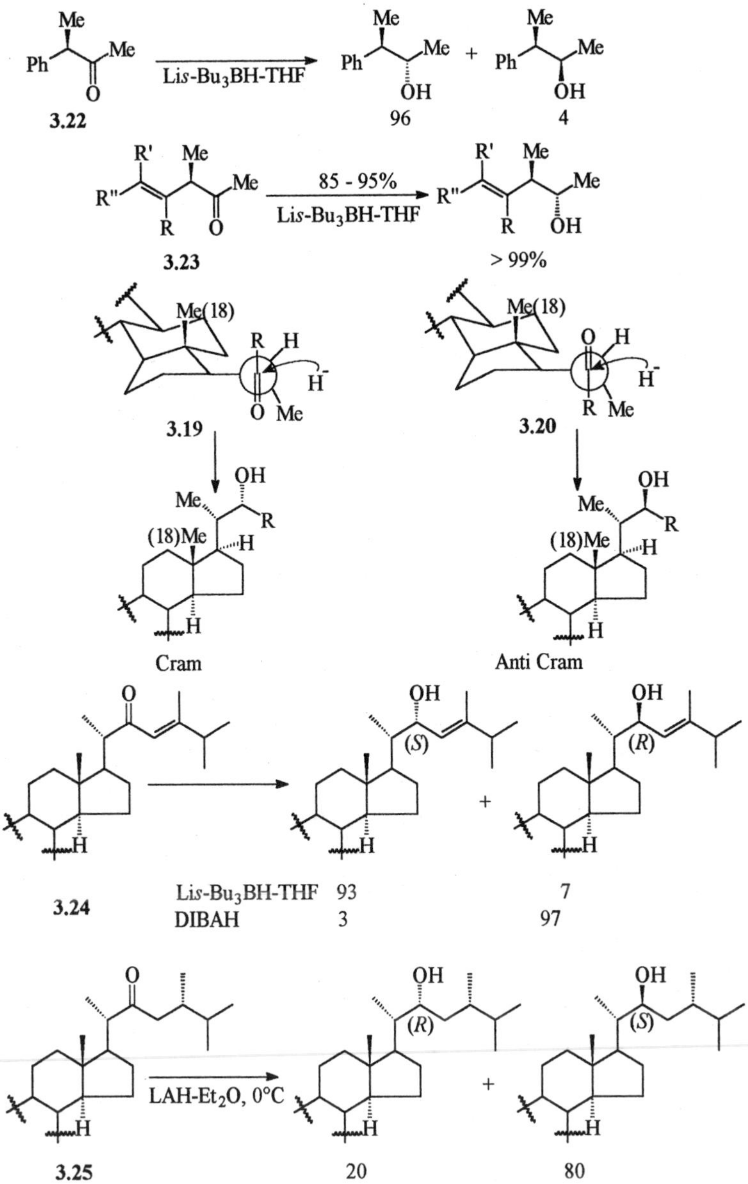

Figure 3.11

disfavors conformer **3.19**, and **3.20** can participate, leading to the other 22-isomer. The reduction of such carbonyl compounds by LAH gives a mixture of (*S*)- and (*R*)-22-alcohols in a 4:1 ratio [PR3] (Figure 3.11).

There has been some debate among the theoretical chemists as to whether the hydride should attack *anti* to the best acceptor or to the best donor group (L or P on

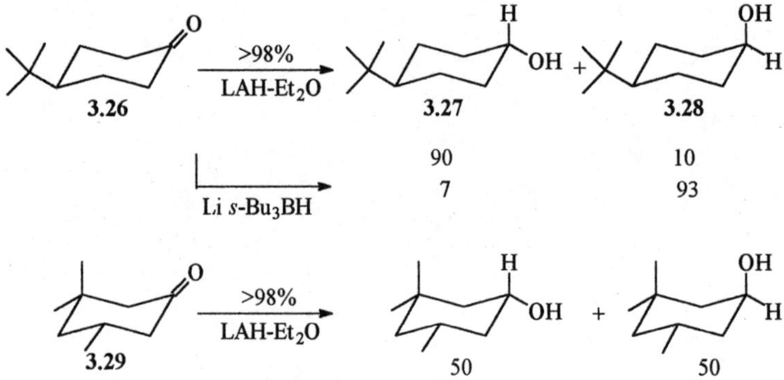

Figure 3.12

conformers **3.19** and **3.20**) [C6, CT3, FK4, LL5, M5, N2, WH1, WT1]. However, this debate only takes into consideration orbital interactions. Taking into account all the different factors including electrostatic, steric, and torsion interactions [WT1], and the position of transition state along the reaction coordinate (early or late) [AM3], it is possible to obtain more reliable trends. Moreover, solvent effects should not be neglected [AM3] (see the following).

The reduction of cyclic ketones can be interpreted by a similar approach [N2, N5, W2]. Stereoelectronic control favors axial attack on the rigid cyclohexanones. But steric interactions, as a result of either the substituents of the molecule or the structure of the reagents, can work against this pathway. The following results illustrate these trends [H3] (Figure 3.12). LAH reduction of 4-*t*-Bu-cyclohexanone **3.26** in diethylether gives mainly the equatorial alcohol **3.27**. Li(*s*-Bu)$_3$BH or Li (*t*-BuEt$_2$O)$_3$AlH [BD2] is very bulky, and it enters from the least hindered face of the molecule to give rise selectively to the axial alcohol **3.28**. LAH reduction of 3,3,5-trimethylcyclohexanone **3.29**, whose axial attack is hindered by the presence of an axial methyl group, gives more equatorial attack than that of **3.26** (Figure 3.12).

In the steroid series, LAH preferentially attacks **3.30** from the axial face of the ketone at the 3-position, while Li(*s*-Bu)$_3$BH does so at the equatorial face. One can thus selectively obtain the 3-cholestanol **3.31** with an equatorial OH or its stereoisomer **3.32** with an axial OH, depending on the reducing agent employed [DA1] (Figure 3.13). Some other examples are described by Ohloff and co-workers [OM2]. In the coprostane series, where the cyclic AB ring junction is *cis,* the same reagents selectively give rise to either **3.33** or **3.34** [OM2] (Figure 3.13). The reduction of 1-oxosteroids has been reviewed recently [WL1].

Similarly, K(*s*-Bu)$_3$BH provides stereoselection in favor of the diaxial 3,7-diol from the starting corresponding dione **3.35**; the ester functional group remains unchanged [TF1] (Figure 3.13). The stereoselectivity of the reduction of polymer-supported 6-ketosteroids by aqueous KBH$_4$ has been examined. This depends on the polymer used and on whether a phase-transfer catalyst is added or not [BH4].

Ring substituents other than alkyl groups can also influence the stereoselectivity of the reduction. The presence of a CN group at the β-position with respect to the

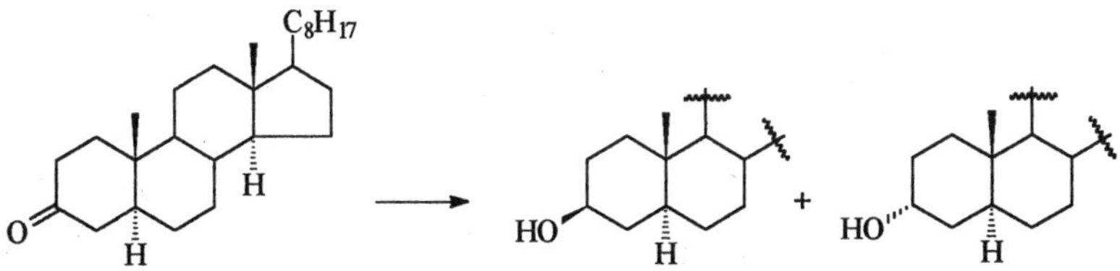

3.30 → 3.31 + 3.32

	3.31	3.32
LAH-Et_2O	88	12
Li *s*-Bu_3BH-THF	20	80

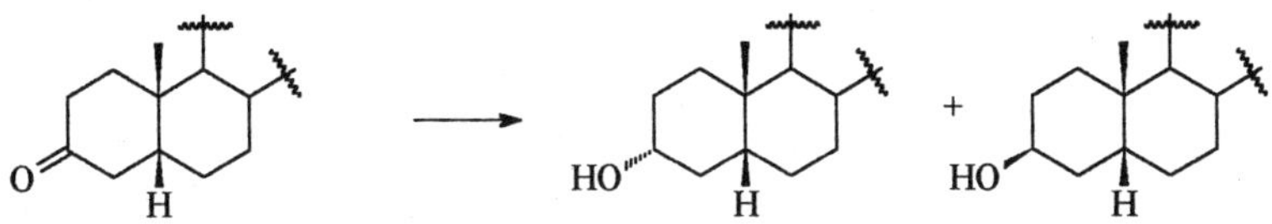

→ 3.33 + 3.34

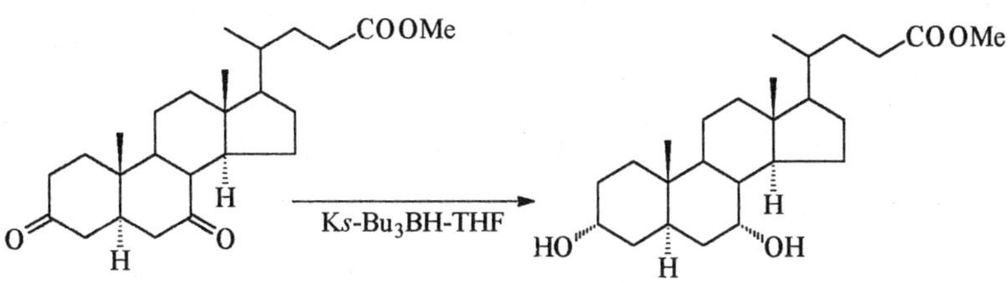

3.35

Figure 3.13

carbonyl on either face of a decalone orients the reduction by LTBA in THF to the preferential generation of the axial **3.36** or equatorial **3.37** alcohols [AF2, CB1] (Figure 3.14). Other related examples have been recorded by Caro and co-workers [CB1]. These results have been ascribed to remote electrostatic effects [WT1].

In the cyclic series, a problem arises due to the relative rigidity of some molecules. The antiperiplanarity requirement between the nucleophilic hydride and the CUL(P) bond as indicated in the Felkin–Anh model is sometimes difficult to attain at the transition state. The flatness and flexibility of the ketone to be reduced have to be considered [HM4, N2, WH3, WH4]. Cyclohexenones, which are flatter than cyclohexanones, give more product of axial attack [CG7, KY3, WH3, YK6] (Section 3.2.9). The reduction of 3-ketosteroids such as **3.30** by $NaBH_4$ results in 80% axial attack (Figure 3.15). Reduction of the 7-keto analogues **3.38** under similar conditions gives only 45% axial attack [WM1]. Kinetic studies indeed showed that this stereoselectivity was due to a decrease in the rate of only axial attack, possibly because the steroid B ring is less flexible than the A ring (Figure 3.15). The reduction of 1,3-dioxane-5-one **3.39** and 1,3-dithian-5-one **3.40** by LAH in Et_2O

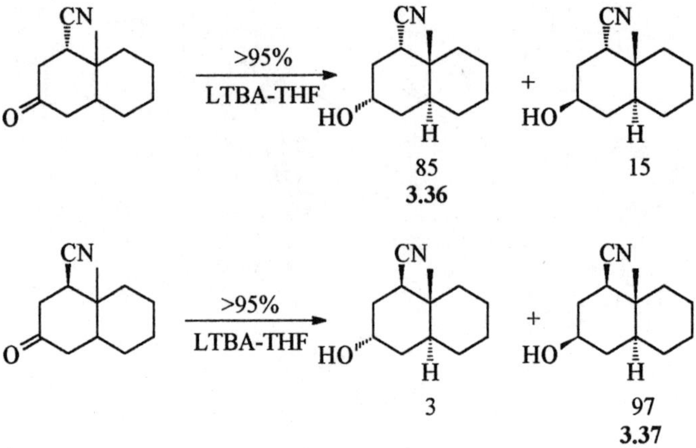

Figure 3.14

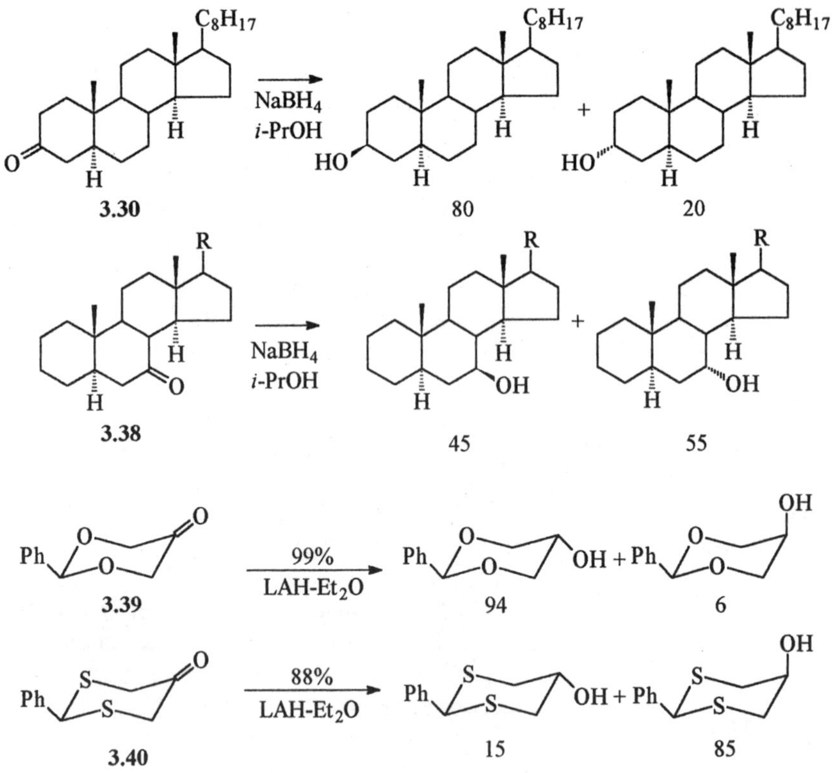

Figure 3.15

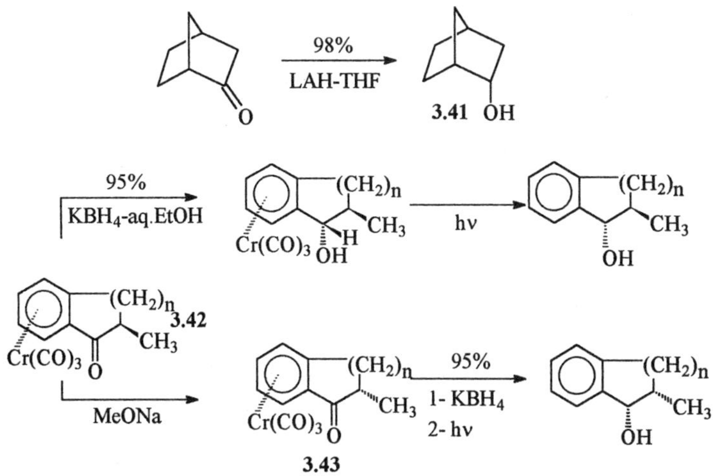

Figure 3.16

gives comparable results [JK1, KL3] (Figure 3.15). In the first case, 94% axial attack is observed, while in the second one only 15% axial attack takes place. This apparent discrepancy has been interpreted in terms of the ring structures of these ketones: The six-membered ring of **3.39** is quite flat, while that of **3.40** is significantly puckered [N2, WH4]. Antiperiplanarity between the incoming bond and the vicinal axial C—H bonds is difficult to attain during the axial attack of **3.40**. Similar results have been observed in reduction of 16-oxa- and 16-thiahomosteroids [TO2].

In the norbornane series, the attack of LAH in Et_2O takes place on the *exo* face, leading to *endo*-norbornanols **3.41**. These are the thermodynamically less stable products. Therefore, the reductions performed by LAH do not follow "product development control," but essentially depend on stereoelectronic factors [AB2] (Figure 3.16). The same *exo* attack is observed in other strained bicyclic systems. In general, Li(*s*-Bu)$_3$BH in THF at $-78°C$ is more stereoselective than LAH in THF [KG2].

The reduction of the chromiumtricarbonyl complexes of indanones and tetralones **3.42** and **3.43** [JM1] are interesting in terms of steric hindrance. The hindered organometallic group blocks the attack of the hydride on the same face of the molecule. Because it is possible to obtain the corresponding ketones in nonracemic form, one has access to enantiomerically pure stereoisomers after decomplexation (Figure 3.16).

***The Houk Model* [CL4, CL5, HP3, HW1, PR1]** The intervention of a favored conformation through which the reduction occurs can depend on the stereoelectronic interactions in a Lewis acid–base complex that is formed between a tricoordinated reducing agent (boranes, DIBAH) and a carbonyl compound. The complex associates in a way that minimizes the different repulsive interactions, and the transfer of the hydride takes place in second stage. Accordingly, the conforma-

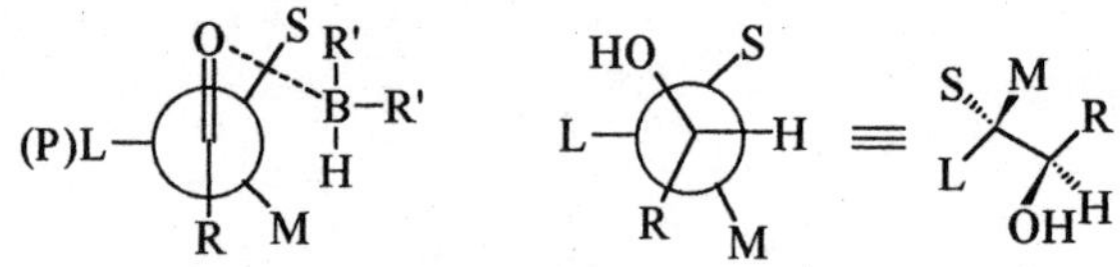

3.44

Houk model

Figure 3.17

tion of the ketone in this complex **3.44** is more favored when the substituents on the boron or aluminum atom are bulkier. As previously mentioned, the hydride transfer takes place in the *anti* position with regard to the most bulky (L) or polar (P) group. Under these conditions, the reduction leads to the "anti-Cram" diastereoisomer (Figure 3.17). In cyclohexanones, the interaction with the axial vicinal C—H bonds is the same as in the previous case [CL5].

In addition to the pioneering work by Midland [MK1], a certain number of reports in the literature show a reversal of stereoselectivity as the reducing agent is varied. For example, the reduction of steroidal prochiral ketones **3.45** and **3.46** either by alumino- and borohydrides or by boranes and DIBAH gives different stereoisomers, enabling the process to be interpreted either by the Felkin–Anh or the Houk model [MK1, SH7, SK1, TO1] (Figure 3.18). Similar results are obtained in the reduction of cyclopropyl ketones either with Li $(s\text{-Bu})_3BH$ or with DIBAH [OS4]. However, when starting from **3.47**, the size of the L substituent is such that a similar selectivity is observed whatever the reducing agent used. If the double bond of **3.47** does not carry a trimethylsilyl ($SiMe_3$) group, reduction by DIBAH gives a greater proportion of the other isomer.

In most of the examples discussed above, stereoselectivity is high either because the structure of the ketone shows two sufficiently different faces in terms of steric hindrance or because the reducing reagent is bulky. In many other cases, the stereoselectivity is decreased because the different factors more or less compensate. Indeed, it is necessary to consider the position of the transition state on the reaction coordinate. Transition states are early with aluminohydrides and late with borohydrides in alcoholic media [CB1,CL4,N2]. The possibility of electrophilic assistance by the alkaline cation is another important factor [N2, S2]. The latter effect especially facilitates the axial attack of cyclohexanones. In the presence of [2.1.1]cryptand, which prevents electrophilic assistance by the Li cation, the amount of axial attack of **3.26** by LTBA decreases [AK1] (Figure 3.19). The size of the reducing agent, the structure of the ketone, and the possibilities of conformational equilibria must be taken into account [GZ1, KW3]. Indeed, the reactive conformer may not be the most stable one [N2, S2, S3]. For example, Table 3.1 indicates the stereoselectivity of the reduction of two rigid ketones, **3.26** and **3.48**, and of a conformationally flexible ketone **3.49**, with different reagents [BD2, BS6, CB1, DL2, FF2, GL2, HF1, M1, PS1, RC2] (Figure 3.19).

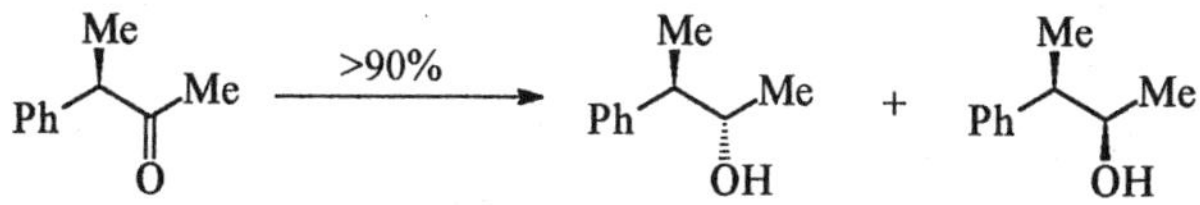

Li s-Bu_3BH-THF	96	4
Sia_2BH-THF	20	80

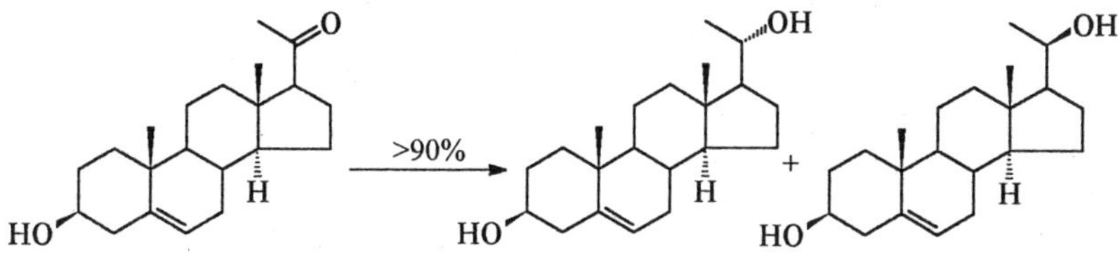

3.45 20-(*S*) 20-(*R*)

Li s-Bu_3BH-THF	1	99
Sia_2BH-THF	90	10

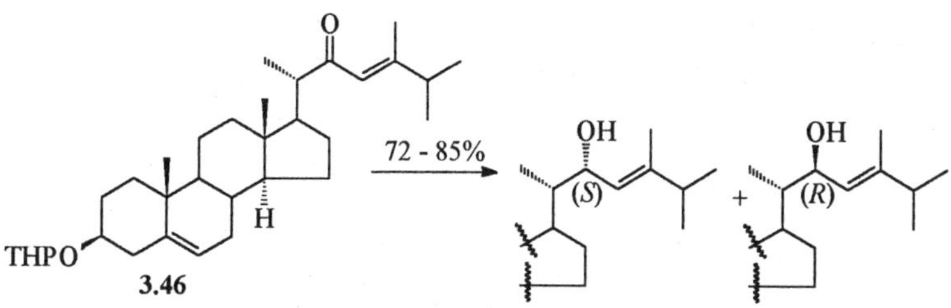

3.46

Li s-Bu_3BH-THF	93	7
DIBAH-toluene	3	97

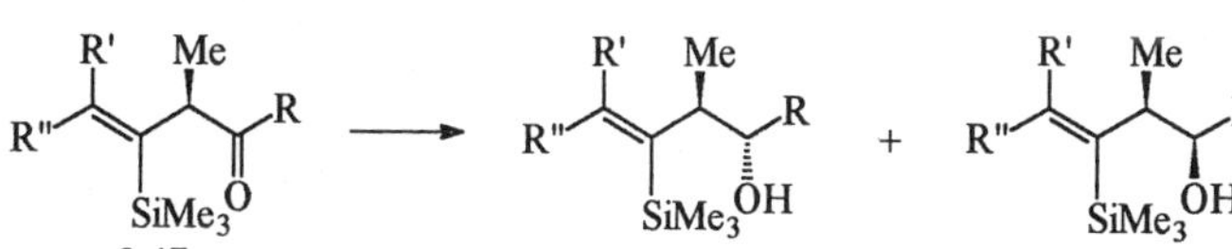

3.47

Li s-Bu_3BH-THF	>99	<1
DIBAH-toluene	>90	<10

Figure 3.18

Adsorption of the ketone on montmorillonite clay enhances the axial attack of $NaBH_4$ reduction to >99% for 4-*t*-butylcyclohexanone **3.26** and 78% for 3,3,5-trimethylcyclohexanone **3.29** [SR1]. Other hindered substituted borohydrides also give higher levels of equatorial attack [CY1]. From the numerous studies to date, it appears that torsional and steric factors are very often predominant, as illustrated by the reduction of eight-membered cyclic taxane derivatives [SH7]. An interesting solvent effect in the reduction of a sugar derivative has been recently shown. The reduction of a substituted rigid six-membered ketone with DIBAH in CH_2Cl_2 or

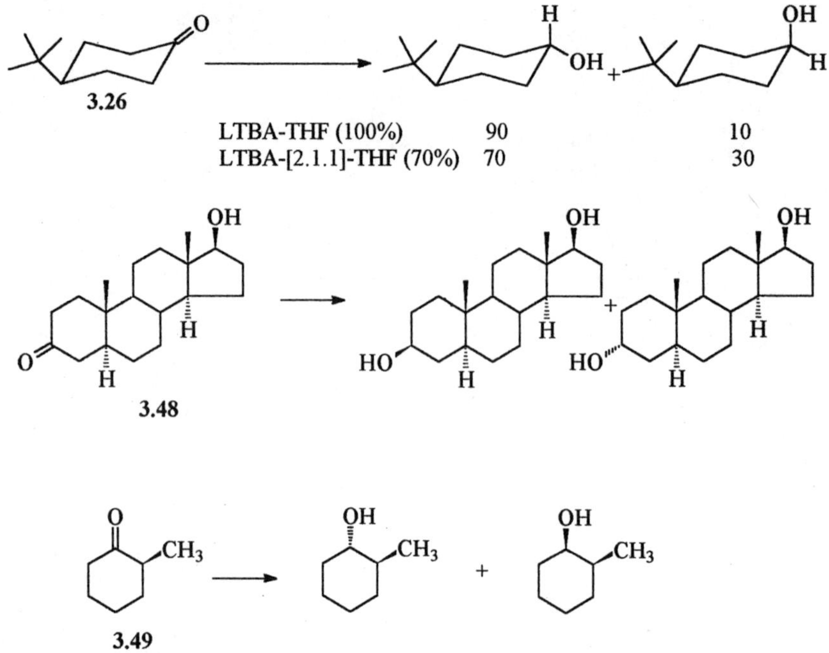

Figure 3.19

TABLE 3.1 Percent axial attack in the reduction of 3.26, 3.48, and 3.49

Reagent	3.26	3.48	3.49
LAH–THF	90	76	76
LTBA–THF	90	98.5	70
Li(*t*-BuEt$_2$O)$_3$AlH	95		>99
LiBH$_4$–THF	86–93	71	70
NaBH$_4$–MeOH	81	81	70
NaBH$_4$–CeCl$_3$–MeOH	94	>95	70
NaBH$_4$–H$_2$O–glycosidic Surfactant	98		
Cp$_2$TiBH$_4$–THF	97		
(*i*-PrO)$_2$TiBH$_4$–CH$_2$Cl$_2$	97		97
Li *n*-Pr$_2$NBH$_3$–THF	99		
LiMe$_2$NBH$_3$–THF	95		64
Li(*s*-Bu)$_3$BH–THF	7		1
K(*s*-Bu)$_3$BH–THF	1	20	8
LiSia$_3$BH–THF	0.5		
AlH$_3$–Et$_2$O	85		
DIBAH–toluene	61		
9-BBN–THF	92		60
ThexylBHCl.Me$_2$S	44		5.5

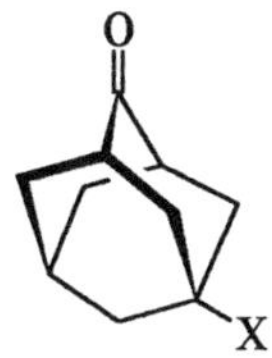

3.50

X = Ar, *t*-Bu, F, Cl, OH, CF_3

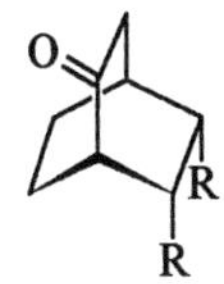

3.51

R = CO_2Me, CH_2OMe, $CH=CH_2$, Et

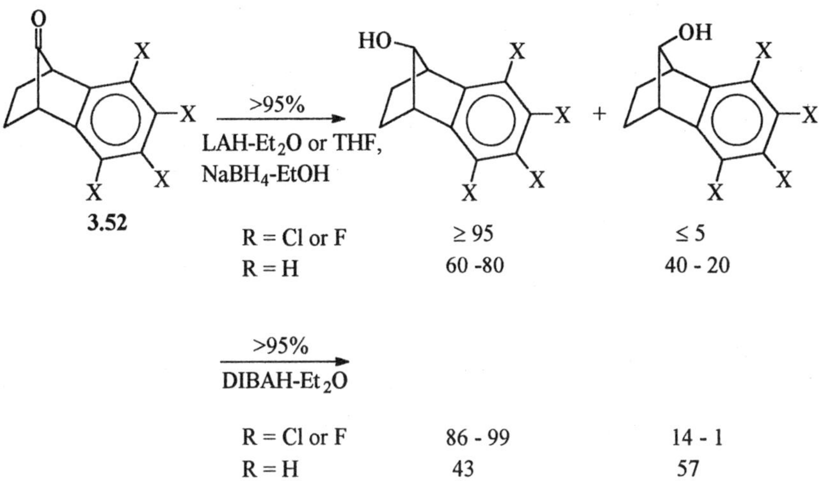

Figure 3.20

pentane is poorly stereoselective. On the other hand, axial attack is highly predominant in THF. This has been explained by the relative bulkiness of the reagent, which is trimeric in CH_2Cl_2 and pentane and monomeric in THF [HP4].

However, the stereoselective reductions of the 5-substituted adamantanones **3.50** [CT2, G5, KA4, LL5], and bicyclo-[2.2.2]-octan-2-one **3.51** [MK5] have been interpreted in terms of Cieplak's model [C6, CT3]. Unfortunately, the observed stereoselectivities are low (de $<40\%$), so that it is difficult to dissect the various parameters involved in such reductions [CH5, N2]. The electronic contribution to the stereoselectivity of the reduction of substituted 9-benzonorbornenones **3.52** is very important [G5, OT1, OT2] (Figure 3.20). Halogen substituents on the aromatic ring favor *syn* attack by all hydrides on those very rigid systems. Electrostatic effects have been invoked to interpret these results [PW2, WT1].

3.2.3 Asymmetric Reductions

The enantioselective reduction of ketones has been the goal of numerous intensive studies in recent years [BD3, BJ1, BR4, DS5, N5, NN1, S3, S4, WM2]. Chiral alkoxy- or aminoaluminohydrides have been used to perform asymmetric reductions

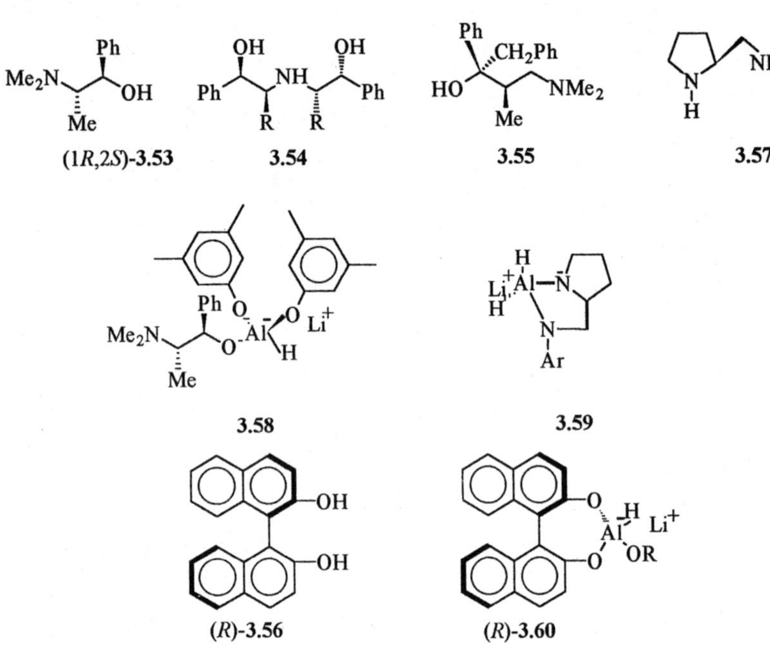

Figure 3.21

of ketones or aldehydes. The most efficient chiral alkoxyaluminohydrides [BP2, GH1, IS3, NN1, S3] are generated in situ from solutions of LAH by addition of (1*R*,2*S*)-N-methylephedrine **3.53** and 3,5-dimethylphenol or 3,5-dimethyl-N-ethylaniline in a 1:1:2 molar ratio. Polymer-grafted ephedrine can be used for a similar purpose [FB2]. Chiral C_2 symmetrical diethanolamines **3.54** also give interesting selectivities as LAH modifiers [VB2], as does Chirald **3.55** or its enantiomer [NN1] or 2-isoindolinylbutan-1-ol [BL4, BL6]. Reaction of (*R*)- and (*S*)-binaphthol **3.56** with LAH and EtOH in a 1:1:1 ratio allows the preparation of (*R*)- or (*S*)-Binal [NN1, BD3]. Proline-derived diamines such as **3.57** [NN1, TK3] (Figure 3.21) are also interesting chiral modifiers. Depending on the additive, the chiral reagent is a rigid mono- or dihydride such as **3.58**, **3.59**, and **3.60** (Figure 3.21), although, due to disproportionations, the reagents must be carefully prepared and used [N5].

Most of these reagents reduce arylalkylketones and some α-enones with a high enantiomeric excess provided that the alkyl group (R) is not too bulky (Figure 3.22). The reagent generated from Chirald **3.55** is useful for the enantioselective reduction of α-ynones [MW2, NN1, PC2, S3], although some limitations are known [MO1] (Figure 3.22).

Compared to the other reagents, (*R*)- and (*S*)-Binal in THF have a broader range of application. Many arylalkylketones, α-enones, and α-ynones are reduced at −78°C by (*R*)-**3.60** to (*R*)-secondary alcohols and by (*S*)-**3.60** to (*S*)-enantiomers [DK2, HM5, NN1, S3, S4]. These reagents also reduce α-deuteroaldehydes such as **3.61** at −100°C to α-deuterated primary alcohols and acylstannanes **3.62** and **3.63**

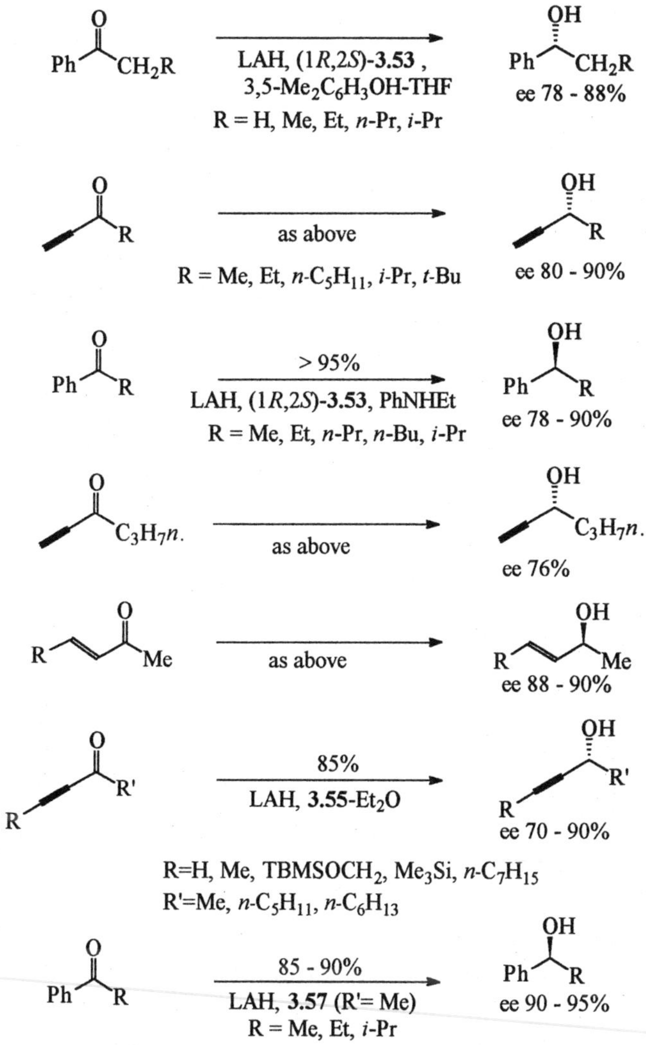

Figure 3.22

give α-stannyl alcohols with excellent enantioselectivity [S3] (Figure 3.23). Reductions of trifluoromethylarylketones occur with poor enantiomeric excesses with the exception of 2,2,2-trifluoromethylanthrone [CM1]. To overcome the disappointing results that are usually obtained in reductions of dialkylketones, crown-ether groups have been introduced on Binal and preliminary results are promising [YU1].

Tartaric acid derivatives (TADDOLs) and a hexamethoxy substituted 1,1′-biphenol give rise to reagents having similar potential [BD3, RM1].

Among the chiral borohydrides, ate complexes generated from α-pinene, such as Alpine-hydride **3.64** and related reagents, reduce 2-octanone or acetylcyclohexane

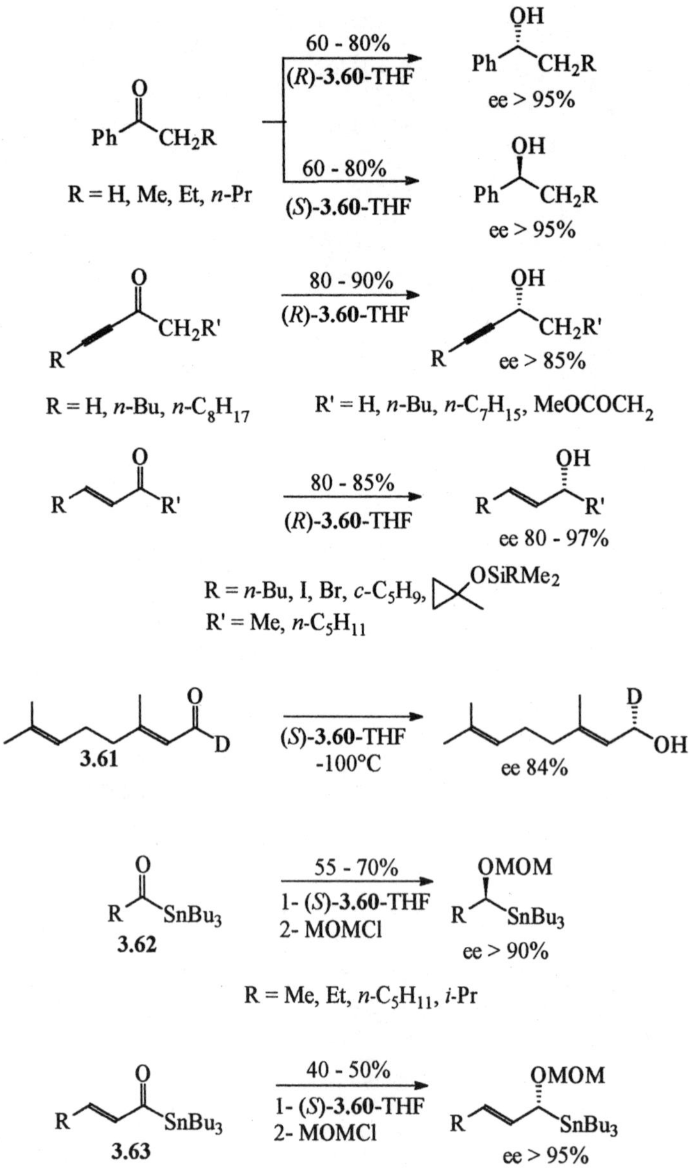

Figure 3.23

at $-100°C$ with a high enantiomeric excess [BJ2, BR4, WR5]. A reagent generated from N-benzoylcysteine **3.65**, *t*-BuOH, and $LiBH_4$ reduces arylalkylketones to (*R*)-alcohols with a 90% ee [NN1, SY2]. Similar selectivities are observed with Li glucoride **3.66** derived from diacetoneglucose [BP1, NN1]. $NaBH_4$–tartaric acid also gives interesting results, but only with ketones functionalized in the α- or β-position with chelating residues such as OR, esters, or amides [YO3]. The reduction of some arylketones with alkali borohydrides in the presence of catalytic amounts of chiral (β-oxoaldiminato)-cobalt complexes gives secondary alcohols with a high ee [NY2]. $NaBH_4$-(*R*)- or (*S*)-*t*-leucine in the presence of AcOH promotes the enantio- and stereoselective reduction of a prochiral ketone, precursor of diltiazem, a drug [YM1].

Borane derivatives have been widely developed for asymmetric reductions [BR4, DS5, N5, NN1, S3, S4]. Although isopinocampheylborane **3.67** and diisopinocampheylborane **3.68** are good asymmetric hydroborating agents, they proved unsatisfactory as chiral reducing agents of prochiral ketones [BJ1, BR4]. While $NaBH_4$ in the presence of β-cyclodextrins gave disappointing results, pyridine–borane under such conditions reduces PhCOMe and $PhCH_2CH_2COMe$ with 90% ee [SI2]. The asymmetric reduction of aralkylketones by BH_3 in the presence of chiral aminoalcohols such as **3.69** was discovered by Itsuno in 1981 [DS5, IN1, IS1, IS2, NN1, S4]. These reductions occur near room temperature, giving ee's greater than 95%. The selectivity is lower with dialkylketones (55–73%). Itsuno and co-workers developed this methodology using, among other things, polymer-supported aminoalcohols [DS5]. In 1987, Corey and co-workers [CB2, CB3] demonstrated that Itsuno's reagent was indeed an ate complex **3.70** of an oxazaborolidine and BH_3. These authors broadened the scope of application of these systems (CBS reagents) designing new oxazaborolidines [DS5, S4, WM2] and showing that they could be used in catalytic amounts with the achiral reducing co-reagents being $BH_3{\cdot}THF$ [CC3, CG6, CK6, CL2, CL10, CR2] or catecholborane [CB4, CH6, CL2, CL6, CL7, CL9]. Corey's oxazaborolines **3.71** are stable, and they bear the (*R*)- or (*S*)-proline skeleton. Several synthetic pathways have been described [CL10, MJ1, S4], as has the preparation of crystalline borane complex of (*S*)-**3.71** (Ar = Ph, R = Me) [MT6]. This complex has been characterized by X-ray crystallography [CA3, MT6].

Chiral reductions are now quite common with these reagents when used either in stoichiometric amounts [CT4, MT6, SC1] or in catalytic amounts in the presence of $BH_3{\cdot}THF$, catecholborane [G6, GB6, QW1, SK5], or $BH_3{\cdot}Me_2S$ as achiral reducing co-reagent [DS6, JM3, N5, S5, SC1, SM6, TA3, TB1, YL4]. The asymmetric reduction of aldehydes, various ketones, α-enones, and α-ynones usually take place with an excellent enantiomeric excess [BB13, CL2, CL6, CR2, CT4, DS5, DK2, GB6, MT6, NN1, PL2, S4, SM6, WM2] (Figure 3.24). The reduction of ketones is carried out between −20°C and r.t., while that of aldehydes requires −126°C for a good asymmetric induction.

Both enantiomeric (*R*)- and (*S*)-oxazaborolidines are available; so it is possible to obtain at will either enantiomeric alcohol. The selectivity depends upon the geometry of the complex formed by coordination of the carbonyl oxygen to the Lewis acidic heterocyclic boron atom, complex **3.72** being favored. The catalytic cycle is shown in Figure 3.25.

The Ar and R substituents in **3.71** that give the best results vary from case to case. Moreover, catecholborane is less reactive than borane; therefore, it must be used when the carbonyl compounds are sufficiently reactive (for example, aldehydes or trihalomethylketones) and the reduction is only performed via the chiral reagent.

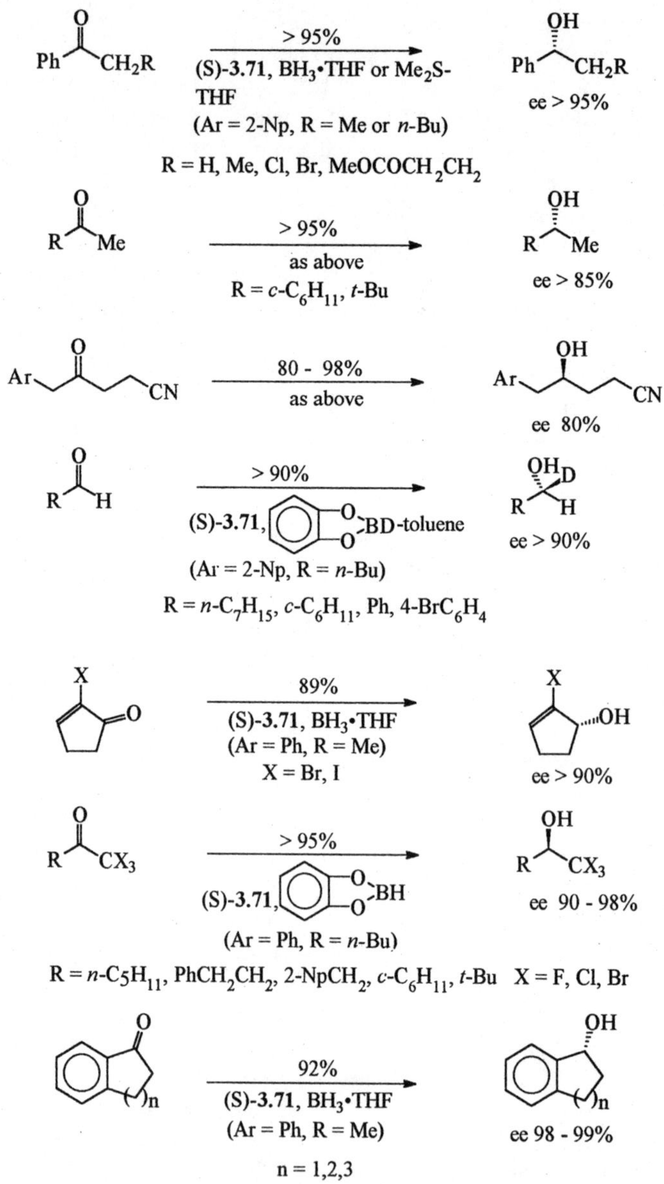

Figure 3.24

Catecholborane, as an achiral co- reagent, has also been recommended for the reductions of ketophosphonates [ML2, ML4], a β-aminosubstituted α-enone [LO1], cyclopropylisopropylketone **3.73**, some cobaltcarbonyl complexed ynones **3.74**, and unsymmetrically substituted benzophenones **3.75** [CH6, CH7] (Figure 3.26). The reduction of 2,2-diphenylcyclopentanone **3.76** by (*S*)-**3.71** (Ar = Ph, R = Me) leads

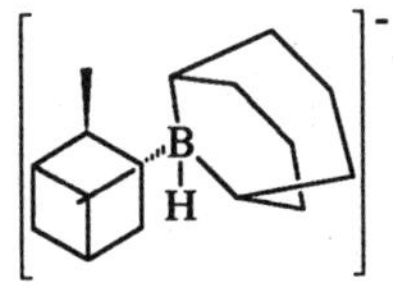

3.64

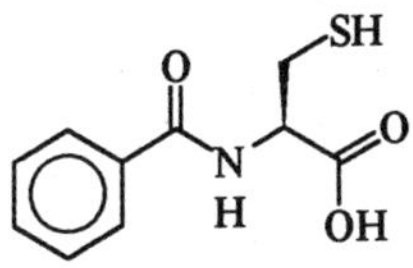

3.65

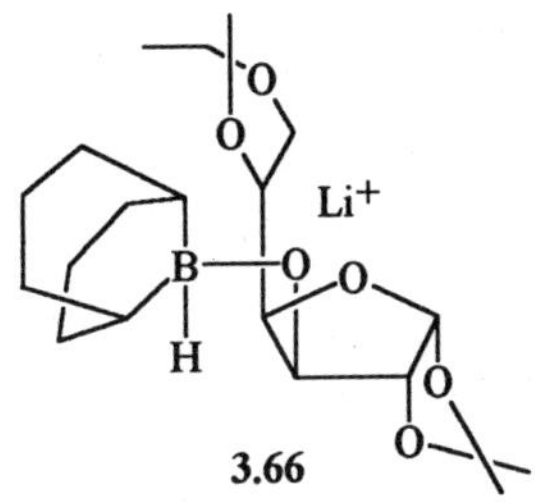

3.66

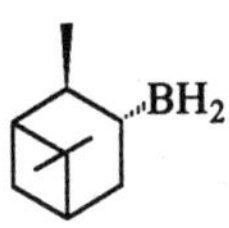

3.67

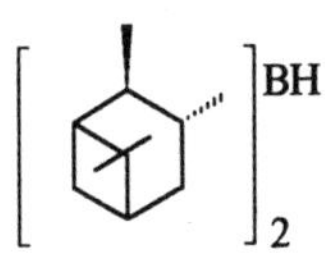

3.68

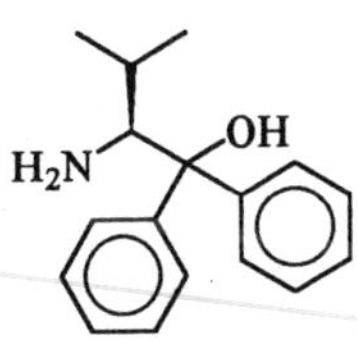

3.69

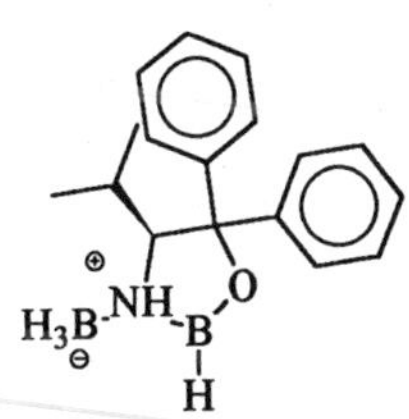

3.70

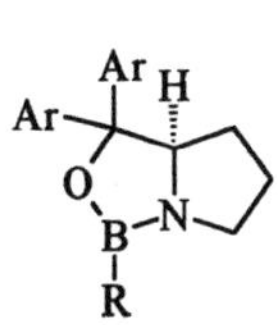

(*R*)-3.71

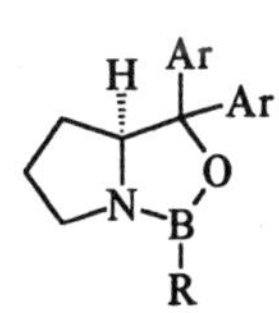

(*S*)-3.71

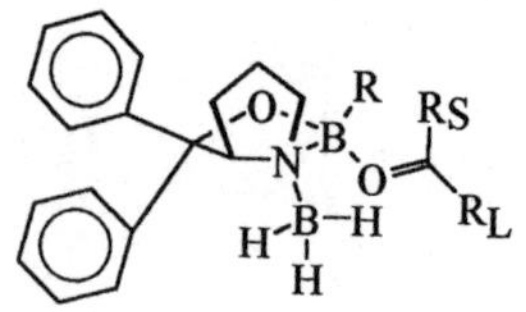

3.72

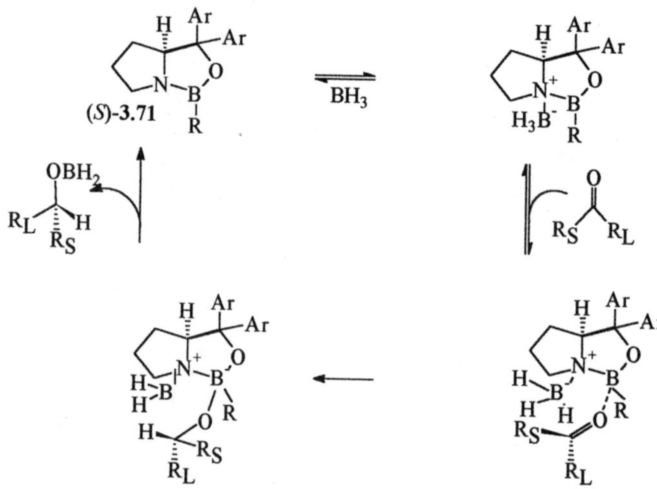

Figure 3.25

to the (*R*)-cyclopentanol, which is itself a useful chiral auxiliary [DS6] (Figure 3.26). Ketones that contain heteroatoms capable of coordinating boranes, particularly nitrogen, can be reduced with high ee's. With compounds such as acetylpyridines, the amount of oxazaborolidines must be increased, but the selectivities are not greater than 80% [QW1]. Ferrocenylketones are also useful substrates [LR3, SK5, WF2]. Benzils are selectively reduced to *syn* diols [PJ2].

The synthetic routes to a number of optically active drugs include, as one step, a reduction of a prochiral keto group by CBS reagents [see, for instance, CH8, CL10, CP4, CT4, HG2, JM3, SC1, TA3, TB1]. From chiral 1-trihalo-2-alkanols, one may prepare chiral α-aryloxy-, α-hydroxy- and α-amino acids [CL7, CL9]. In some cases, the presence of Et_3N in the reaction medium increases the enantioselectivity [CT4].

The influence of temperature on the selectivity of the reduction of acetophenone and cyclohexylmethylketone by various oxazaborolidines **3.71** (Ar = Ph, R = Me, Bu, Ph) has been examined [S5]. The use of polymer-bound **3.71** has also been proposed [FS3]. Some mechanistic [DT1] and theoretical investigations of the reductions have also been carried out [DL1, LS6, N4, NU2, QB1].

Other amino alcohols have been transformed into oxazaborolidines that have been tested as precursors of CBS reagents [DS5, MS9, NN1, PM2, RW1, S4, SM7, WM2]. Analogs of **3.71** in which the proline ring has been either enlarged, dimin-

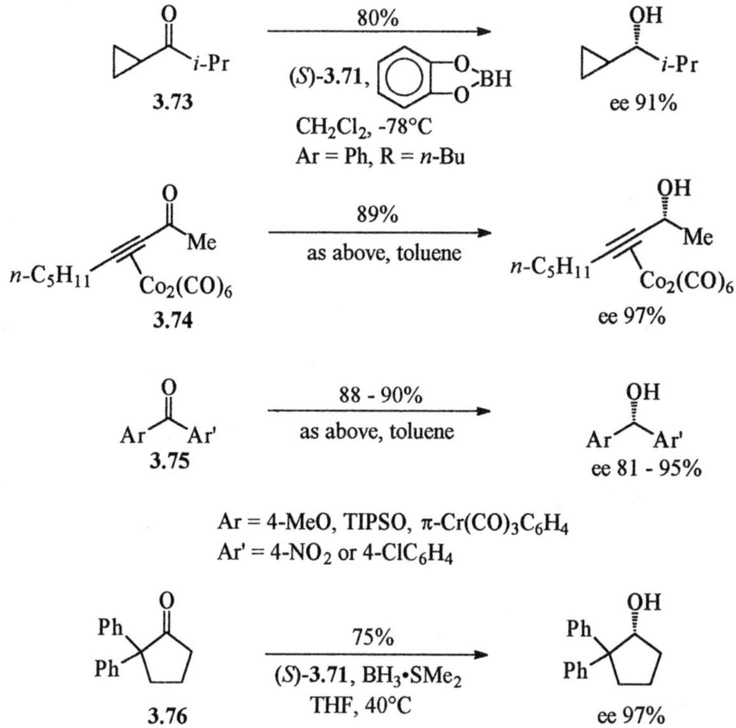

Figure 3.26

ished [WD2, WM2], or rigidified by fusion of an aromatic ring [WM2] have been proposed. Diethanolamines gave disappointing results [DF2], as did (1*R*,2*S*)-norephedrine grafted on polymer [CC6, CE3, DL3]. Reductions of arylalkylketones with reagents generated from (*S*)-prolinol **3.77** [BM5], *cis*-1-amino-2-indanols **3.78** [DL3, DS7, HG2] and indolinemethanols **3.79** [KP3] give selectivities similar to those obtained in reductions with reagents generated from **3.71**. From methionine or cysteine, sulfur-substituted amino alcohols **3.80** and **3.81** have been prepared [MM6, MM7]. The corresponding oxazaborolidines, generated in situ with BH_3·THF, also reduce arylalkylketones with an excellent ee. Interesting reagents were obtained from (1*S*,2*R*)- and (*1R*,*2S*)-2-amino-1,2-diphenylethanol **3.82**. The corresponding oxazaborolidines **3.83**, (R = Me, *n*-Bu, or Ph) were prepared in the usual fashion. Used in catalytic amounts in the presence of $BH_3 \cdot Me_2S$ at room temperature, these reagent effect the reduction of various ketones with useful ee's [QW2] (Figure 3.27). Reagent **3.83** (R = H) is prepared in situ. One advantage of such reagents is the good enantiomeric excess obtained in the reduction of acetylpyridines, although higher amounts of BH_3·THF must be used (Figure 3.27). (*R,S*)-**3.83** has also been used in stoichiometric amounts in the presence of $BH_3 \cdot Me_2S$ to reduce symmetrical diketones **3.84** to (*S,S*)-diols, which are formed along with about 15% of one *meso* isomers [QK1] (Figure 3.27). (*R*)- and (*S*)-1-

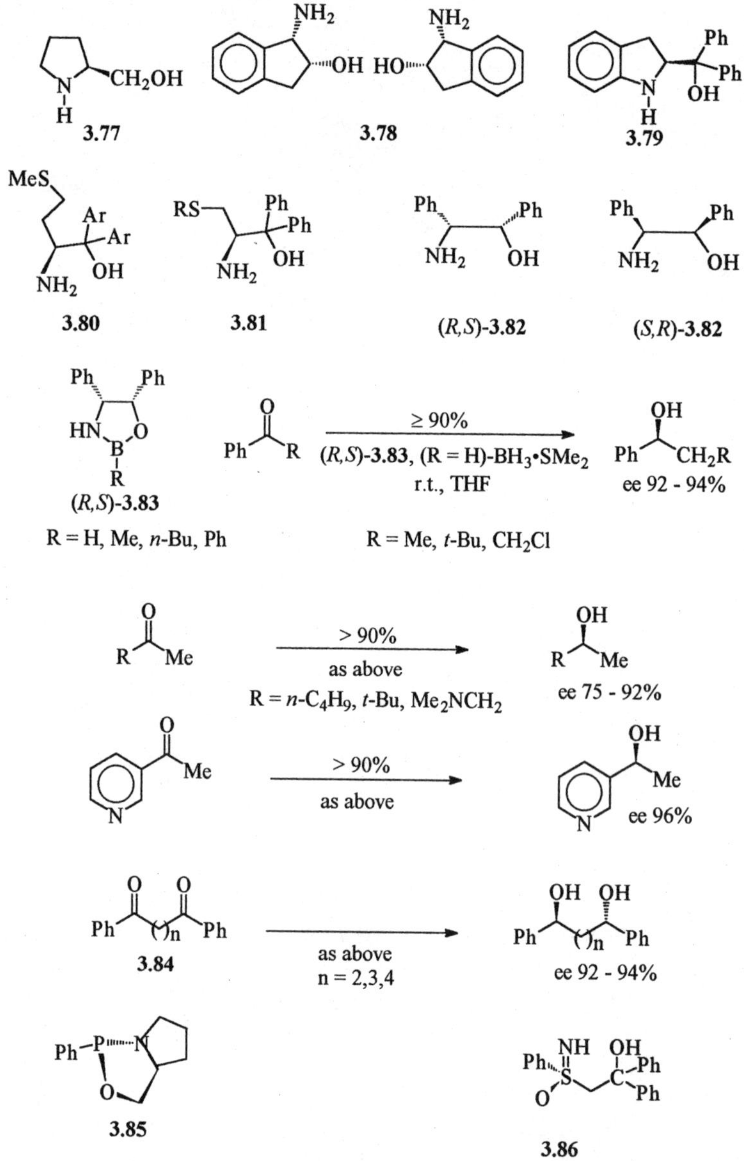

Figure 3.27

diphenyl-2-phenylaminoethanol have also been used to generate oxazaborolidines. These reagents give high selectivities in the reduction of α-enones [BB8]. The stereoselectivity of the reduction of 17-oxosteroids is increased in the presence of CBS reagents [RM2].

Oxazaphospholidines such as **3.85** [BP6, CF1, PM1] and β-hydroxysulfoximines **3.86** [BF3] have been proposed as chiral additives in borane reductions. With **3.86**,

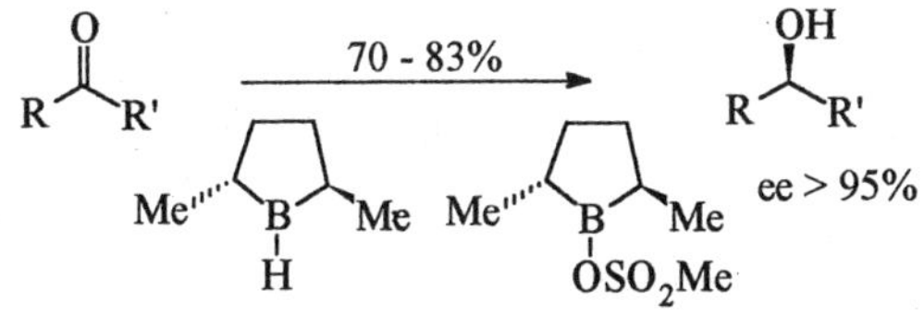

3.87 **3.88**

R = Me, n-C_5H_{11}, n-C_6H_{13}
R' = $PhCH_2$, t-$BuCH_2$, i-Pr, i-Bu, c-C_6H_{11}, t-Bu

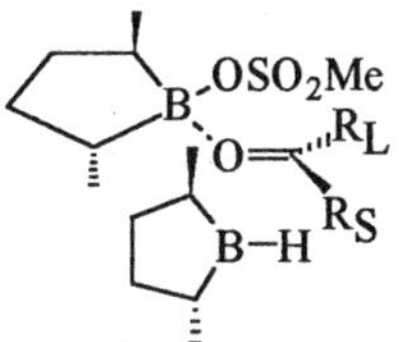

(R,R)-**3.87**

Figure 3.28

borane is generated in situ from $NaBH_4$ and Me_3SiCl [BS7]. Under various conditions, the results are not better than those obtained with **3.71**. N-Sulfonyloxazaborolidines give lower enantiomeric excess [OI1].

Chiral titanates have also been used as additives in the reduction of acetophenone by catecholborane. At best, the ee was 80% [GD1].

Another method of asymmetric reduction of unsymmetrical dialkylketones is to use chiral borane **3.87** [IT2, NN1, S3, S4] (Figure 3.28). As shown by Masamune and co-workers, asymmetric induction occurs by coordination of the carbonyl group of the ketone to mesylate **3.88**, which is present in catalytic amounts. This is followed by hydride transfer in such a fashion that steric interactions are minimized (Figure 3.28). However, the chiral reducing agent has to be used in stoichiometric amounts.

Borane in the presence of (R)-binaphthol-La(O-i-Pr)$_3$ reduces various ketones but with a low ee [ZY1]. Other chiral boranes can be used to carry out enantioselective reductions [AC1, BP2, BR4, DS5, M2, NN1, S3, S4], but the transferred hydride comes from the carbon skeleton; therefore, these reagents are outside the scope of this book.

3.2.4 Functionalized Aldehydes and Ketones

If heteroatoms are present in the vicinity of the carbonyl group, the formation of chelates around an alkali, another cation, or the aluminum or boron atom can influence the course of the reduction by playing the role of a Lewis acid. The Lewis acid–base interaction can be quite strong depending on the nature of the heteroatom and its substituents, on the ligands attached to boron and aluminum, and finally on

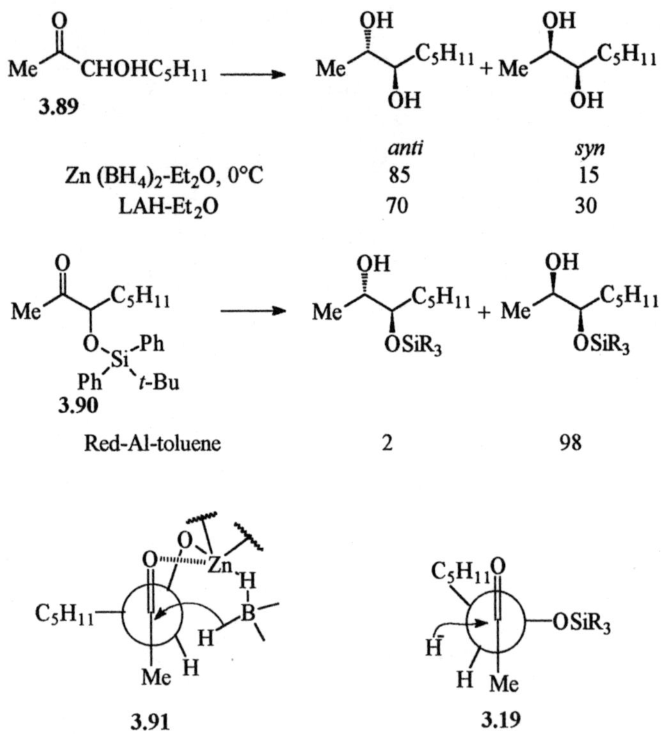

Figure 3.29

the solvent. In this way, the stereoselectivity of the reduction can be modulated [EG1, G5, N5, R4, S3] (see the following).

The stereoselectivity of the reduction of α-hydroxyketones [ON1, PR4], α-ketoethers [B2, CN2, E1, KT1, MK2, RO1], and α-epoxy ketones [B2, BB6, E1, ON1, RC3, TF2] has been reexamined, and conditions have been described that allow the highly stereoselective synthesis of each diastereomeric alcohol. Thus the reduction of an α-ketoalcohol **3.89** by $Zn(BH_4)_2$ in Et_2O leads to the *anti* diol with good selectivity [ON1]. When one uses LAH in Et_2O, this selectivity is lower. On the other hand, the reduction of the corresponding silyl ether **3.90** by Red–Al in toluene gives predominantly the silyl ether of the corresponding *syn* isomer [ON1], which, in the presence of n-Bu_4NF, generates the *syn* diol. (Figure 3.29).

In the first case, chelation between the carbonyl group and the zinc atom of the alcoholate that is formed facilitates a cyclic transition state **3.91** (Cram cyclic model), the hydride being subsequently transferred to the side bearing the least bulky substituent H. In the second case, one can propose a Felkin–Anh transition model **3.19**, with the most polar group (P) being $OSiR_3$ and the smallest one (S) being H (Figure 3.29). Again, according to steric hindrance and conformational effects, the reductions can be more or less stereoselective [SK4].

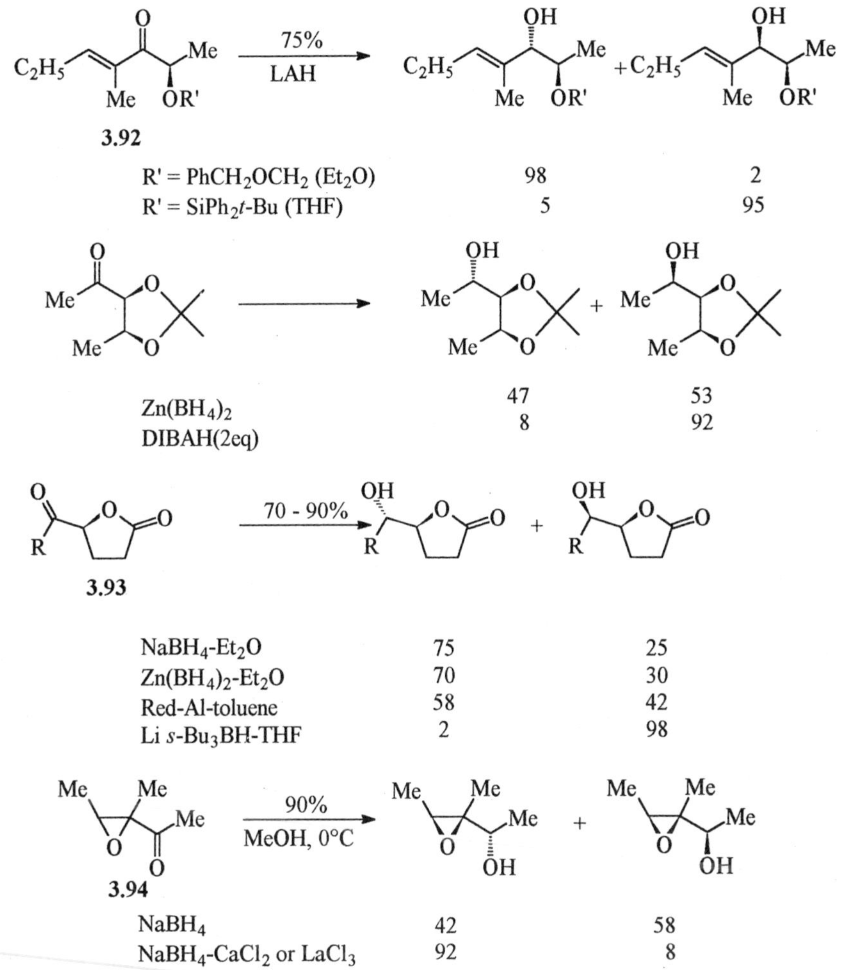

Figure 3.30

The possibility of chelation at the transition state can also be envisioned during the reduction of the α-alkoxyketones RCOCH(OR′)R″. This depends above all on the nature of the R′ group. If this group is $PhCH_2OCH_2$ or $MeOCH_2$, chelation may take place. On the other hand, as seen in the reduction of **3.90** when R′ = $SiPh_2$-*t*-Bu, chelation does not take place and the Felkin–Anh model applies. If R′ = alkyl, the reductions are slightly stereoselective. The stereoselectivity of the reduction of **3.92** according to R′ is a striking example of this dichotomy [OM1] (Figure 3.30). Similar results have been obtained starting from the alkynylketones [TM1]. In certain cases, the stereoselectivity is low in the presence of $Zn(BH_4)_2$, but this can be enhanced by using an excess of DIBAH, which is a stronger Lewis acid and a bulkier reagent (Figure 3.30). Comparable results have been obtained with the nonracemic acyl-4-butanolides **3.93** [LL2] (Figure 3.30) and other chiral α-alkoxyketones [GB4].

In contrast, 2-methoxy-1,2-diphenylethanone is reduced to the *anti* α-methoxy alcohol with good stereoselectivity whatever the reducing agent (LAH, DIBAH, $NaBH_4$, or K(*s*-Bu)$_3$BH [FH4]. Epoxy ketones such as **3.94** are also reduced highly stereoselectively by $NaBH_4$–$CeCl_3$ [BB6], $NaBH_4$–$CaCl_2$, or $LaCl_3$ [TF2] (Figure 3.30). $NaBH_4$-$SmCl_3$ can also be used to reduce α-alkoxyketones or their precursors [YN1]. A chelate is formed around the metal and reduction occurs from the least hindered side.

The reduction of aminoketones by LAH, LTBA, and borohydrides has been well studied [N5, T2]. Highly stereoselective reductions have been observed in certain cases [BL1, E1, KL5, T2]. As early as 1972, it was shown that the reduction of α-aminoketones by LAH in Et_2O could be highly stereoselective. According to the size of the nitrogen substituent of **3.95**, the reduction takes place either with or without chelation control [DD3] (Figure 3.31). Similar results were observed with α-aminocyclanones [T2].

The reduction of 2-acyl-1,3-oxathianes such as **3.96** (X = S) or of 2-acyl-3-oxa-N-benzylpiperidines **3.96** (X = NCH_2Ph) can also take place with or without chelation control [E1, EF1, EH2, KE1, KF4]. In cases of chelation control, the oxygen atom of the heterocycle participates in the chelation process (Figure 3.32). When the reaction is carried out with Li(*s*-Bu)$_3$BH in the presence of LiI as an additive, the reduction occurs under chelation control. However, when using two equivalents of DIBAH, each of them coordinates to a one basic site, and no chelation takes place. The use of these chiral auxiliaries allows the synthesis of nonracemic α-hydroxyaldehydes or α-hydroxyesters with a high enantiomeric excess [NN1, S3].

This method can also be applied to α-phenylthioketones **3.97**. Their reduction by Li(*s*-Bu)$_3$BH leads very selectively to the *syn* isomer, whereas $Zn(BH_4)_2$ preferentially gives the *anti* isomer with lower stereoselectivity [SM1] (Figure 3.32). The problem of the reduction of 2-methylthiocyclohexanones has also been examined [CD1].

Other additives can induce the formation of chelates. α-Phosphinyloxyketones **3.98** undergo a stereoselective reduction to *anti* alcohols with $NaBH_4$–$CeCl_3$ in

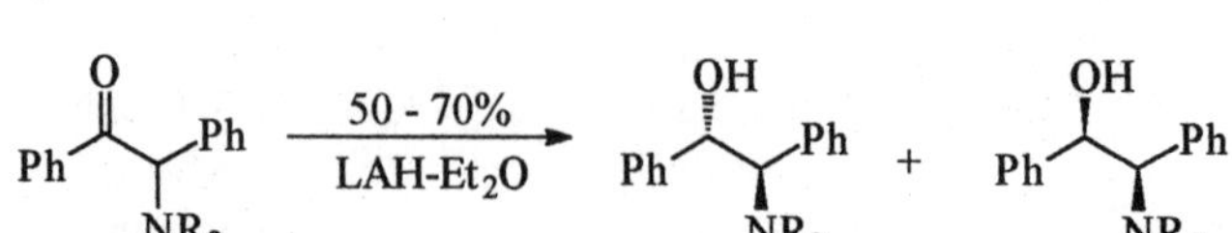

3.95

NR_2		
$NR_2 = NMe_2$, N(pyrrolidinyl)	> 98	< 2
NPr_2	38	62
N*i*-Bu_2	< 2	> 98

Figure 3.31

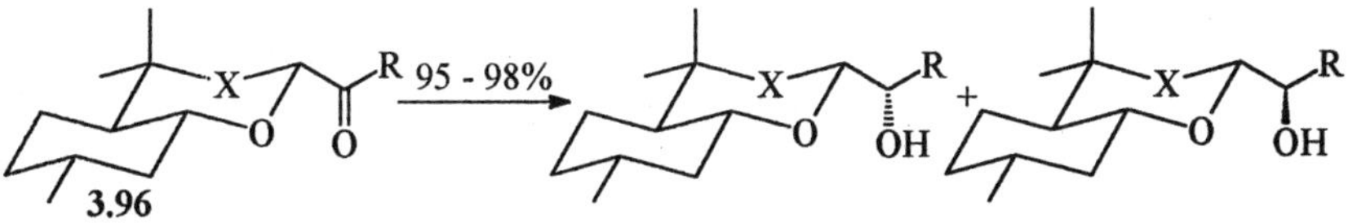

3.96
(X = S)

Li s-Bu_3BH LiI, toluene, -78°C	91 - 98	9 - 2
DIBAH 2 eq. toluene, -78°C	9 - 11	91 - 89

R = n-$C_{10}H_{21}$, Ph, c-C_6H_{11}

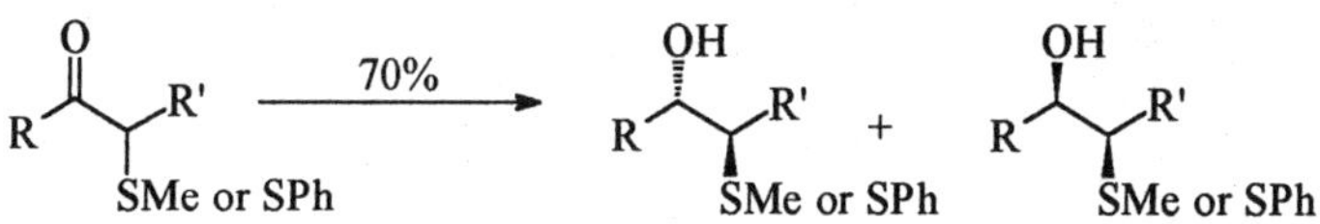

3.97

Figure 3.32

MeOH or $LiBH_4$-$TiCl_4$ [BB11] provided that R′ is bulky, while *syn* alcohols are obtained without $CeCl_3$ [CW3, EW1, HJ5] (Figure 3.33). Hydroxy substituents on R′ can induce good stereoselectivity [CW3, GW2]. α′-Phosphinyloxy-α-enones **3.99** do not suffer a stereoselective reduction by $NaBH_4$–$CeCl_3$ unless the R″ group is bulky enough (i-Pr, cyclohexyl). In these reactions the *anti* isomer is predominant, in agreement with a chelated transition state [CW3, EH1] (Figure 3.33). The extension of this method to α-phosphinoxyketones bearing an oxazolidine substituent has

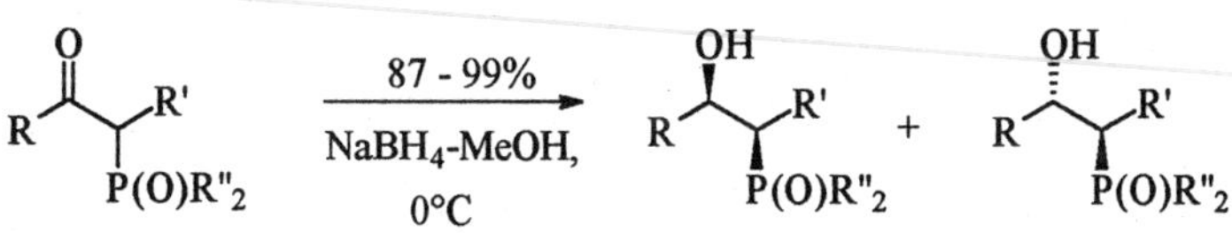

3.98

-	97 - 60	3 - 40
+$CeCl_3$	16 - 3	84 - 97

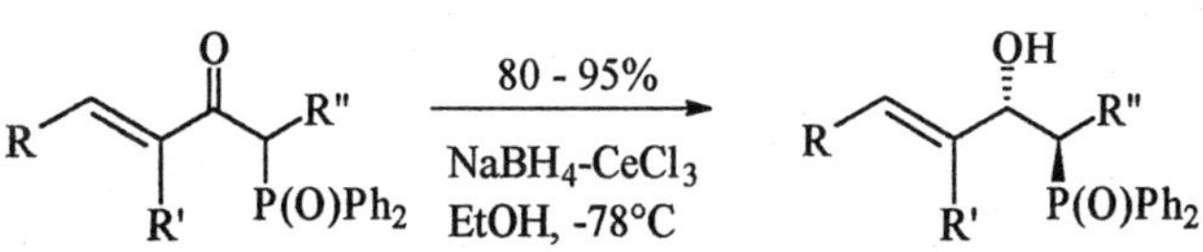

3.99

de > 90%

Figure 3.33

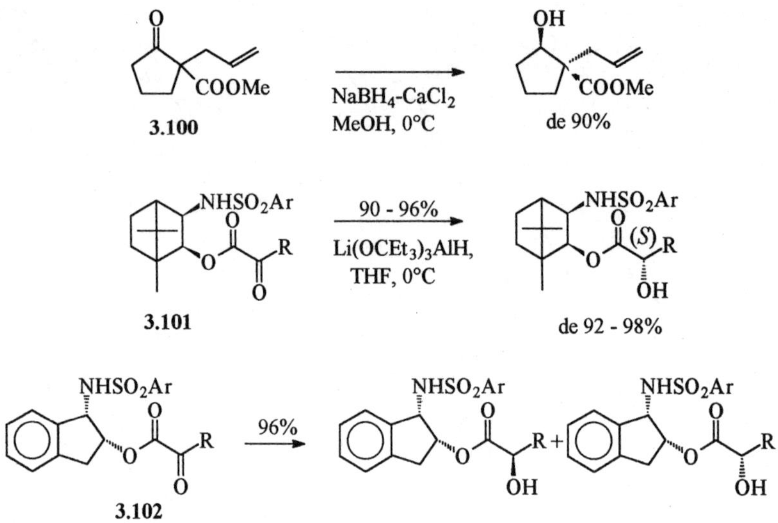

Li s-Bu_3BH-$ZnCl_2$, -78°C	> 99	< 1
K s-Bu_3BH-18-crown-6, -78°C	5	95
$Li(OCEt_3)_3AlH$, 0°C	95	5

Figure 3.34

been examined [OW3]. β-Ketophosphonamidates can also be reduced under chelation control with $NaBH_4$–$CeCl_3$ with a high selectivity [DA3].

In the case of polysubstituted derivatives, such as α,β-dialkoxycarbonyl compounds, Li(s-Bu)$_3$BH and $Zn(BH_4)_2$ lead each to a different isomer, but the stereoselectivity is lower than in the former cases because of competition among the different association sites [FK3, IY1, YK3, YK4]. Analogous problems also arise in the chemistry of sugars [MT2].

The stereoselective reduction of α-ketoesters and α-ketoamides can also be performed. α-Ketoesters may form five-membered chelates that suffer reduction from their least hindered side. For example, ketoester **3.100** is reduced by $NaBH_4$–$CaCl_2$ to the *syn* hydroxyester with a good selectivity, while $NaBH_4$ or Li(s-Bu)$_3$BH gives a mixture of stereoisomers [FB3] (Figure 3.34). Similarly, $LiAl(OCEt)_3H$ in THF at 0°C reduces the chiral ester **3.101** to the (S)-α-hydroxyester under chelation control [XS1] (Figure 3.34). Li(s-Bu)$_3$BH–$ZnCl_2$ also promotes the reduction of **3.102** via chelation control, while the other isomer is formed in the presence of K (s-Bu)$_3$BH-18-crown-6, where chelation is impeded [GC1] (Figure 3.34). K (s-Bu)$_3$BH reduction of phenmenthyl phenylglyoxylate follows the same pathway [SB4].

The reduction of the chiral ketoamide **3.103** by $LiBH_4$ in THF in the presence of

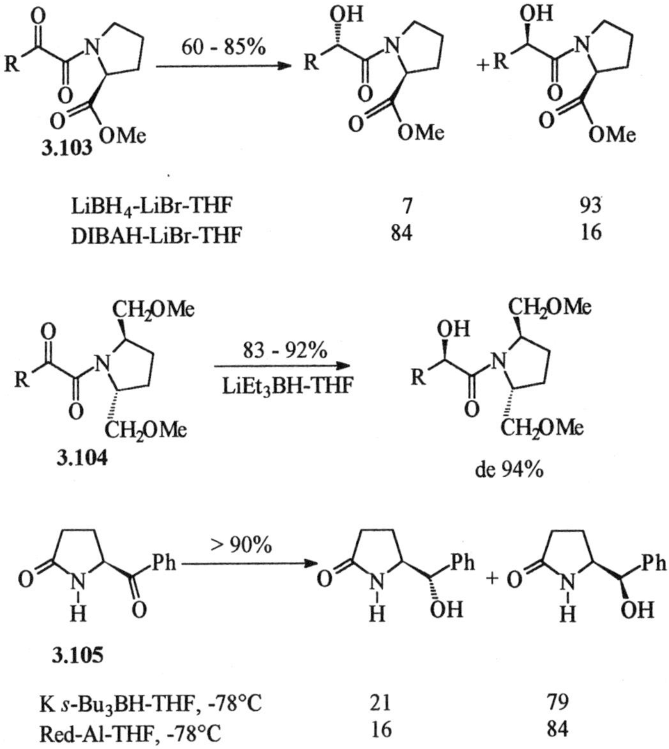

Figure 3.35

LiBr leads selectively to one of the diastereomers [SI1], but with DIBAH, the reduction appears to be less selective (Figure 3.35). The most interesting results are obtained if the heterocycle carries two methoxymethyl [KF3] or CH_2OMEM groups [KF5] such as in **3.104** (Figure 3.35). The reaction of $Zn(BH_4)_2$ is not stereoselective, whereas the use of $NaBH_4$ or KBH_4 in *i*-PrOH is clearly less interesting than that of $LiEt_3BH$. If one employs a secondary cyclic amide such as **3.105**, the stereoselectivity is low [OS1] (Figure 3.35). When the reductions are carried out at low temperature or at 0°C, the amides are not affected (Section 3.2.8).

Stereocontrolled aldol reactions have been the subject of many recent studies [H4, KW4, P2], and the reduction of β-hydroxyketones or their ethers to 1,3-diols has been extensively studied. This reduction can be extremely stereoselective. Thus, from β-hydroxyketones, one obtains *syn* diols very selectively either by the action of DIBAH in THF or Et_2O at low temperature [KK4, M6, PC4, SL2] or with $LiAlH_4$–LiI in Et_2O at −100°C [MK4]. *Syn* diols may also be obtained after the initial formation of an alkyl borate **3.106** with *n*-Bu_3B, Et_3B, BJ6, KP1, NP1] or better yet Et_2BOMe [CG4, CH1, HH4, PC3], and subsequent reaction with $NaBH_4$ in THF (Figure 3.36). The diol is obtained after H_2O_2–NaOH treatment. An improvement to this method is the use of *n*-Bu_2BOMe in THF–MeOH, followed by

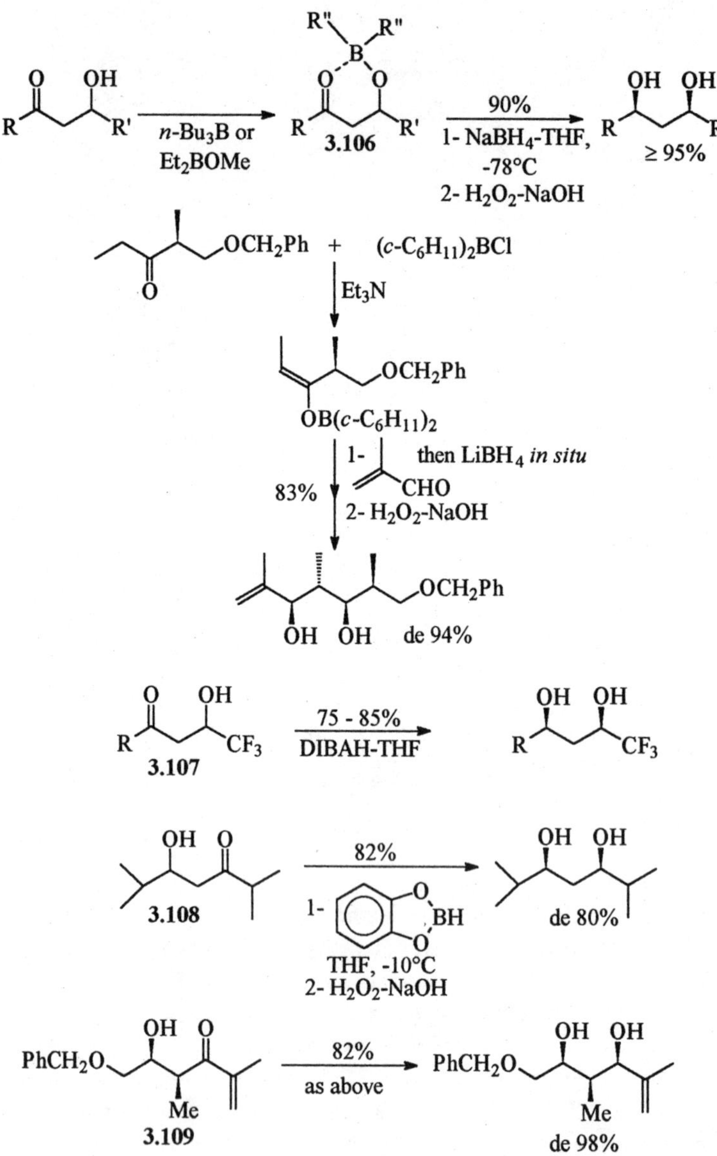

Figure 3.36

reduction with $LiBH_4$ at $-78°C$ [PC4, PP5]. In some cases, the reaction is poorly stereoselective so that the aldol borate **3.106** (R″ = c-C_6H_{11}) obtained by reaction of an aldehyde with a boron enolate is directly reduced in situ by $LiBH_4$, leading thus to a higher selectivity [PC4, PP5] (Figure 3.36). The reduction of trifluoro β-ketoalcohols **3.107** by DIBAH also leads very stereoselectively to *syn* diols [LY1] (Figure 3.36). Provided that the ketone is substituted by large groups, such as **3.108** or

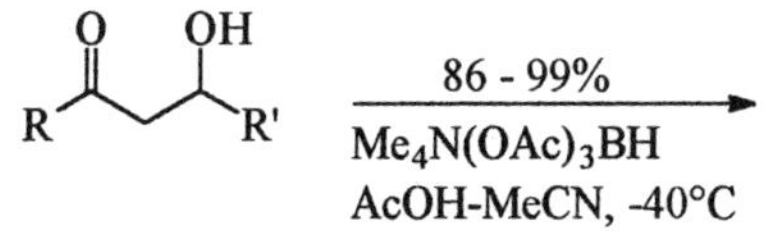

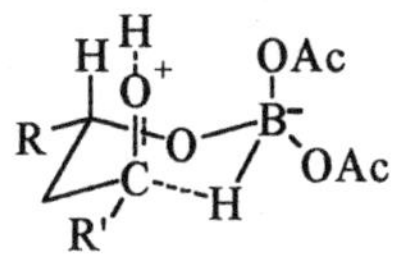

3.110

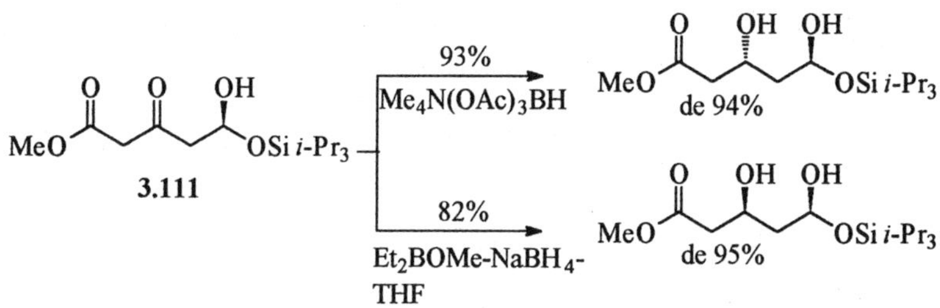

Figure 3.37

3.109, excess catecholborane at $-10°C$ also reduces β-hydroxyketones to *syn* 1,3-diols, most likely via an intermediate borate [EH3]. In some cases, catalysis by $Rh(PPh_3)_3Cl$ improves the selectivity. *Syn* 1,3-diols can also be formed by $TiCl_4$ or BCl_3-mediated reduction of β-hydroxyketones [SC2] via chelation control. Depending to the substituents, one or the other co-reagent is recommended.

The *anti* diols are obtained in a stereoselective manner by reaction with $Me_4N(AcO)_3BH$ in MeCN–AcOH [EC1, EC2, LY1] (Figure 3.37). The reduction takes place by an intramolecular hydride delivery via the least hindered chairlike transition state **3.110** (Figure 3.37). These methods are compatible with functional groups such as esters or amides. For instance, from polyfunctional β-hydroxyketone **3.111**, either diastereoisomeric diol is obtained with a high selectivity [EG2]. In some cases, $Na(AcO)_3BH$ can be used instead of the ammonium salt.

When the β-hydroxyketones carry an alkyl substituent in the α-position, the relative stereochemistry of the R″ group and the hydroxyl group define the stereoselectivity [NP1]. If these two groups are in a *syn* relationship, one obtains selectively the *syn,syn* 1,3-diols by reaction with n-Bu_3B followed by $NaBH_4$ in THF (Figure 3.38). The *syn,anti* isomers can be obtained by action of $Me_4N(AcO)_3BH$ in AcOH–MeCN [EC1, EC2] (Figure 3.38). Many applications of these reductions to the synthesis of polyketide natural products have been published [ER1, H5]. Cyclic β- or γ-hydroxyketones such as **3.112** or **3.113** also experience a highly stereoselec-

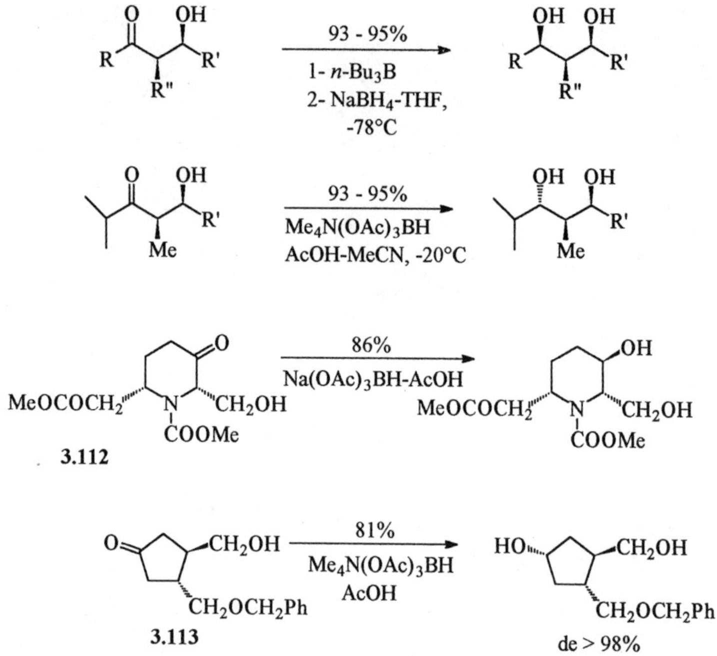

Figure 3.38

tive reduction by Na or $Me_4N(AcO)_3BH$ [HW3, NS5, RS2, SL1, TM2, TY3] (Figure 3.38).

On the other hand, if the OH and R″ groups are in an *anti* relationship, one obtains a mixture of stereoisomers in the acyclic **3.114** as well as in the cyclic **3.115** series [NP1, TM2] (Figure 3.39).

With regard to the reductions involving the intermediate alkylborates, chelated chairlike intermediates **3.116** and **3.117** can be suggested (Figure 3.39). The course of the reduction depends on the relative configurations of R″ and the hydroxyl group. When R′ and R″ are on the same side of the plane of the chelate **3.116**, the hydride approaches the carbonyl on the side opposite to the substituents, and the reduction is stereoselective. On the other hand, when R′ and R″ are on opposite sides of this plane as in **3.117**, either approach of the hydride is constrained, and stereoselectivity is no longer observed (Figure 3.39).

Furthermore, when the R′ and R″ groups are too bulky, chelation becomes unlikely. With either Li(s-Bu)$_3$BH or DIBAH, the reaction yields the product expected from the Felkin–Anh model [SS2], as shown in Figure 3.40. Similar results are observed in the reduction of 2-fluoro-2-trifluoromethyl-3-hydroxy-ketones [IY3].

Similarly, the Felkin–Anh products are formed when chelation is prevented by the formation of t-BuMe$_2$Si ethers of β-hydroxyketones **3.118** and **3.119**. This allows access to *syn,anti* or *anti,anti* α-alkylated diols, depending on the configura-

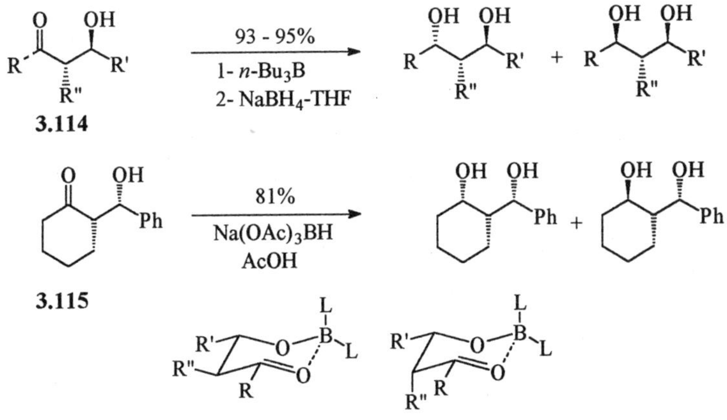

3.114 3.115 3.116 3.117

Figure 3.39

tion of the starting β-hydroxyketone [BG3] (Figure 3.41). The stereoselectivity is poor when R′ = H [BG3]. A similar stereoselection is observed in the reduction of substituted 2-cyclobuten-1-yl methyl ketones with LTBA [HW3], or in reduction of 1-alkoxy-2-phenylalkan-3-ones with Li(s-Bu)$_3$BH [GB8]. If suitable conditions are used, other functional groups can remain intact. The reduction of **3.120** by $LiBH_4$ at room temperature transforms the benzoate into the corresponding alcohol [PW1], while at −78°C, the ester group remains untouched (Figure 3.41).

When less bulky ethers (Me or $PhCH_2$) of β-hydroxyketones are reduced, LTBA

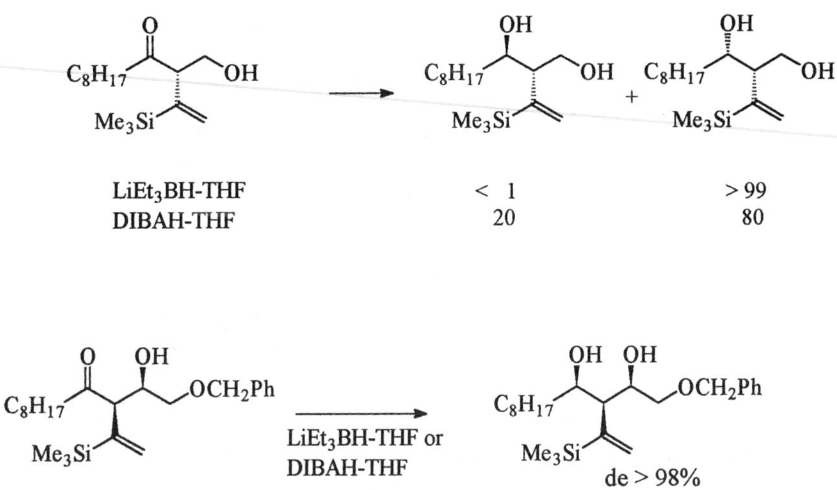

Figure 3.40

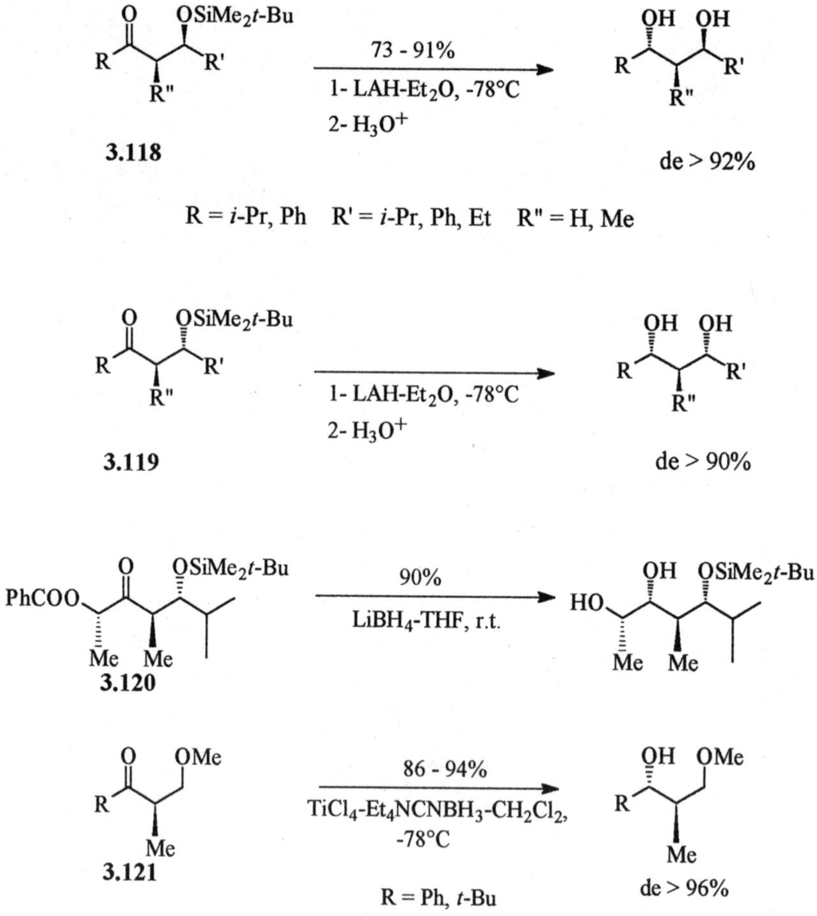

Figure 3.41

or DIBAH promotes reduction with chelation control, while Li(s-Bu)$_3$BH does not. Moderate stereoselectivities are observed [ED2, MK6] even when using β-keto-1,3-oxazolidinones [PP3]. A refined model to interpret these results has been proposed by Evans and co-workers [ED2].

In order to promote chelation control, addition of $TiCl_4$ to β-methoxyketones **3.121** prior to reduction has been recommended [SG2]. The best reducing system is tetraethylammonium cyanoborohydride in CH_2Cl_2 at −78°C (Figure 3.41). Benzyl or MOM ethers give interesting selectivities as well, but disappointing results are obtained when R = Me. Other β-benzyloxyketones are stereoselectively reduced by Li(s-Bu)$_3$BH–$MgBr_2$·Et_2O in CH_2Cl_2 [TC1].

The reduction of the 1,3-diketones **3.122** and **3.123** to diols can also be stereoselective and can lead to either *syn,syn* 1,3-diols or to *syn,anti* 1,3-diols, depending

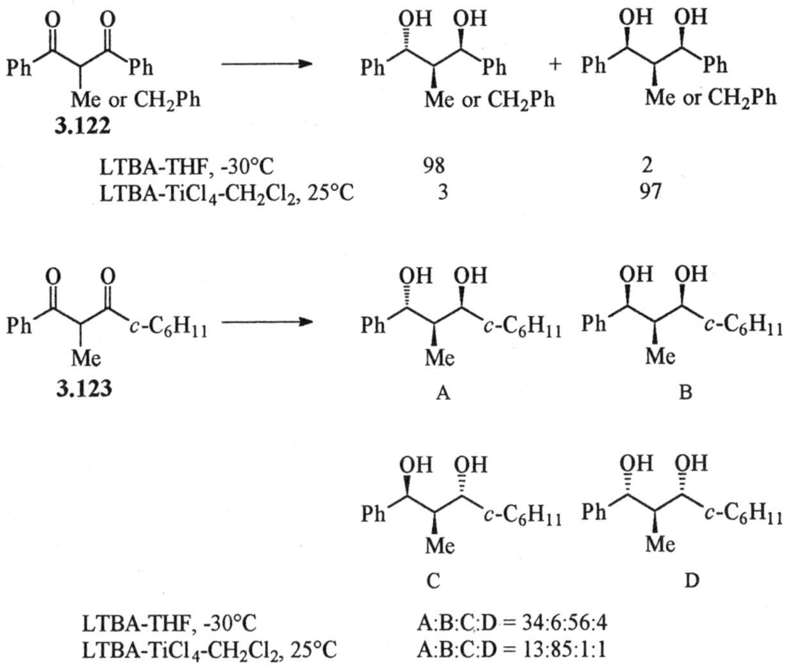

Figure 3.42

on chelation. Chelation control is obtained by adding $TiCl_4$ after the formation of the alcoholate that results from the initial reduction [BR2, MS7]. The selectivity is higher when the environment of the two carbonyl groups is similar (Figure 3.42). A very stereoselective reduction of a bicyclic [3.1.0] diketone by $NaBH_4$–$CeCl_3$ at low temperature has been described [KS6]. $LiBEt_3H$ gives the other stereoisomer, while reductions by LAH and $LiBH_4$ are poorly stereoselective.

β-Aminoketones can also be reduced in a stereoselective fashion [N5, T2]. For instance, the reduction of **3.124** by LAH in THF is stereoselective when R = Ph (Figure 3.43). However, when R = Me or CH_2Ph, the selectivity is poor. The reduction of **3.125** by LAH in THF is stereoselective as well, but better results are obtained in the presence of $TiCl_4$ [BO1] or by reduction with DIBAH–$ZnCl_2$ at −78°C with **3.126** [BV2] (Figure 3.43). When the reduction is carried out at reflux in CH_2Cl_2, the corresponding amides can be used [BA3]. A highly selective reduction of ketone **3.127** by $Zn(BH_4)_2$ has been described [GM3] (Figure 3.43). However, when the Me group is located on the other side of the carbon skeleton, the selectivity is low (40%). The reduction of triazolylketones with n-Bu_4NBH_4 gives the isomer predicted by the Felkin–Anh model, while chelation control operates when the reduction is carried out in the presence of $TiCl_4$ [TS2].

The reduction of β-ketoesters such as **3.128** or β-ketoamides such as **3.129** by $Zn(BH_4)_2$ is extremely stereoselective in favor of the *syn* β-hydroxy isomer, while KBH_4 or n-Bu_4NBH_4 in EtOH or, better, K(s-Bu)$_3$BH, lead selectively to the *anti*

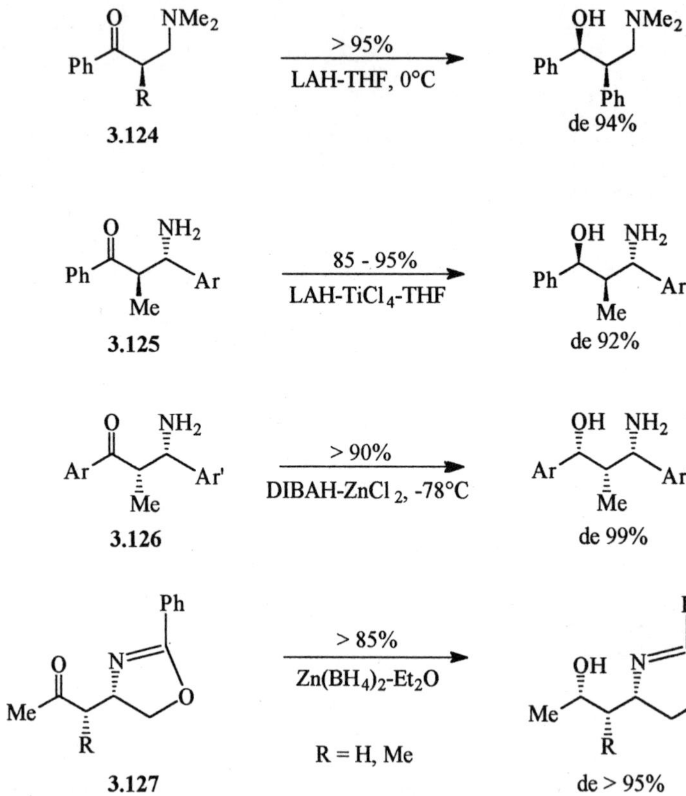

Figure 3.43

isomer [IK1, ON1] (Figure 3.44) from racemic or nonracemic series. Similarly, β-keto-N-acyloxazolidinones **3.130** are stereoselectively reduced to *syn* alcohols by $Zn(BH_4)_2$ in CH_2Cl_2 [NF1] (Figure 3.44).

As before, when the cation associated with the reducing agent or when the reagent itself is a strong Lewis acid, one can consider a chelated transition-state model **3.131** (Figure 3.45). In the absence of chelation, a Felkin–Anh transition-state model **3.19** is envisioned. The attack of the hydride takes place on the face opposite to the most polar (P) group, in this case, the ester or amide group (Figure 3.45). When the reduction is carried out with $LiEt_3BH$, the results (low stereoselectivity) show that the two possible transition-state models can be considered. Chelation control operates during the reduction of cyclic ketoamides such as **3.132** by Li $(s\text{-Bu})_3BH$ in THF at low temperature. The observed stereoselectivity is opposite to that using $K(s\text{-Bu})_3BH$ in Et_2O at 20°C (Figure 3.45). Other reagents are less selective. Chelation is favored by entropic effects due to the lowering of the temperature [PA1]. Similar results have been obtained for related cases [SK2].

Other substituents in the vicinity of the ketone group can also influence the stereochemical course of the reduction. An OMe group or a fluorine atom at the

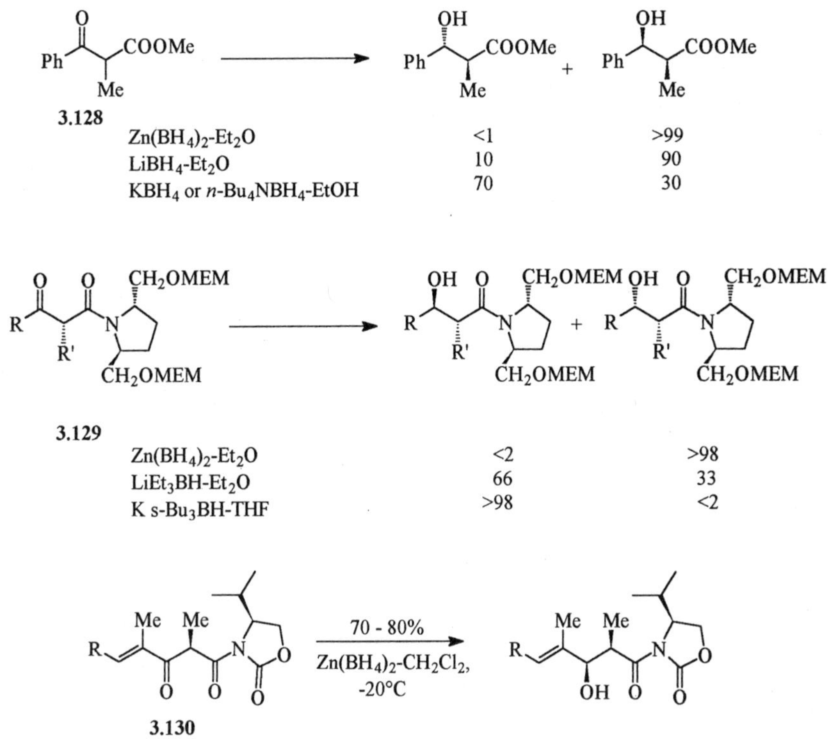

Figure 3.44

ortho position in an arylketoester **3.133** induces an opposite stereoselectivity depending on the reagent employed [BF1] (Figure 3.45). The presence of a fluorine atom on the α-alkylated carbon diminishes the stereoselectivity of the reduction of β-ketoesters **3.134** by $Zn(BH_4)_2$ [KK7] (Figure 3.45).

Finally, the introduction of additives may allow the stereoselectivity of the reductions to increase. Thus the addition of $ZnCl_2$ to $Zn(BH_4)_2$ or the coordination of the carbonyl group by a bulky Lewis acid such as diisobutylaluminum 2,6-di-*t*-Bu-4-methylphenolate (BHT) induces high and opposite stereoselectivities from chiral β-ketoesters **3.135** (Figure 3.46). In the first case, chelation is strengthened, and the reduction involves a cyclic transition state. In the second case, chelation is disfavored, and the other isomer is formed [TD1]. Chelation may also be promoted in reductions of β-ketoesters or amides by addition of $TiCl_4$ [SG2] or $MnCl_2$ in catalytic amounts [FO1].

The reduction of the 14-membered ring β-ketolactone **3.136** by Li(s-Bu)$_3$BH in THF or by $NaBH_4$–$MnCl_2$ in MeOH is also highly stereoselective, while $NaBH_4$ gives a mixture of stereoisomers [NO1] (Figure 3.46). The poor selectivity in the absence of chelation has been ascribed to conformational factors. This problem has also been raised in some other cases [W7].

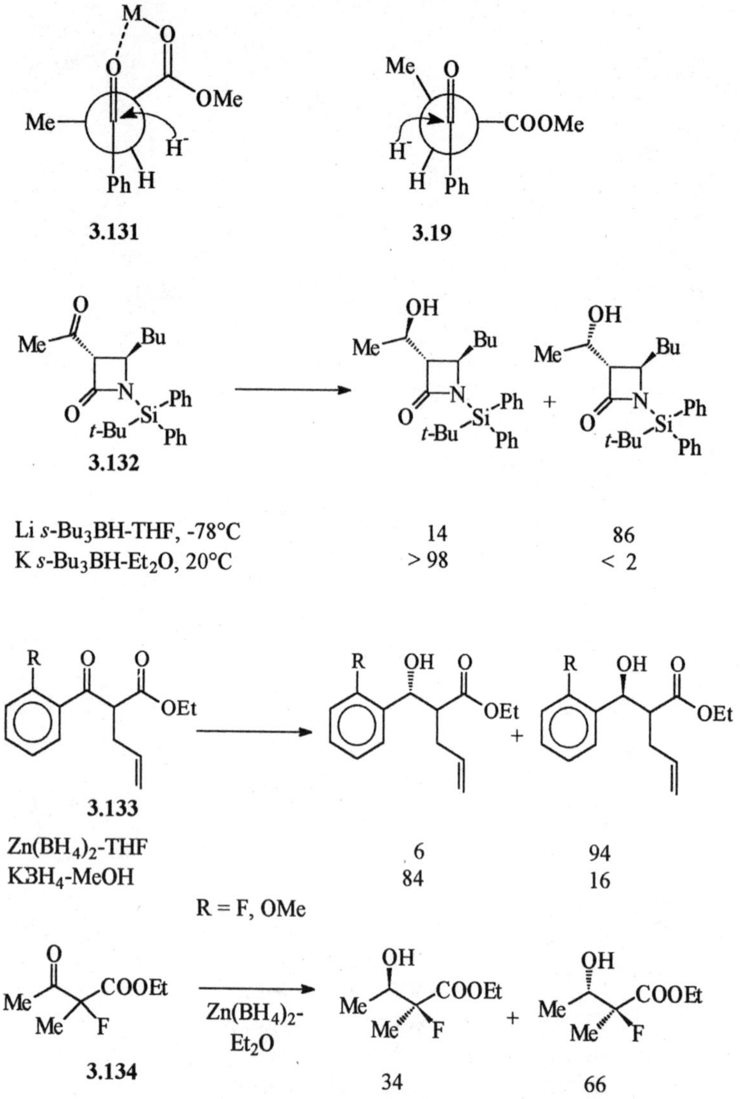

Figure 3.45

When the chelating group is located in the γ-position, stereoselective reduction can also be observed. For example, γ-oxo-γ-phenylbutanoic acids **3.137** are reduced to *syn* γ-hydroxyacids by DIBAH–$ZnCl_2$. The other reducing agents are far less stereoselective [FK2] (Figure 3.47). N-Acylamides such as **3.138** have been reduced with a high stereoselectivity by $NaBH_4$–$CeCl_3$. Chelation takes place between the keto group to be reduced and the BOC residue [AC3] (Figure 3.47). Similar results were observed with other amides such as **3.139** [WH5]: One stereoisomer was generated under chelation control and the other one without chelation (Figure 3.47).

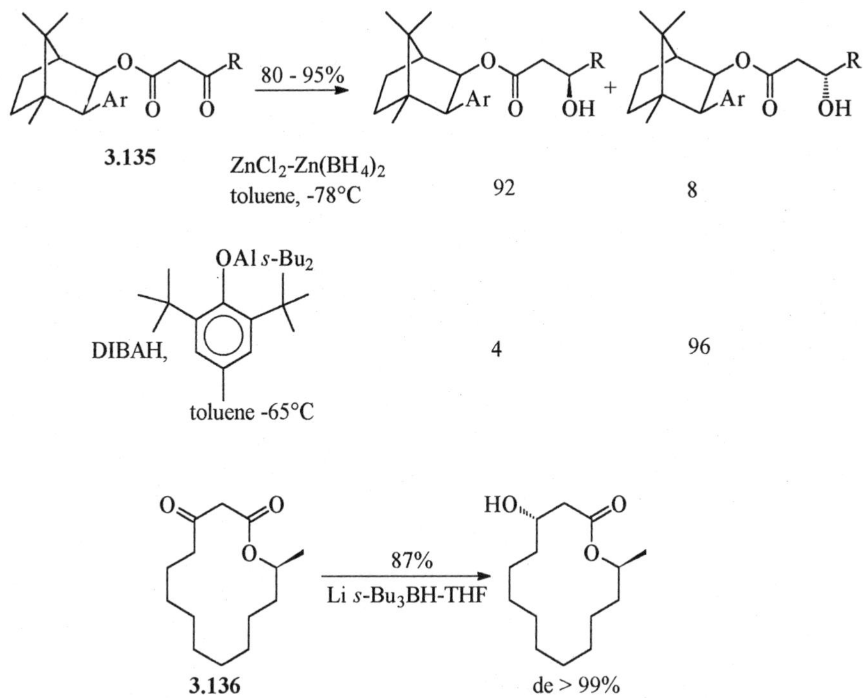

Figure 3.46

Other examples were described in similar systems [GP2, KK12, R5] and with N-sulfamides, where chelation involves one of the SO residues [MP3].

β-Ketosulfoxides **3.140** can be reduced in a stereoselective fashion and, depending on the conditions chosen (i.e., DIBAH or DIBAH–$ZnCl_2$), they can lead to either of the two possible diastereoisomers [CD1, CG5, KK3, SD1, SG1, SS6] (Figure 3.48). The transition-state models **3.141** and **3.142** for this reduction are chelated, and the hydride attacks the carbonyl on the face opposite to the tolyl group (Figure 3.48). $ZnCl_2$ can be used in catalytic amounts [SS6]. The interest in these reductions lies in the access to chiral alcohols after the cleavage of the C—S bond by aluminum amalgam. Starting from these types of compounds, one can also obtain chiral epoxides and lactones. This methodology has often been used in the synthesis of drugs or natural products [BL3, S6, SA2, SS7]. It has also been applied to the stereoselective reductions of different β-sulfoxides [CC7, LT2].

The reduction of ketosulfoxides by BH_3·THF, $Zn(BH_4)_2$ in Et_2O–THF, LTBA in Et_2O, or $NaBH_4$ in EtOH is poorly stereoselective [KK3, GP1], as is the reduction of chiral ketosulfoximines [JS3]. The stereoselective reduction of chiral β-ketosulfones has been described [BB12].

Long-range chelation control has been observed in the reduction of γ-ketoboronates **3.143** and **3.144** [CM2, MB3, MB4] (Figure 3.49). Among the various

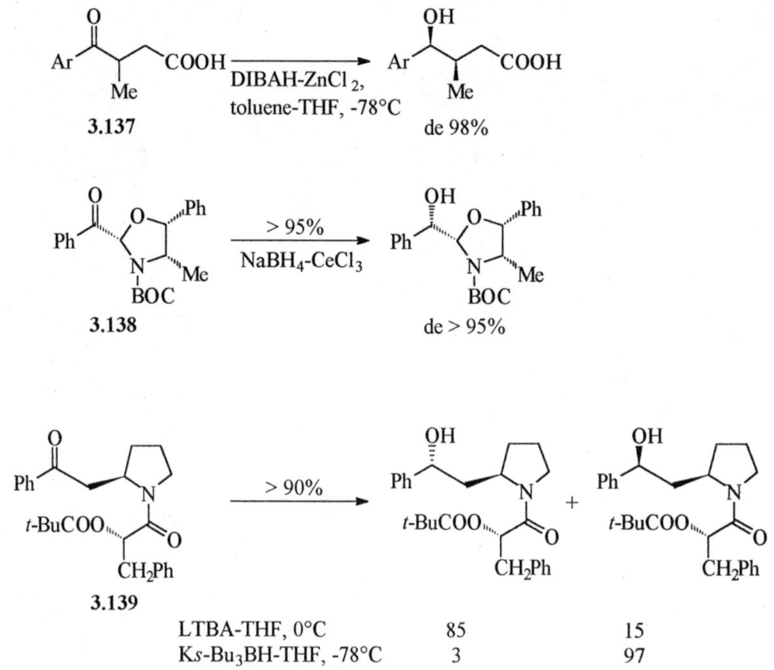

Figure 3.47

reagents tried, $BH_3{\cdot}SMe_2$ gave the best selectivities. These results have been interpreted in terms of intermolecular hydride delivery on a boron chelate [CM2], although the hypothetical intermediate **3.145** could not be observed by IR or NMR spectroscopy [MB3] (Figure 3.49). The formation of a bicyclic metal chelate species has also been proposed to interpret the stereoselectivity of the reduction of **3.146** by (*R*)-Alpine hydride or $Zn(BH_4)_2$ [ZC1, ZH2] (Figure 3.49).

When the ketones to be reduced are substituted by several groups capable of chelating the reducing agent or the associated cation, the reactions are only slightly stereoselective unless one of the groups bears a substituent that disfavors chelation. The reduction of aminoketoesters **3.147** is poorly selective except if R′ = CH_2Ph [R4] and if the reduction is performed in a slightly acidic medium [GB1, GB3, RD1] (Figure 3.50). Chelation with nitrogen is prevented in acid, and reduction takes place preferentially on a rigid chelate system with hydrogen bonding involving the ketone and the ester groups. An analogous example is compound **3.148**, whose reduction is not stereoselective in neutral media [OF1] (Figure 3.50). A similar stereoselectivity towards the *syn* isomer has been observed in the reduction of 2-methyl-2-thiophenyl-β-ketoesters by $Ca(BH_4)_2$ in MeOH–THF [SS4]. Similarly, only $Li(s\text{-}Bu)_3BH$ in THF selectively reduces the multifunctional compound **3.149**, wherein the transition state concerned is of the Felkin–Anh type [MT1]. Similar results are obtained for other aminoketones **3.150** that lead predominantly to *syn* isomers with Li or $K(s\text{-}Bu)_3BH$ in THF and to *anti* isomers with $LiBH_4$ in *i*-PrOH

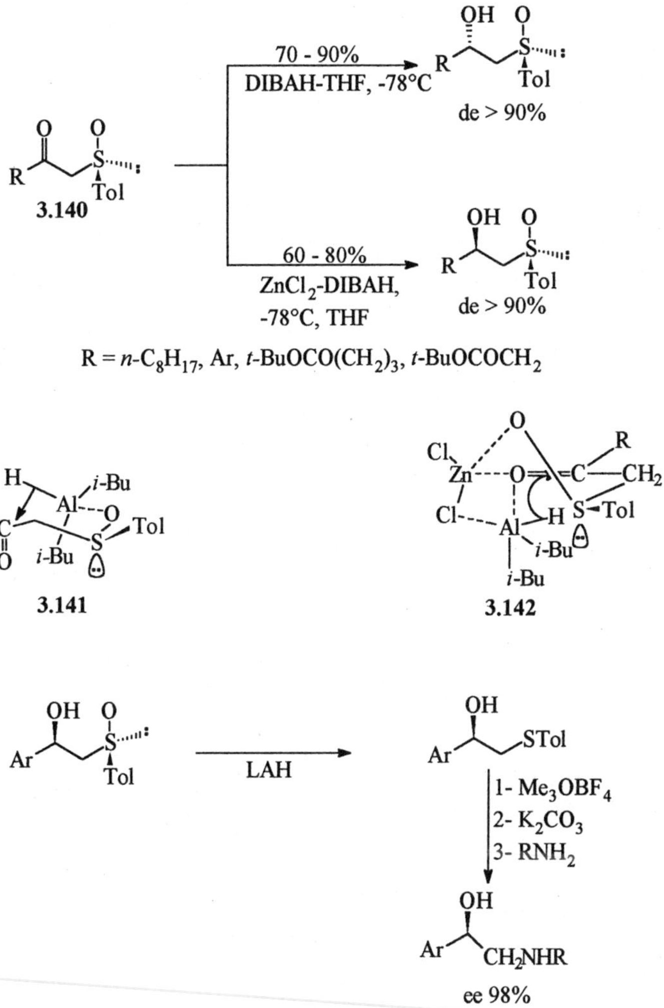

Figure 3.48

[RR1] (Figure 3.50). The use of unsymmetrical ketodiethers such as **3.151** also allows the stereoselective reduction of the carbonyl group under chelation control. Their reduction by DIBAH–$MgBr_2$ in diethylether leads selectively to one diastereoisomer **3.152** because chelation does not take place with the bulky $OSiPh_2$-*t*-Bu group. The best results are obtained when R′ is 4-$MeOC_6H_4CH_2OCH_2$ [BG4, GB7] (Figure 3.50).

Similar observations have been made concerning the reduction of β-pyridylketosulfoxides [GP1]. Some stereoselective reductions have been performed when substituents able to induce chelation are located on each side of the carbonyl group [HJ4, MS4, MT3].

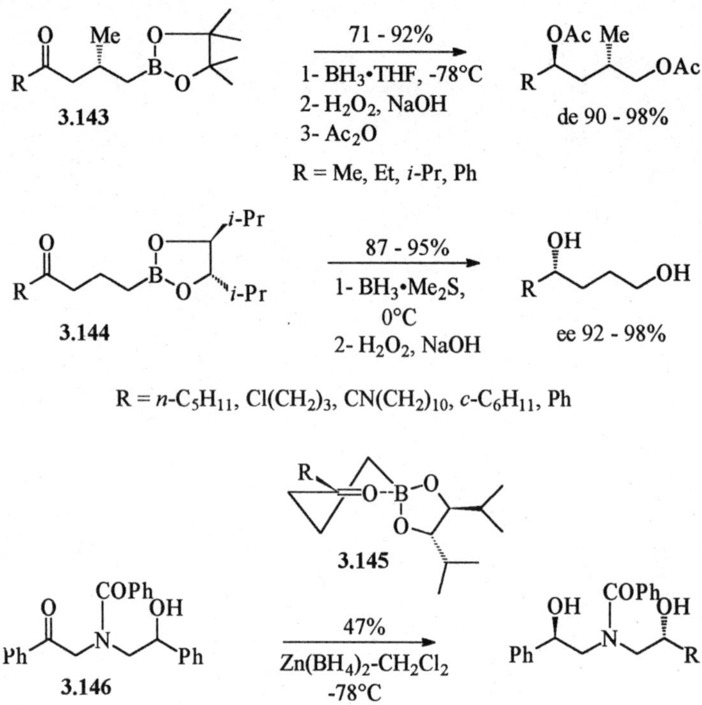

Figure 3.49

3.2.5 Esters, Lactones, and Thiolesters: R–C(=O)OR′, R–C(=O)SR′

Esters and lactones can be partially reduced with the participation of a single hydride. Starting from esters, aldehydes are formed via the corresponding hemiacetals. Starting from lactones, the products are generally lactols (Figure 3.51).

These transformations are still difficult to realize in practice. LTBA in THF at 0°C sometimes allows phenyl esters to be transformed to aldehydes with good yields, while the reaction of aliphatic esters is much slower [BK5, C5, M3] (Figure 3.52). Aluminum bis(N-methylpiperazino)hydride in THF at 25°C allows this transformation to take place starting from alkyl esters [C5, H3, M3]. The use of LAH in the presence of diethylamine in pentane has been proposed for monoreduction of esters [CK1, CK3]. $Na(Et_2N)_3AlH$ seems more useful; alkyl esters are transformed into the corresponding aldehydes in THF at −78°C [CK5] (Figure 3.52). Sodium diethylpiperidinohydroaluminate reduces ethyl esters of aromatic acids to aldehydes in good yields at 0°C in THF–toluene. However, esters of aliphatic acids give mixtures of aldehydes and alcohols, while lactones are not transformed into lactols under these conditions [YA2].

DIBAH in toluene at low temperature in stoichiometric amounts is often recommended for carrying out the reduction of saturated esters to corresponding al-

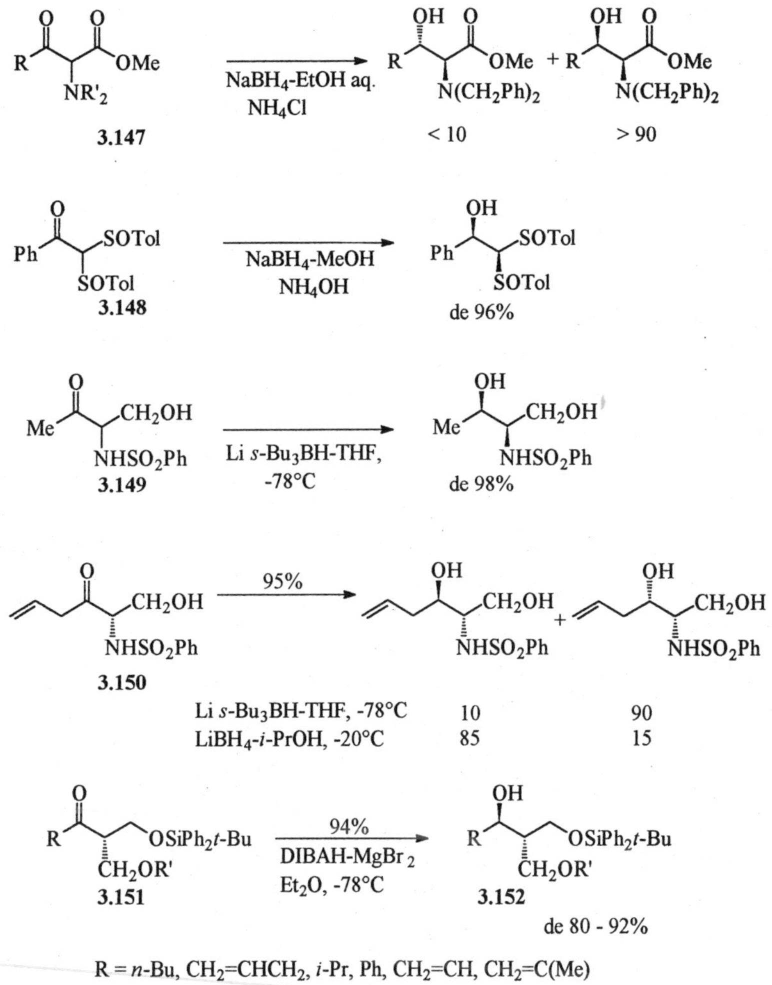

Figure 3.50

dehydes [C5, K2, W1, YG1] (Figure 3.52). In the presence of *o*-anisidine, yields are improved [KK9]. The reduction is compatible with an α-SePh substituent, whose absolute configuration in a chiral molecule is retained [DD1]. Use of *i*-Bu_2AlD leads to deuterated aldehydes [KW1] (Figure 3.52). The intermediate hemiacetal may, in some cases, be reacted in situ with another reagent [DR1]. For example, protected α-amino esters **3.152** react sequentially with DIBAH or *i*-Bu_3Al–DIBAH and then with PhMgBr. This process provides the corresponding α-amino alcohols in excellent chemical yield with good stereoselectivity [PP2] (Figure 3.52). A similar two-step process can also be realized by reduction of ethyl esters by limited amounts of $LiBH_4$ in the presence of an organomagnesium reagent [CH2] (Figure 3.52).

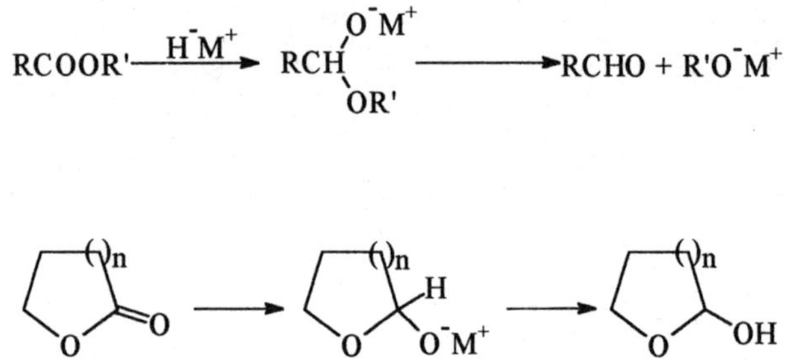

Figure 3.51

The reaction of DIBAH with ketones is, however, more rapid. The selective reductions of ketoesters (Section 3.2.4) have already been described. Moreover, α,β-unsaturated esters give rise to allylic alcohols, even if a less than stoichiometric amount of reagent is used (Section 3.2.9).

Reductions of lactones, under the same conditions, or in Et_2O, lead to the corresponding lactols [BG5, DT2, M3, SY4, W1], as shown in Figure 3.53. One of Corey's early syntheses of prostaglandins involves the reduction by DIBAH of lactone **3.152a** [W1], followed by a Wittig reaction, which can be carried out directly on the lactol. The reduction of multifunctional lactone **3.153** respects the integrity of the other groups [WV1] (Figure 3.53). Thiolesters are also reduced by DIBAH in toluene at low temperature into aldehydes [GM2, GW4]. Under these conditions, the 1,3-oxazoline functionality of **3.154** remains untouched (Figure 3.53). Sia_2BH reduces lactones to lactol borates, which, after hydrolysis, give γ-hydroxyaldehydes [BK5] (Figure 3.53).

Esters and lactones are reduced to alcohols or diols by numerous reagents (Figure 3.54). LAH in ethers and on silica gel converts these groups to alcohols resulting from a double reduction. At low temperature, a mixture of aldehydes and alcohols is usually obtained [BK5, H3, KH2, M3]. Reduction of ethyl *trans*-3,4-epoxycyclopentanecarboxylate to the corresponding alcohol, without epoxide ring opening can be performed with 0.5 equiv. LAH in THF at −78°C [MN2]. The reduction of esters by LAH may allow alcohols to be prepared from acetates. This is a useful method for acetate cleavage when hydrolysis would induce side reactions, such as epimerization (Section 2.2). An example is given in Figure 3.54 for the reduction of **3.154** [AK2].

$NaAlH_4$ [CB5] or LAH–N-methylpyrrolidine [FS1] is also an efficient reducing agent, as is excess $AlH_3{\cdot}Et_3N$ [CB7]. All these reductions are run in THF at 0°C or at r.t. Ate complexes formed from DIBAH and *n*-BuLi in THF–hexane at r.t. also give alcohols from esters [KA1]. Red-Al in petroleum ether allows the selective reduction of long-chain bromoesters to bromoalcohols [W8].

Since LTBA reacts slowly with esters at low temperature, selective reductions of ketones can be performed [KW5, M3]. AlH_3 also reduces esters to alcohols [BK5],

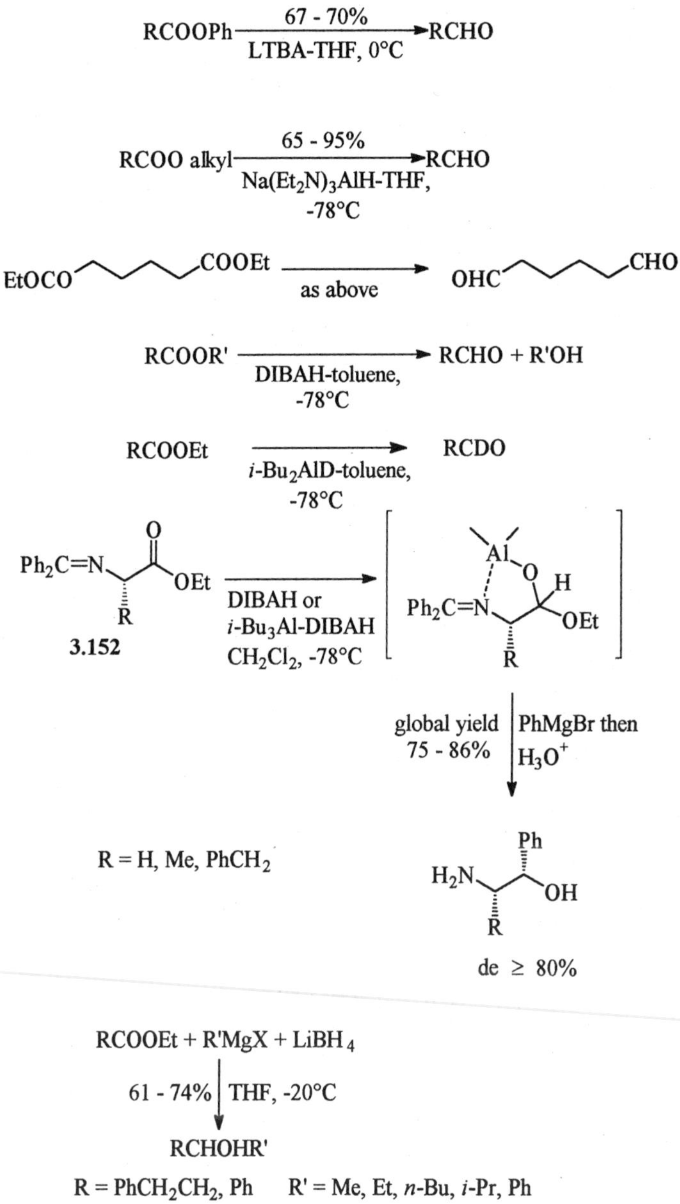

Figure 3.52

as does $LiBH_4$ in hot DME or $LiBH_4$ in refluxing Et_2O or in THF at r.t. [BK5, BS1, PW1]. The reduction requires electrophilic assistance since it is faster with $LiBH_4$ when the solvent is not a good solvating agent of the cation: Et_2O > THF ~ DME > *i*-PrOH [BN1]. It is therefore possible to perform the selective reduction of the keto group of α-ketoester **3.155** by $LiBH_4$ in THF at −78°C, while both functional

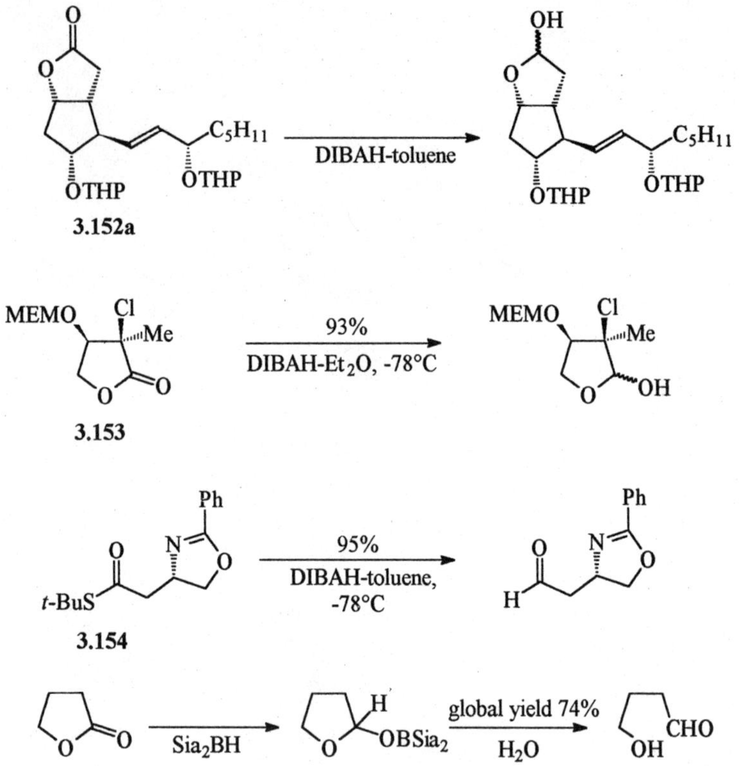

Figure 3.53

groups of **3.156** are reduced at r.t. [PW1] (Figure 3.54). Other examples have been given in Section 3.2.4. As acids are inert to $LiBH_4$, it is possible with this reagent to reduce selectively the ester group of hemiesters of diacids [BG2, BK5, OK1]. $LiBH_4$ can be generated in situ from $NaBH_4$–LiCl in THF–EtOH [HI1, HS4]. With this system, chiral amino esters can be reduced to amino alcohols without epimerization [JF2]. $Ca(BH_4)_2$, formed from $CaCl_2$ and $NaBH_4$ in aqueous ethanol, can also be used for this purpose [BR3, HC2, LR1]. This reagent does not reduce alkali carboxylates; so it is possible to perform selective reduction of the ester group of **3.157**, which is followed by lactonization in situ [LR1]. Similarly, N-(ethoxycarbonyl)amino esters are reduced to N-(ethoxycarbonyl)amino alcohols [LM3]. The selective reduction of an ester functionality in the presence of a secondary tosylate can be performed with $LiBH_4$–$LiBEt_3H$ in Et_2O–THF at 0°C [AS1].

The reduction of esters by $LiBH_4$ can be accelerated by the addition of other Lewis acids such as Et_3B [YP2], $B(OMe)_3$ [BN3], or methanol [SO3]. In the last case the reduction is carried out under reflux in Et_2O in the presence of 4 equivalents of MeOH. Acid, Cl, and NO_2 groups remain unchanged under these conditions.

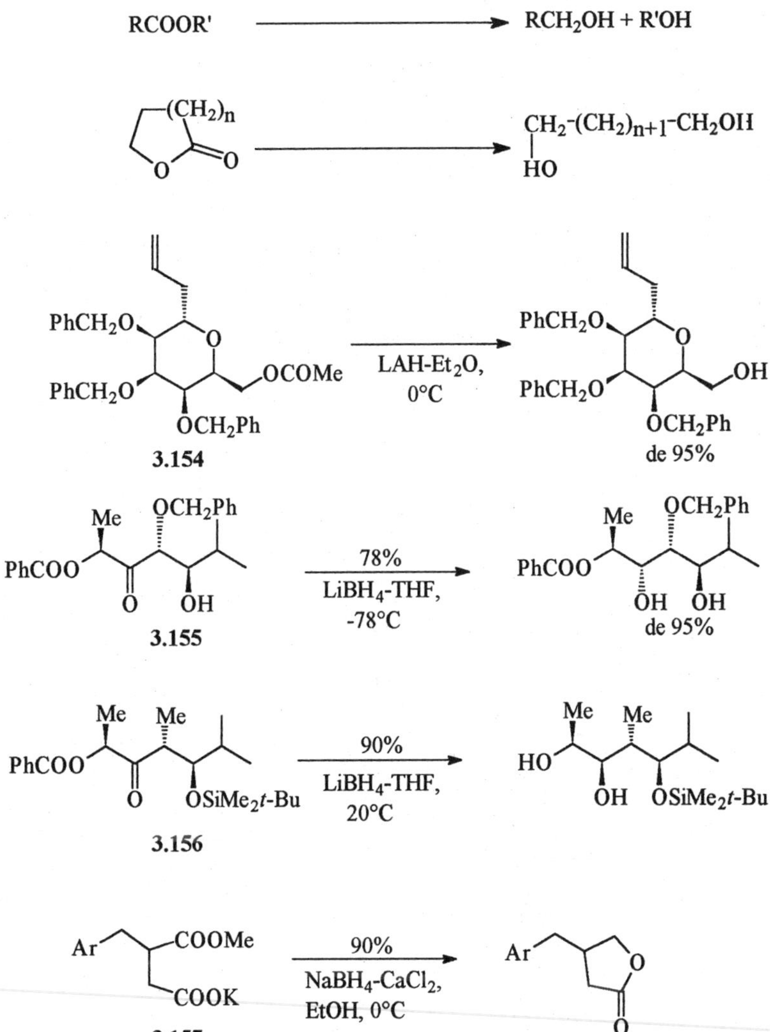

Figure 3.54

$LiBuBH_3$ reduces esters in Et_2O at 0°C, but leaves them untouched in toluene–hexane at −78°C [KM2]. Lithium trialkylborohydrides as well as lithium aminoborohydrides in THF transform esters to alcohols and lactones to diols [BK5, FH5]. Steric hindrance around ester groups can allow the regioselective reduction of the least hindered functional group of an unsymmetrically substituted diester by $LiEt_3BH$ in THF at 0°C [FR1]. An ester group can be reduced in the presence of an amide or a carbamate [TY3]; the reduction of **3.158** illustrates this compatibility (Figure 3.55). KEt_3BH in THF at r.t. reduces esters quickly enough to leave epoxides, amides, and nitriles unchanged [YY1]. Moreover, K(*s*-Bu)$_3$BH in THF at 0°C

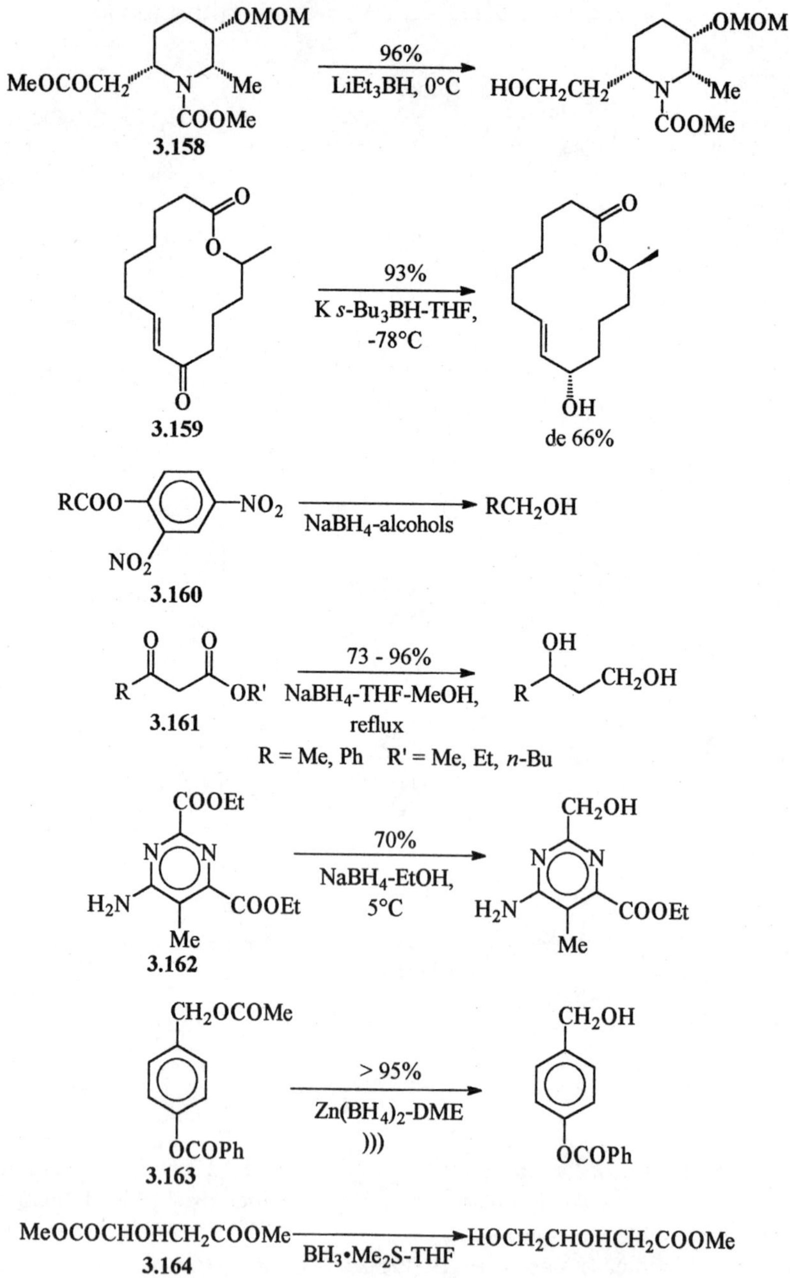
OMOM
MeOCOCH2
N
Me
COOMe
3.158
96%
LiEt3BH, 0°C
HOCH2CH2
OMOM
N
Me
COOMe
O
O
O
3.159
93%
K s-Bu3BH-THF,
-78°C
O
O
OH
de 66%
RCOO
NO2
NO2
3.160
NaBH4-alcohols
RCH2OH
O
O
R
OR'
3.161
73 - 96%
NaBH4-THF-MeOH,
reflux
OH
R
CH2OH
R = Me, Ph R' = Me, Et, n-Bu
COOEt
N
N
H2N
COOEt
Me
3.162
70%
NaBH4-EtOH,
5°C
CH2OH
N
N
H2N
COOEt
Me
CH2OCOMe
OCOPh
3.163
> 95%
Zn(BH4)2-DME
)))
CH2OH
OCOPh
MeOCOCHOHCH2COOMe
3.164
BH3•Me2S-THF
HOCH2CHOHCH2COOMe

Figure 3.55

reduces lactones to diols faster than PhCOOEt or ethyl caproate [YH2]. However, at $-78°C$, lactones are left unchanged, thereby allowing the selective reduction of the keto group of 9-oxo-7-tetradecen-13-olide **3.159** by K(*s*-Bu)$_3$BH [KW5] (Figure 3.55).

n-Bu$_4$NBH$_4$ in CH_2Cl_2 does not react with esters [RG1], and reductions by $NaBH_4$ in alcohols or on alumina are very slow (Section 3.2.1). However, 2,4-dinitrophenyl esters **3.160** are easily reduced to alcohols [PS1] (Figure 3.55). This reduction takes place in the presence of ethylene glycol oligomers at 80°C [SF1]. Reduction can also be accomplished in the presence of additives. The action of $NaBH_4$ on esters in THF or refluxing *t*-BuOH in the presence of MeOH [SO1], or in refluxing EtOH [OS3], or even in water [BP3] leads to the corresponding alcohols. Under these conditions, primary amides, acids, and NO_2 groups remain inert. Thus, starting from ketoesters **3.161**, one can obtain diols [SO2] (Figure 3.55). $NaBH_4$ in EtOH allows the selective reduction of the less sterically hindered ester group of diester **3.162**, while $LiBH_4$, DIBAH, or Selectrides are unable to differentiate them [BH6] (Figure 3.555).

Moreover, α-cyano-α-epoxy esters are easily reduced to α-cyano-α-epoxy alcohols by $NaBH_4$ in aqueous THF [MR4]. Ethanedithiol can also be an additive for the reduction of esters by $NaBH_4$, except *t*-butyl esters. Nitrile groups remain unperturbed under these conditions [GE1]. Methyl benzoate is reduced to benzylalcohol by $NaBH_4$–$ZrCl_4$ in THF [IS1].

Esters are much less sensitive than ketones to $Zn(BH_4)_2$ or cyanoborohydrides [PS1], and the selective reduction of the ketone groups of α- and β-ketoesters can be accomplished without problems (Section 3.2.4). Moreover, $Zn(BH_4)_2$ in DME under sonication reduces acetates or cyclohexanecarboxylates while benzoates are left untouched [R3]. The chemoselective reduction of the acetate residue of **3.163** can be performed under these conditions (Figure 3.55).

BH_3·THF or BH_3·SMe_2 reacts very slowly with esters in THF at room temperature even in the presence of amino alcohols [BK5, IW1, L2, PS1]. Under reflux, BH_3·SMe_2 reduces esters to alcohols [BC1], as does the in situ generated borane [BB7]. Nevertheless, α-hydroxyesters can be reduced at room temperature by BH_3·SMe_2 in THF in the presence of a catalytic amount of $NaBH_4$. The selective reduction of **3.164** is an example of this reaction [SH1] (Figure 3.55). γ-Carboxyesters also undergo reduction of the ester group by BH_3·SMe_2 [FC2]. The reduction of esters to alcohols by catecholborane also takes place in refluxing THF [KB7].

Most acyloxyboranes [MM1] and aminoboranes [A1] do not react either with esters or with lactones. However, Ph_2NH·BH_3 reduces aliphatic esters [CU1]. Substituted boranes are more efficient. 9-BBN reduces esters under reflux in THF [PS1], while ThexBHCl gives rise to alcohols with heating [BN5]. Finally, the ate complex Li 9-BBNH reduces esters to alcohols and lactones to diols. Acids, amides, nitriles, and halogenated derivatives remain intact under these conditions [BM1].

It has clearly been emphasized that ketones are reduced more rapidly than esters (Section 3.2.4). It is nevertheless possible to reduce ketoesters such as **3.165** to corresponding ketoalcohols [BH2, KF2] by forming the corresponding lithium enolates in a first step (Figure 3.56). The limitation of the method is the stability of the

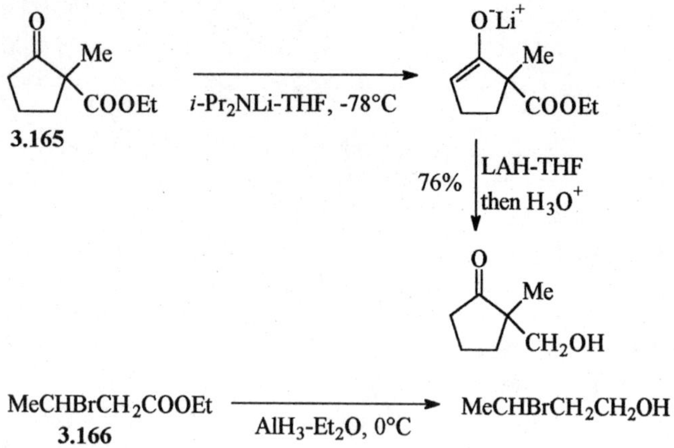

Figure 3.56

enolates formed and the need for the absence of labile hydrogen at the α position of the ester group. Competitive enolization decreases the yield for reduction of malonates to 1,3-diols. In this case, good reagents for avoiding this side reaction are electrophilic hydrides such as AlH_3 or, better, DIBAH in THF [CE2]. The reduction of bromoesters **3.166** to bromoalcohols by AlH_3 in Et_2O leaves the carbon–halogen bonds unchanged [BK5, E2] (Figure 3.56).

Thiolesters (RCOSEt) are reduced to alcohols by $LiBH_4$ in Et_2O or by an excess of n-Bu_4NBH_4 in refluxing $CHCl_3$, but are inert in the presence of borane [KH5, LL4]. Dithioesters and thioxoesters are reduced to thiols by $BH_3 \cdot Me_2S$ [JS4].

3.2.6 Carboxylic Acids, Acid Anhydrides: RCOOH, RCOOCOR′

Again, reduction can lead to aldehydes or alcohols, and the choice of reducing reagent and reaction conditions dictate the formation of one or the other functional group.

The reduction of acids to aldehydes may be accomplished by aluminum bis(N-methylpiperazino)hydride in THF in excellent yields, starting from both aliphatic and aromatic acids [C5, H3, HE1, MM3]. On the other hand, the use of DIBAH on a preparative scale does not appear to give satisfying results [YG1]. Likewise, Thex-$BHCl \cdot Me_2S$ in CH_2Cl_2 as well as 9-BBN in excess also give aldehydes [BC4, BC5, C5, CO1], and these reductants are compatible with aliphatic and other halogenated substituents and with NO_2, CN, and ester groups (Figure 3.57). Another method consists of treating acylboranes obtained by the action of 9-BBN on acids with 1 equivalent of Li 9-BBNH [CK2], the reduction being compatible with the same groups as before.

LAH, SAH and AlH_3 in ethers, LAH–N-methylpyrrolidine complex, $AlH_3 \cdot Et_3N$, $Li(MeO)_3AlH$, or Red–Al in C_6H_6 at 80°C can be used to reduce acids and anhydrides to the corresponding alcohols [BK5, BY1, CB5, CB7, FS1, H3, M3]. Cyclic anhydrides are transformed into diols (Figure 3.58). Sodium diethyl-

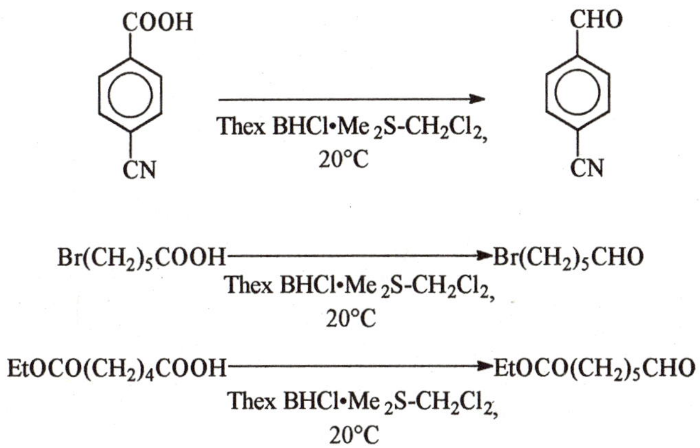

Figure 3.57

piperidinoaluminate does not reduce acids [YA2]. With a limited amount of LAH, cyclic anhydrides can be reduced to lactones at low temperature [M3] (Figure 3.58). Reductions with $LiBH_4$ in THF at 25°C [N1], $NaBH_4$ in THF at 25°C in the presence of methanol added dropwise, DIBAH in the presence of *n*-BuLi [KA1], or lithium trialkylborohydrides in THF [BK6] also lead to lactones. The enantioselective reduction of *meso* anhydrides such as **3.167** can be performed with Binal [MI2] (Figure 3.58).

The reduction of unsymmetrical anhydrides is regioselective: Hydride attack takes place on the carbonyl group that is vicinal to the most substituted carbon, as

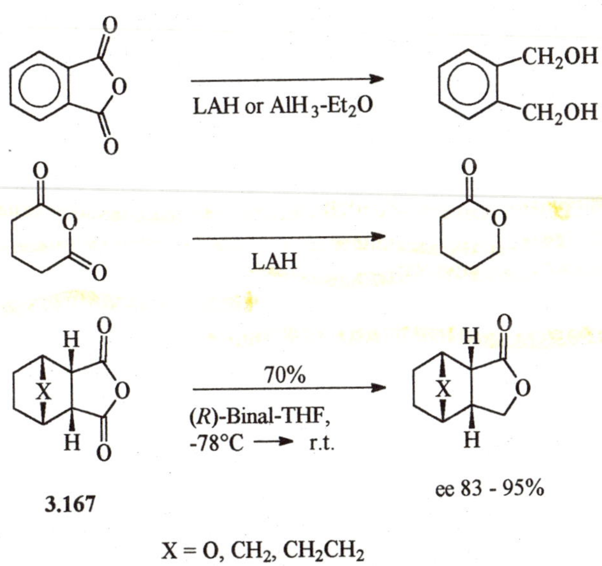

Figure 3.58

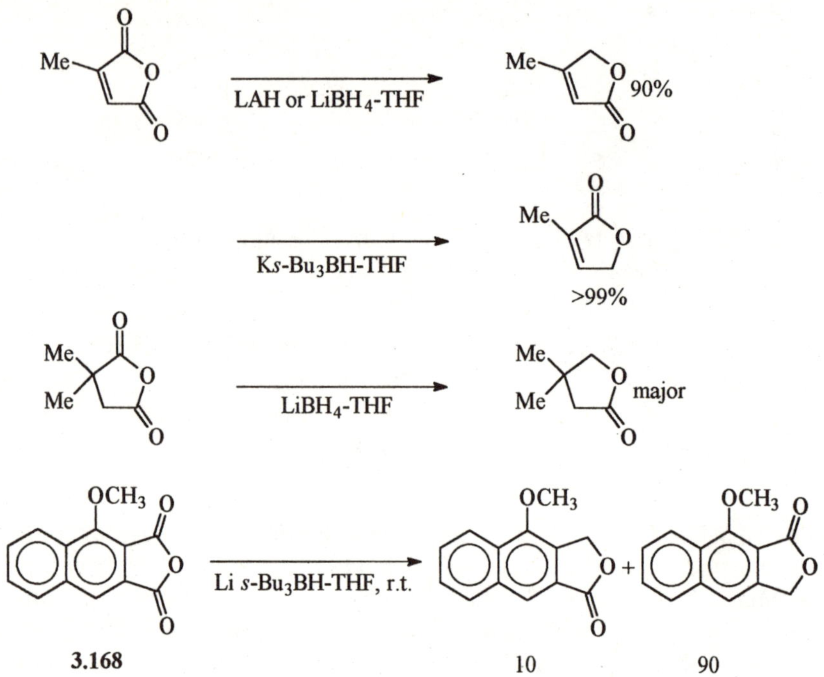

Figure 3.59

shown in Figure 3.59 [KM1, M3, VN1]. Similar results are also obtained with aspartic anhydride [MH4]. $NaBH_4$ in DMF may lead to the same regioisomer [BJ2]. However, bulky complex metal hydrides such as K or Li(*s*-Bu)$_3$BH lead to a complete reversal of regioselectivity [M3] (Figure 3.59). For example, alkoxy-substituted phthalic anhydride **3.168** lead regioselectively to the lactone resulting from reduction of the carbonyl group that is away from the OMe group (Figure 3.59). Contrary to another report [MM4], $NaBH_4$ and $LiBH_4$ in THF are poorly regioselective in this case.

An easy and clean method of reducing acids to alcohols consists of transforming them into mixed carboxylic–carbonic anhydrides by reaction with ClCOOEt–Et_3N. These are easily reduced by $NaBH_4$ in THF, sometimes in the presence of MeOH or by borohydride exchange resin-Ni(OAc)$_2$. This method does not affect double bonds or NO_2, CN, $CONH_2$, and COOR groups, as shown in Figure 3.60 [BM9, IK2, SY1]. It can be applied to N-protected anhydrides of amino acids, leading thus to N-protected 1,2-amino alcohols [FC3, K5, RL1]. In the presence of MeOH, the reduction takes place at 10°C, and aromatic halides are not affected [SY1] (Figure 3.60).

In certain cases, reduction of mixed anhydrides of unsaturated acid such as **3.169** by $NaBH_4$ can be delicate [JU1]. $NaBH_4$–$CeCl_3$ in MeOH does not appear to be efficient, as the formation of the methyl ester hinders the reduction. The use of $NaBH_4$–SmI_3 in THF leads to the expected allylic alcohol (Figure 3.60). Some

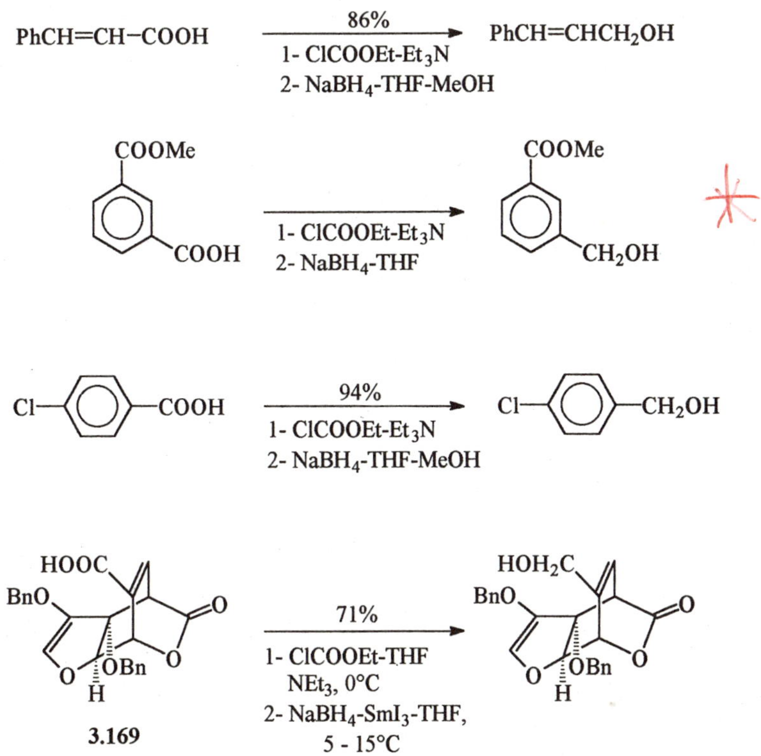

3.169

Figure 3.60

authors have recommended the use of mixed carboxylic phosphoric anhydrides, but the yields are not very high [KY1].

A method involving the in situ transformation of carboxylic acids into acylfluorides that are reduced by $NaBH_4$ in MeOH has been recently proposed [KN4].

Carboxylic acids are not reduced by alkali borohydrides [BK5], aminoborohydrides [FH5], trialkylborohydrides [BK5], 9-BBN, or Sia_2BH at room temperature [BK5], or by the acyloxyboranes [HM1, GN1]. The activation of $LiBH_4$ by MeOH in refluxing THF induces the reduction of carboxylic acids to alcohols [SO3], but in the presence of $B(OMe)_3$, the reduction is incomplete [BN3]. $NaBH_4$–$ZrCl_4$ in THF also reduces benzoic acid to benzyl alcohol [IS1], while $NaBH_4$–$TiCl_4$ in DME [KT2] or $(i\text{-PrO})_2TiBH_4$ in CH_2Cl_2 [RC2] transforms all acids into alcohols. The reduction of acids into alcohols can be performed by $Zn(BH_4)_2$, either in DME in the presence of $(CF_3CO)_2O$ [R3] or in refluxing THF [NM3], although there are some limitations to these methods.

Methods for performing the reduction of acids to the corresponding alcohols use an excess of BH_3·THF or catecholborane [BK5, KB7, PS1]. These reagents react more rapidly with aliphatic acids than with aromatic ones. The reaction process

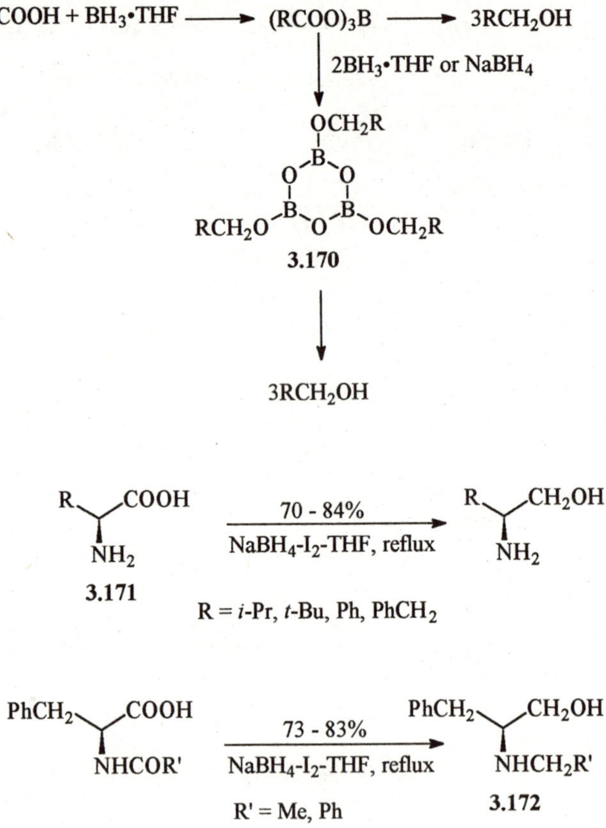

Figure 3.61

(Figure 3.61) implies the formation of a triacycloxyborane. This is reduced by excess of $BH_3 \cdot THF$ to a cyclic borate **3.170**, which is hydrolyzed to the corresponding alcohol. The intermediate triacyloxyborane obtained by reaction of 3 equivalents of acid with $BH_3 \cdot THF$ can also be reduced to the corresponding alcohol by $NaBH_4$ in an alcoholic medium (Figure 3.61). Borane can be generated in situ by action of Me_3SiCl, I_2, H_2SO_4, or $MeSO_3H$ on solutions of $LiBH_4$ or $NaBH_4$ in THF or else $NaBH_4$-$BF_3 \cdot Et_2O$ in THF [AM2, BB7, DA2, DC2, GS2, JJ2, MB5, MM2]. This methodology has been applied to the reduction of chiral amino acids **3.171** to related amino alcohols without epimerization (Figure 3.61). This transformation can also be run with $NaBH_4$–$ZnCl_2$ in THF. If the free amine is transformed into an N-acetyl or N-benzoyl group, N-alkylamino alcohols **3.172** are formed [AM2, MM8] (Figure 3.61). However, $NHCOOCH_2Ph$ and NHTs residues remain unaffected [AM2]. The reduction of NHBOC-protected amino acids can be performed with $(i\text{-PrO})_2TiBH_4$ in CH_2Cl_2 [RC2].

When using boranes, esters, halogen derivatives, nitriles, amides, and nitro compounds are inert [BF2, BK5, HC1, HI1, OK1, YP1]. This makes possible the

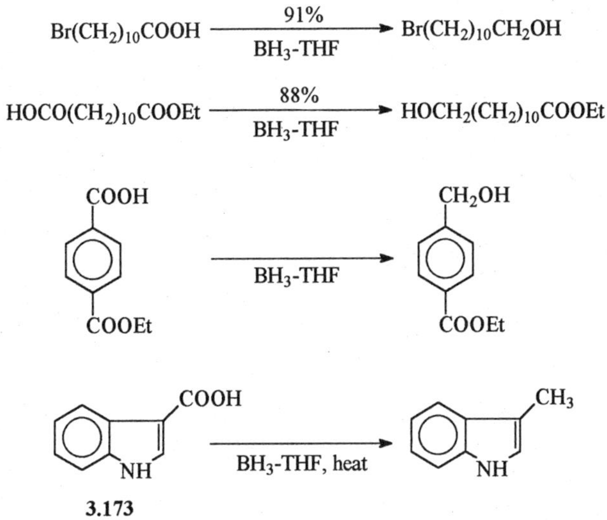

Figure 3.62

selective reductions shown in Figure 3.62. In some cases, the reduction may proceed directly to the hydrocarbon, as shown in the case of **3.173** [PS1] (Figure 3.62). Such is also the case for cyclane-substituted carboxylic acids, which lead to ring-expanded cycloalkanes by action of $NaBH_4$–HOTf in Et_2O at −78°C [OW2].

When starting from substituted malonic acids, the reduction by BH_3 in THF does not give good yields because of competing formation of an enolborate. This can be prevented if the reduction is run at −20°C [CE1]. However, reduction of a chiral malonic acid monoester **3.174** by $BH_3 \cdot Me_2S$ takes place on the ester group [FC2] (Figure 3.63).

$BH_3 \cdot Me_2S$ or amine–boranes induce the same reductions of acids to alcohols; linear anhydrides are reduced by $Ph_2NH \cdot BH_3$ in THF to alcohols, while succinic and phthalic anhydrides remain intact [CU1].

The special reactivity of carboxylic acids allows the following selective reductions shown in Figure 3.63:

- β-Chloropropionic acid is reduced by AlH_3 in Et_2O to β-chloropropanol [BK5].
- The nonracemic hemiester **3.175**, obtained by the action of pig liver esterase (PLE) on the corresponding dimethyl diester, may be transformed into two nonracemic enantiomeric lactones. This occurs either through reduction of the methyl ester by $LiBH_4$, which leaves the carboxylic acid group unattacked, or else through action of $BH_3 \cdot THF$, or, better, via transformation of the acid group to the mixed carboxylic–carbonic anhydride followed by reduction by $NaBH_4$, which does not reduce the ester [BG2] (Figure 3.63).

Thioacids are also reduced to thiols by $BH_3 \cdot Me_2S$ [JS4].

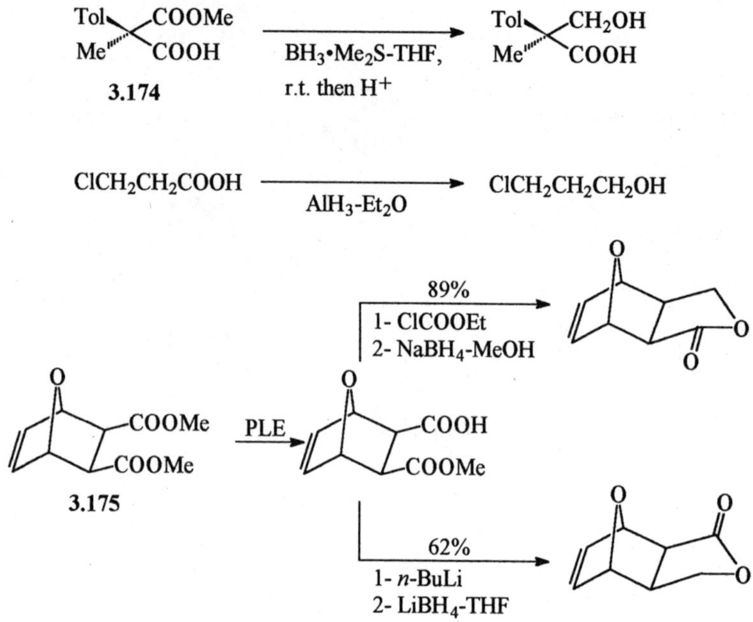

Figure 3.63

3.2.7 Acid Chlorides: RCOCl

The reduction of acid chlorides is particularly easy; however, by carefully adjusting the reaction conditions, the reduction process can be controlled. Starting from acid chlorides, aldehydes can be obtained by four different methods [C5, M3]:

- By the action of LTBA or Na(t-BuO)$_3$AlH in diglyme at -78°C, the aldehyde formed is not reduced. Due to the basicity of the reaction medium, the yields are good only if the aldehydes are not easily enolized. Aromatic or α,β-unsaturated acid chlorides are good substrates; however, Na(t-BuO)$_3$AlH reduces some aliphatic acid chlorides [BK3, CB6] (Figure 3.64). The method is compatible with ester, nitrile, and nitro groups, and it has been applied to the transformation of N-protected amino acid chlorides **3.176** into the corresponding aldehydes [Z1] (Figure 3.64).
- By the action of $NaBH_4$ in DMF–THF in the presence of pyridine at 0°C [B1]. This reduction generates borane, which coordinates to pyridine, forming a complex that precipitates under these conditions (Figure 3.64). The reduction leaves the halide functional group intact (Figure 3.64).
- By the action of $NaBH_4$ in the presence of $CdCl_2$ in DMF. This method is compatible with aliphatic chlorides, esters, nitriles, NO_2 groups, and double bonds [EB3].
- By the action of complex borohydrides $(Ph_3P)_2CuBH_4$ [DF1,W4] or $(Ph_3P)_3CuCNBH_3$ [HM2]. These reductions take place at room temperature in acetone, and only acid chlorides are sensitive under these conditions.

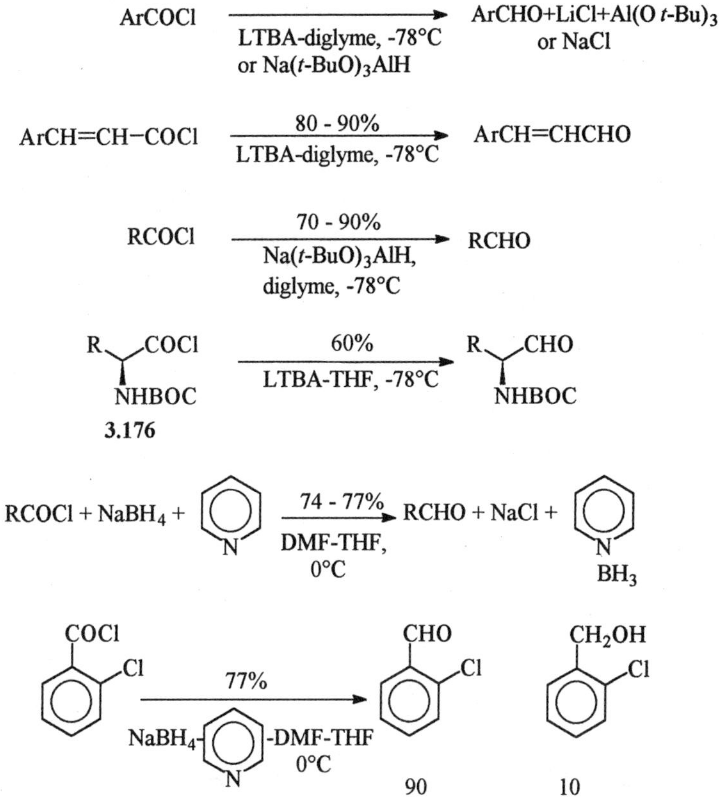

Figure 3.64

Other reducing reagents transform acid chlorides into corresponding alcohols. These include: LAH or SAH in THF, on silica gel or complexed to amines [BK5, CB5, FS1, KH2], AlH_3 in Et_2O, $AlH_3{\cdot}Et_3N$ [BK5, CB7, MP2], DIBAH [YG1], $NaBH_4$, and $LiBH_4$ in THF, dioxane, DME, or on alumina or in the presence of polyethylene glycol [BK5, PS1, SF2], lithium aminoborohydrides [FH5], n-Bu_4NBH_4 in CH_2Cl_2, $(i\text{-PrO})_2TiBH_4$ in CH_2Cl_2 [RC2], $Zn(BH_4)_2$–TMEDA in Et_2O [KU3], 9-BBN in the cold [PS1]. The reductions by $BH_3{\cdot}$THF and Sia_2BH are nevertheless relatively slow [BK5].

The selective reduction of acid chlorides in the presence of esters by 9-BBN in cold THF is possible because esters are reduced only under reflux in this solvent [PS1]. Reduction by $Zn(BH_4)_2$–TMEDA in Et_2O leaves Cl, NO_2, ester groups, and conjugated double bonds unchanged [KU3].

3.2.8 Amides and Imides: $RCONR_2'$, $(RCO)_2NR'$

The attack of the amide carbonyl group by a hydride occurs through a tetracoordinated intermediate, which can proceed either by the breaking of the C—N bond

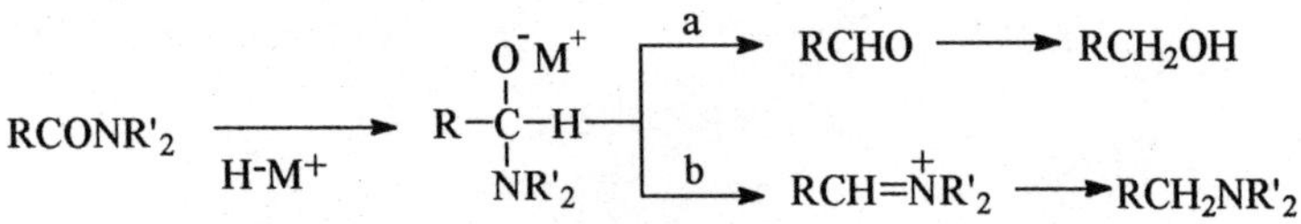

Figure 3.65

(path a), leading thus to an aldehyde (which can eventually be reduced to an alcohol), or by the breaking of the C—O bond (path b), producing an iminium salt (which is the precursor of an amine) (Figure 3.65). Depending on the nature of the reducing agent and the nitrogen substituents, the reaction follows one pathway or the other.

Path a: Access to aldehydes and alcohols.

N-Dimethylamides **3.177** are transformed into aldehydes by reaction of 0.25 equivalent of SAH in THF at 0°C [CB5], by one equivalent of Na diethylpiperidinoaluminate in THF–toluene at 0°C [YA2], or by the ate complex DIBAH–*n*-BuLi in THF–hexane at the same temperature [KA1] (Figure 3.66).

The synthesis of aldehydes [C5] can also be accomplished by controlled reduction of acylaziridines **3.178** [BK5] or of acylimidazoles **3.179** [W3] by LAH in Et_2O at −10°C, by LTBA or by Red–Al in C_6H_6 [H3, M3]: R can be aliphatic or aromatic (Figure 3.66). The N-methoxy-N-methylcarboxamides **3.180** are also cleanly reduced to aldehydes by LAH in excess in THF at low temperature or by DIBAH in THF at 0°C. In many cases, the latter reagent does not lead to formation of alcohols as byproducts resulting from a subsequent reduction of the aldehyde [NW1]. This behavior can be understood by the stabilization through chelation of the lithium or aluminum intermediate (Figure 3.66). α,β-Unsaturated aldehydes may also be prepared by this method, using DIBAH in THF [BS8, NB1].

The method can be applied to N-protected amino acid derivatives **3.181** or to peptides **3.182** [FC1, FH3] without racemization (Figure 3.67).

LTBA or, better, LTEA in Et_2O, reduces all tertiary amides to aldehydes at 0°C. Tertiary amides coordinate the Li cation better because their carbonyl is more basic [BK5, C5, M3] (Figure 3.68). The reduction of precursors of chiral aldehydes **3.183** by LTEA does not cause any epimerization [MY2] (Figure 3.68). Treatment of tertiary amides by $LiEt_3BH$ in THF at 0°C leads, via triethylborates, to aldehydes, which can be reduced to alcohols by an excess of reagent [BK3, BK5] (Figure 3.68).

N-Dimethylamides react with EtOTf to give iminium salts, which can be selectively reduced to aldehydes by Li(*s*-Bu)$_3$BH in THF at −78°C [TR2]. This method can be applied to α,β-unsaturated amides and is compatible with isolated double bonds, nitriles, and esters.

The other alkylborohydrides, 9-BBN and Sia_2BH, also transform tertiary amides to alcohols [BK5, PS1]. Alcohols are obtained via the action of $LiBH_4$ in MeOH–hot diglyme on some tertiary amides [SO3] or by the controlled reduction with Li pyrrolidinoborohydride in THF [FF1] (Figure 3.69). This latter method, however, has some limitations. The reduction of **3.183** by Li pyrrolidinoborohydride or better by $LiNH_2BH_3$ obtained from borane·ammonia and *n*-BuLi does not promote any racemization [MY2, MY4] (Figure 3.69).

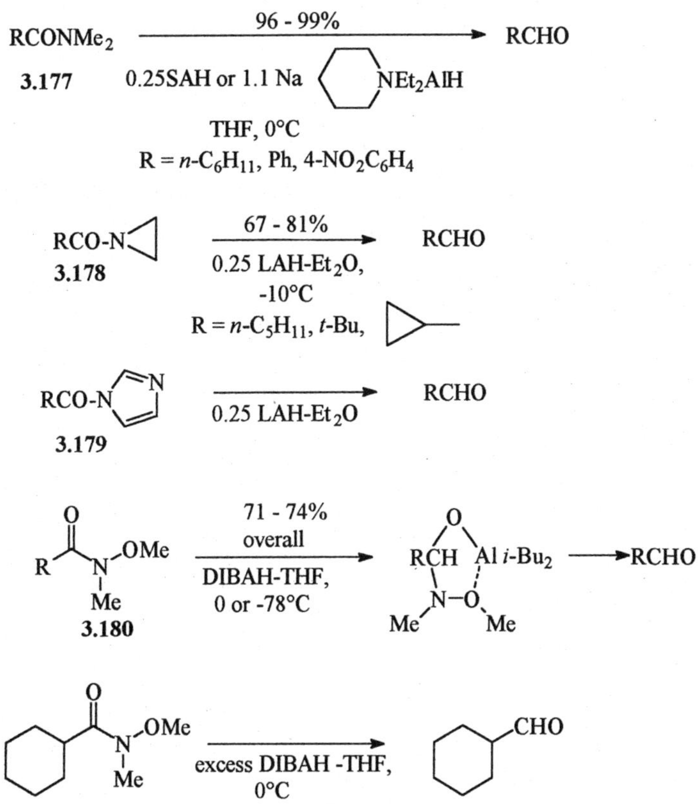

Figure 3.66

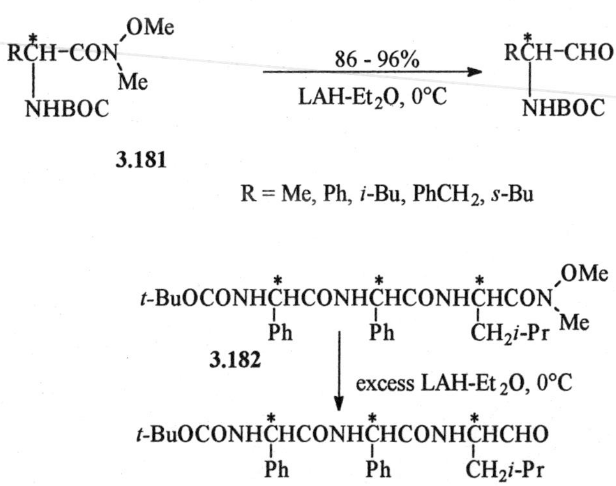

Figure 3.67

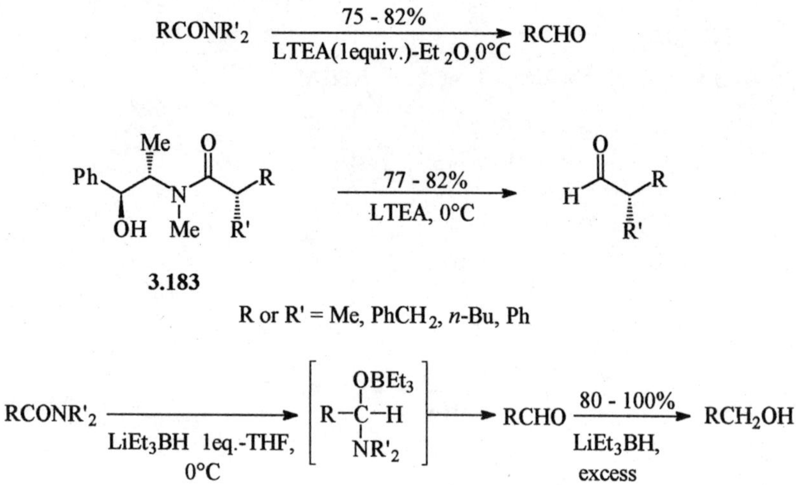

Figure 3.68

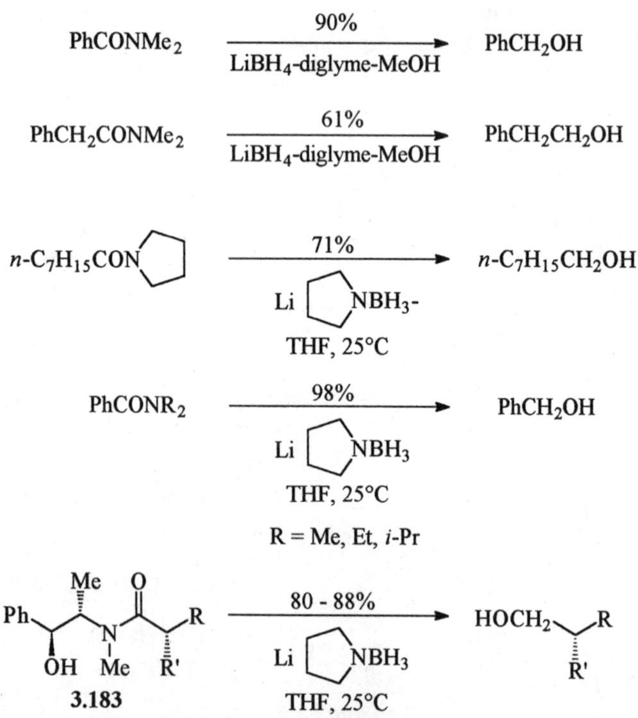

Figure 3.69

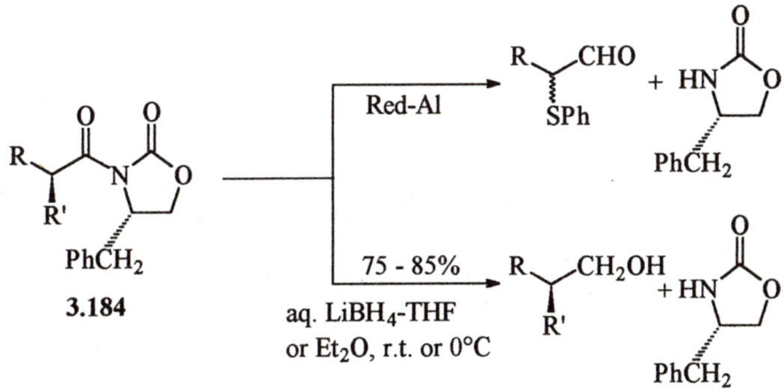

3.184

$R = Et, n\text{-}Pr, i\text{-}Pr, R' = PhS$
$R = Ph_3C, Ph_2MeC, PhMe_2C, R' = N_3$

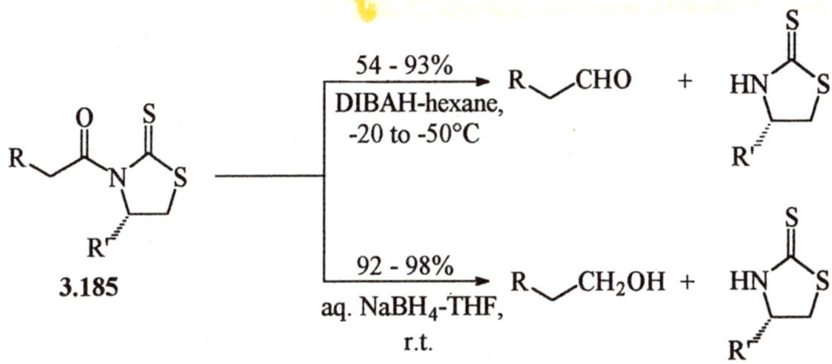

3.185

$R' = H, i\text{-}Pr, COOMe \quad R = n\text{-}C_5H_{11}, n\text{-}C_9H_{17}$, PhCH=CH, Ph or chiral residue

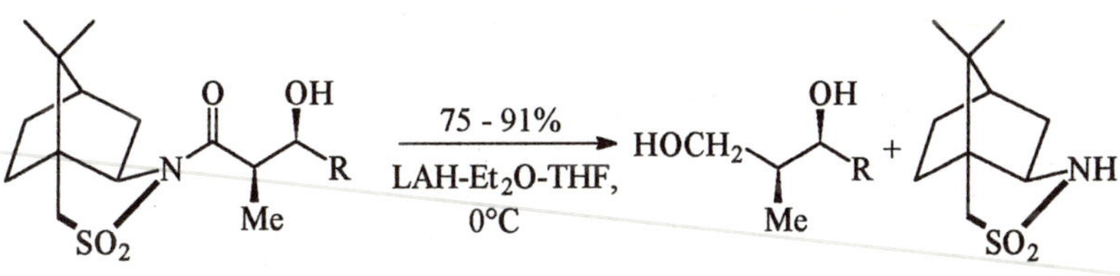

3.186

R = Me, *i*-Pr, Ph

Figure 3.70

Secondary amides remain intact in the presence of $LiBH_4$–MeOH [SO3], while at lower temperature or in the absence of MeOH, reduction of tertiary amides seldom takes place. However, an exception has been found with fused xanthines [CK4].

The reductive cleavage of the amide residue of many chiral auxiliaries is recommended for recovery of chiral compounds and auxiliary regeneration [S3]. Evans's acyloxazolidinones **3.184** have been transformed into aldehydes by DIBAH or Red–Al at low temperature [CW2, EB5, MS8], but in the case of R′=SPh, some epimerization occurs (Figure 3.70). DIBAH has also been proposed to transform N-acylthiazolidinthiones **3.185** [NK1] into the corresponding aldehydes (Figure

3.70). Related chiral auxiliaries [AM4] suffer the same transformations. If the reductive cleavage of **3.184** is carried out with LAH in Et_2O, $LiBH_4$ in THF [EG2, EE1, EG3, ES2], or, better, $LiBH_4$ in the presence of water [CW2, DN1, IA1, PD1], alcohols are formed. When R′ = PhS, N_3, or CF_2Br, no racemization is observed [CW2, DN1, IA1] (Figure 3.70). From **3.185**, $NaBH_4$ in aqueous THF gives the best results [NK1] (Figure 3.70). Similar reductive cleavage of unsymmetrically substituted chiral amides by LAH or $LiBH_4$ in EtOH without racemization have been described [AM4, DD4]. Chiral sultams **3.186** can be cleaved in the same fashion [OB2] (Figure 3.70).

Path b: Access to amines.

LAH in ethers (Et_2O, THF), as well as its complex with N-methylpyrrolidine, Red–Al in C_6H_6, AlH_3 in Et_2O, $AlH_3 \cdot Et_3N$, $BH_3 \cdot THF$, or $BH_3 \cdot SMe_2$ reduce most types of amides to amines at room temperature. Primary amides, however, are reduced by $BH_3 \cdot THF$ only under reflux in THF [BC11, BH1, BK5, BN2, CB7, FS1, L2, M3, PS1]. Borane can be generated in situ from $NaBH_4$ and I_2 in THF [BB7], or in the presence of Me_3SiCl [GS2]. Since $BH_3 \cdot THF$ does not reduce esters, nitro derivatives, or nitriles under these conditions, selective reductions can be run [BK5] (Figure 3.71). Selective reductions of tertiary amides to amines in the presence of secondary amides can be carried out when the secondary amides are protected as lactim ethers [WB2]. Sulfonamides are reduced by $BH_3 \cdot Me_2S$ only in refluxing THF [BF2]. Tertiary amides are reduced to amines by BH_3·aminoethanol [IW1], but no reduction takes place if the aminoalcohol is grafted onto a polymeric support [IW1].

DIBAH reduces tertiary amides well and appears to be more selective than LAH with α,β-unsaturated derivatives. LAH in Et_2O induces the partial reduction of the double bond of **3.187** [W1] (Section 3.2.9) (Figure 3.71). The reduction of amides by LAH in Et_2O is compatible with the presence of an SePh group in the molecule, as shown in the case of **3.188** [TT1] (Figure 3.71).

The alkali borohydrides in an ether medium do not reduce amides at room temperature [BK5], but under reflux of THF secondary and tertiary amides are reduced to amines [PS1]. Lithium diisopropylaminoborohydride in THF reduces aliphatic and aromatic N,N-dialkylamides to the corresponding amines [FF1] (Figure 3.72). $NaBH_4$–$ZrCl_4$ reduces $PhCONMe_2$ to the corresponding amine [IS1], while $NaBH_4$–$TiCl_4$ in DME reduces all amides and lactams to the corresponding amines [KT2]. Tertiary amides are transformed into the corresponding amines by n-Bu_4NBH_4 in refluxing CH_2Cl_2 [WI1] or by $NaBH_4$ in refluxing pyridine [KI1]. In the presence of an organic acid and under reflux, $NaBH_4$ reduces all amides to amines [GN1, UI1]. Under these conditions, a diarylketone can remain unaffected, as shown in the reaction of **3.189** (Figure 3.72).

The activation of $NaBH_4$ by ethanedithiol allows access to primary amines with nitriles remaining intact [GE1]. The same reduction can be accomplished either by $LiBH_4$–diglyme–hot MeOH starting from primary aliphatic and aromatic amides [SO3] or by $NaBH_4$ in an alcohol medium in the presence of $CuCl_2$ (for aromatic amides only) [W4]. $(CF_3COO)_2BH$ leaves the amides unchanged [MM1].

An alkylation method for primary and secondary amines consists of treating them

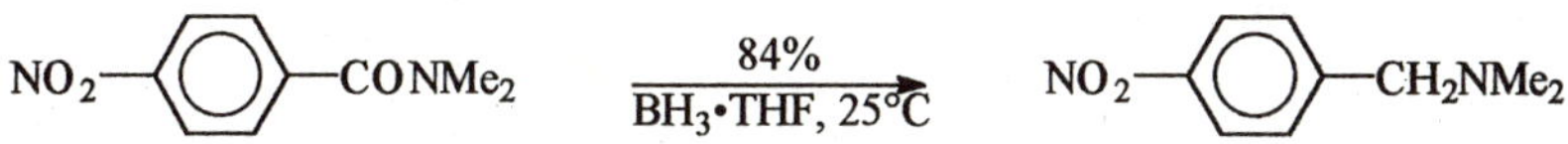

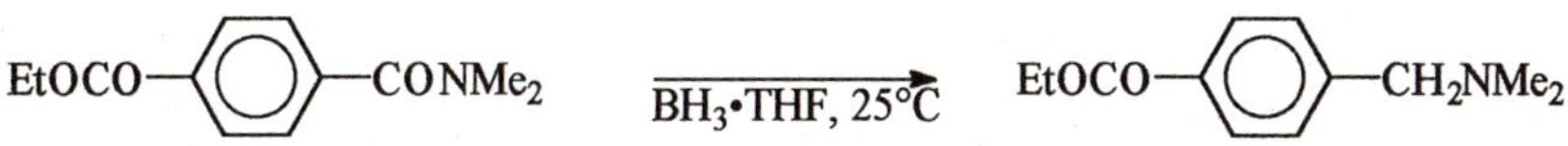

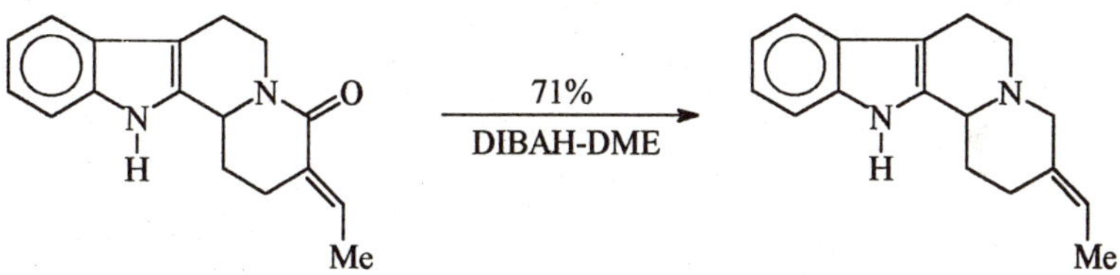

3.187

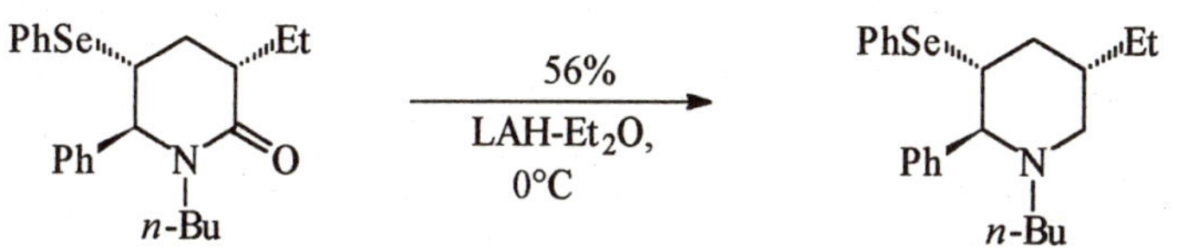

3.188

Figure 3.71

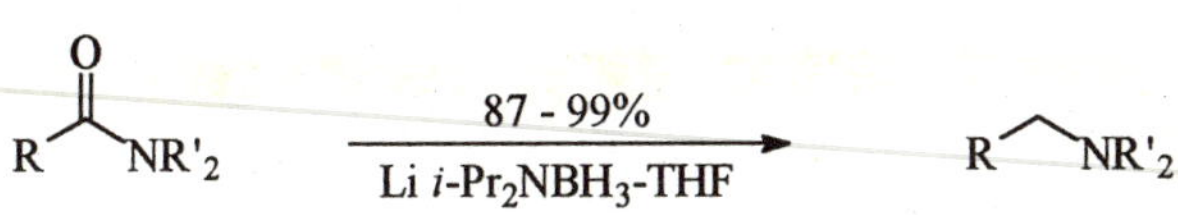

R = n-C_7H_{15}, n-C_9H_{19}, Ph, 3-MeC_6H_4,

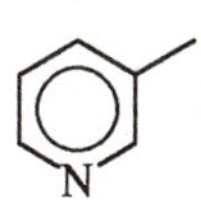

R' = Et, i-Pr,$(CH_2)_4$

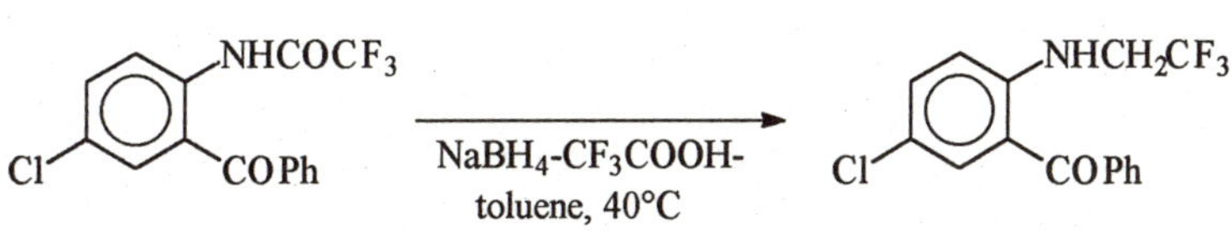

3.189

Figure 3.72

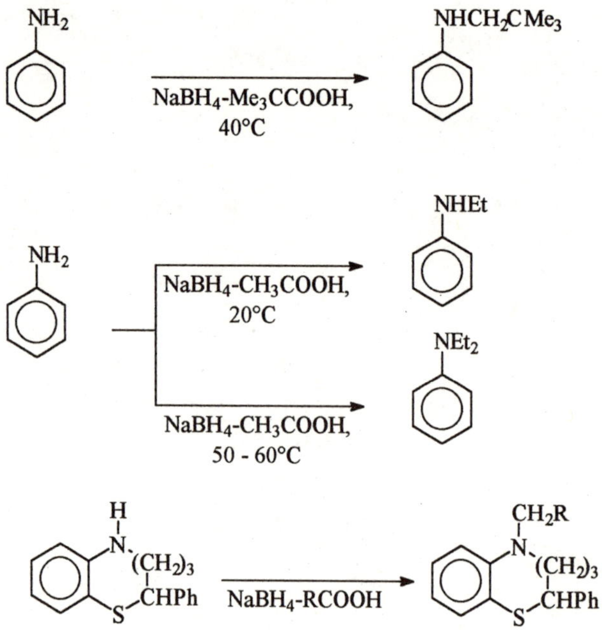

Figure 3.73

with $NaBH_4$ in an organic acid medium. In the cold, monoalkylation of primary amines occurs, while at high temperature, the secondary amines are further alkylated. The mechanism of this reaction has not been elucidated [GN1]. The reaction is better with aromatic amines and is compatible with OH, COOEt and $CONR_2$ groups, and heterocycles (Figure 3.73). With formic acid, the reaction is not always easy to run. In the presence of CF_3COOH, or if $NaBH_4$ is replaced by $NaCNBH_3$, this alkylation does not take place (Section 3.3.1).

Another facile method of methylation of amines consists of their transformation into carbamates, which are reduced in situ either by LAH [RE1] or by $NaBH_4$ in the presence of AcOH or CF_3COOH in dioxane or THF [GN1]. However, under the latter conditions, the *t*-butyl carbamates react poorly (Figure 3.74).

Lactams such as **3.190** or **3.191** are reduced under the same conditions as linear amides, as shown in Figure 3.75 [BK5, BM6, LH1, WM2]. However, LAH on silica gel reduces esters, but it does not reduce lactams (Figure 3.75). Under suitable conditions with LAH, it is possible selectively to reduce a lactam bearing a sulfone group that remains unchanged [TG1]. The reduction of the lactams **3.192** to cyclic tertiary amines has been accomplished through reaction with $NaBH_4$ in refluxing *t*-BuOH in the presence of MeOH added dropwise [MG1, MG2] (Figure 3.75). Partial reduction of carbamate-protected five-membered lactams such as **3.193** to hemiaminals can be performed with DIBAH or $LiEt_3BH$ [KN3, LA1, PE1] (Figure 3.75). The stepwise reduction of lactams such as **3.194** into hemiaminals followed by dehydratation to iminium salts, which suffer a second reduction, gives better

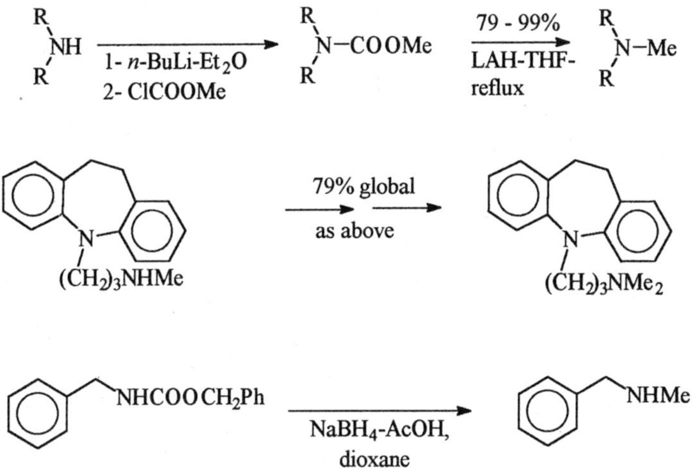

Figure 3.74

yields in the expected pyrrolidine than the one-step reduction by $BH_3{\cdot}Me_2S$ [LR2, PE1] (Figure 3.75). Esters are not affected under such conditions. 2-Pyrrolidinone **3.195** (X = H) as well as its N-carbamoyl derivative **3.195** (X = $CONH_2$) remain unchanged in the presence of $NaBH_4$ in MeOH [KM3]. However, an amino alcohol **3.196** (X = Ts) is obtained from **3.195** (X = Ts) under these conditions, probably via the corresponding hemiaminal. Ring opening takes place from **3.195** (X = $CONH_2$) only in the presence of K_2CO_3 [KM3] (Figure 3.75).

A problem that has received some attention in relation to the chemistry of β-lactam antibiotics is the reduction of the azetidin-2-ones **3.197** [YO2]. While AlH_3 or LAH in Et_2O and $BH_3{\cdot}THF$ cleave the N-alkylazetidinones to 3-aminopropanols, *cis*-3-benzyloxy-1,4-diphenylazetidinone **3.198** is cleaved only by LAH or lithium trialkylborohydrides in THF (Figure 3.76). The reduction of **3.198** by DIBAH in a hexane–THF mixture under reflux or better by AlH_2Cl or $AlHCl_2$ in Et_2O preserves the heterocycle and selectively provides the substituted azetidine **3.199** (Figure 3.76). When the β-lactam carries a methyl ketone functional group at the 3-position, such as **3.200**, the selective reduction of the ketone group by $NaBH_4$ in THF, $Zn(BH_4)_2$ in Et_2O, or Li and K(s-Bu)$_3$BH in THF leaves the β-lactam unchanged [PA1] (Section 3.2.2) (Figure 3.76).

Cyclic imides undergo a reduction with a regioselectivity that is comparable to that of cyclic anhydrides (Section 3.2.6). $NaBH_4$ in EtOH in the presence of HCl or in MeOH partially reduces the imides to α′-hydroxyamides, the carbonyl adjacent to the most substituted carbon being preferentially reduced [PS1, SH2] (Figure 3.77). The reduction of bicyclic imides bearing CN or COOEt groups must be performed with $NaBH_4$–$CeCl_3$ in order to prevent overreduction [DR2]. Chelation can direct the reduction of one carbonyl, as in the case of **3.201** [GK2] (Figure 3.77). DIBAH in toluene at low temperature also brings about this reduction [HT2, SH2, W1], but

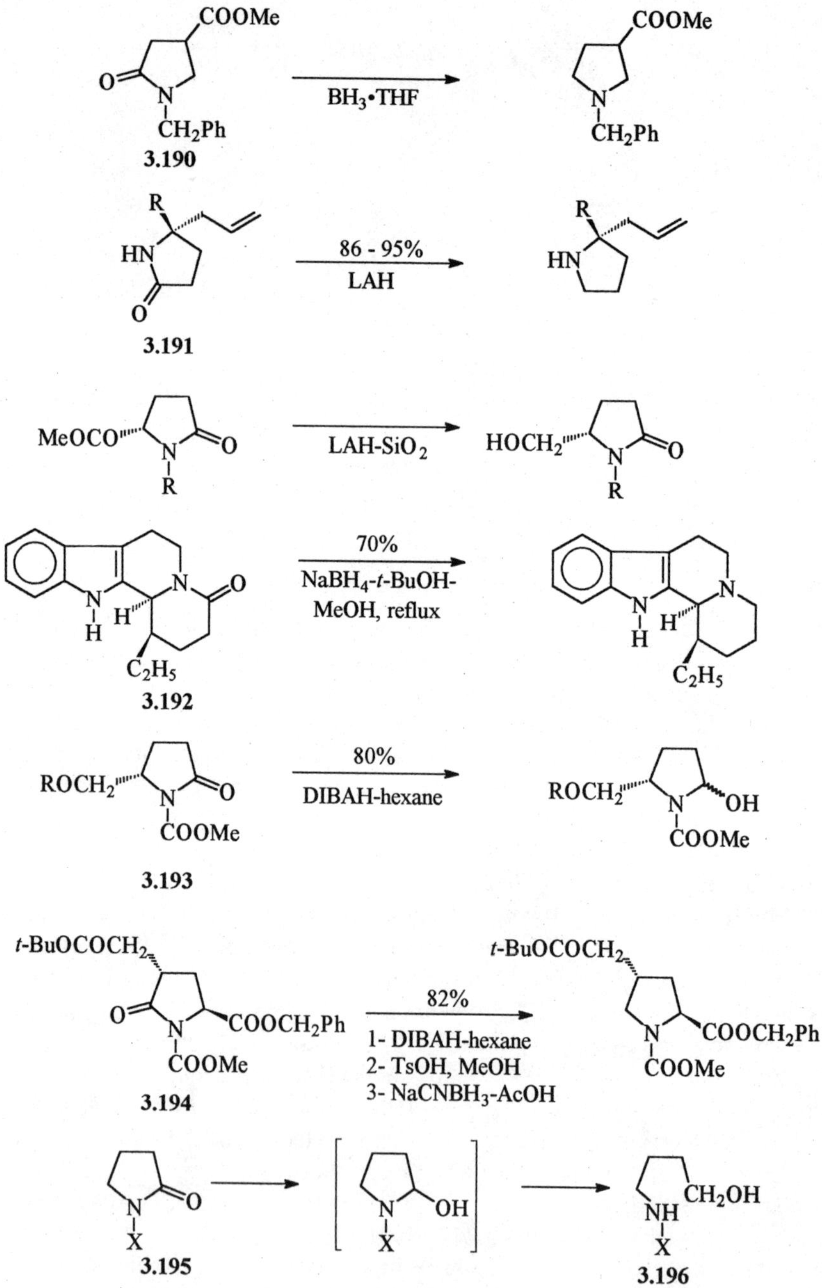
COOMe
O
N
CH₂Ph
3.190
BH₃•THF
COOMe
N
CH₂Ph
R
HN
O
3.191
86 - 95%
LAH
R
HN
MeOCO
N
O
R
LAH-SiO₂
HOCH₂
N
O
R
N
N
H
H
O
C₂H₅
3.192
70%
NaBH₄-t-BuOH-
MeOH, reflux
N
N
H
H
C₂H₅
ROCH₂
N
O
COOMe
3.193
80%
DIBAH-hexane
ROCH₂
N
OH
COOMe
t-BuOCOCH₂
O
N
COOCH₂Ph
COOMe
3.194
82%
1- DIBAH-hexane
2- TsOH, MeOH
3- NaCNBH₃-AcOH
t-BuOCOCH₂
N
COOCH₂Ph
COOMe
N
O
X
3.195
N
OH
X
CH₂OH
NH
X
3.196

Figure 3.75

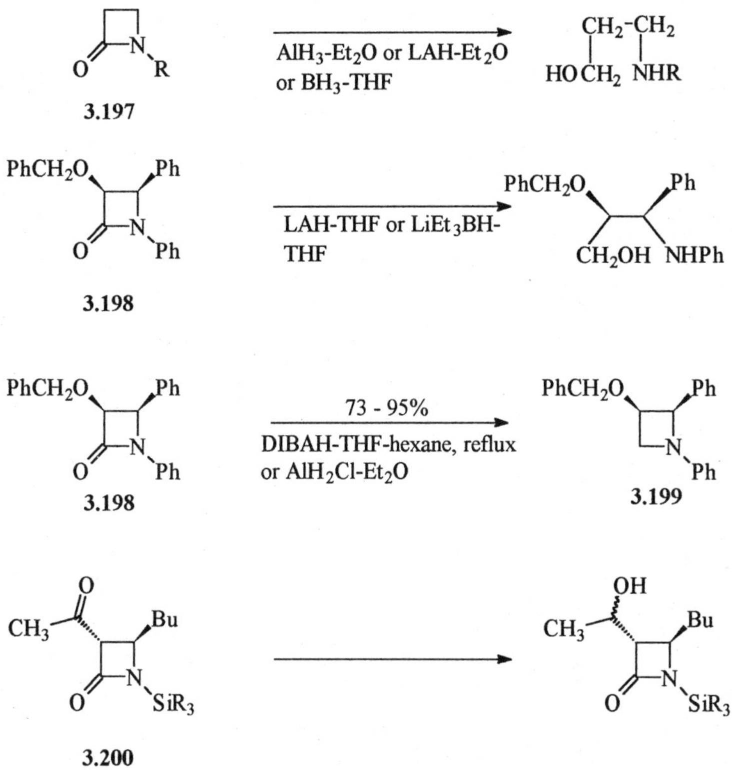

Figure 3.76

the regioselectivity is inverted. Asymmetric reduction of *meso* imides by boranes associated to chiral co-reagents has been described [KL4, RR2].

The reduction of N-methylglutarimide to the corresponding lactamol by DIBAH gives very poor yields, but this transformation can easily be performed with $LiEt_3BH$ in CH_2Cl_2 at low temperature [TR1]. The reduction of chiral N-(1-phenethyl)glutarimide with $LiEt_3BH$ is highly stereoselective [PB3]. This is also the best reagent to reduce pyrrolizinediones **3.202** to the corresponding lactamols [TR1] because $NaBH_4$ in acidic conditions induces ring cleavage and $Zn(BH_4)_2$ gives lower yields (Figure 3.78). The highly stereoselective formation of *cis*-substituted α'-hydroxylactams via the auxiliary controlled reduction of imides has been carried out from chiral imides **3.203** [MC3]. Reduction of the free alcohol **3.203** (R = H) by $Me_4N(AcO)_3BH$ or of the related silyl ether **3.203** (R = $SiMe_2$-*t*-Bu) by Li-Selectride selectively gives each *cis* enantiomer (Figure 3.78). Red-Al is the most efficient reagent to perform the selective reduction of chiral imide **3.204** derived from *meso cis*-caronic anhydride [MY3] (Figure 3.78).

Red-Al reduces N-methylsuccinimide to N-methylpyrrolidone [H3] (Figure 3.78). Reduction of succinimides to pyrrolidines can be carried out by $NaBH_4$–

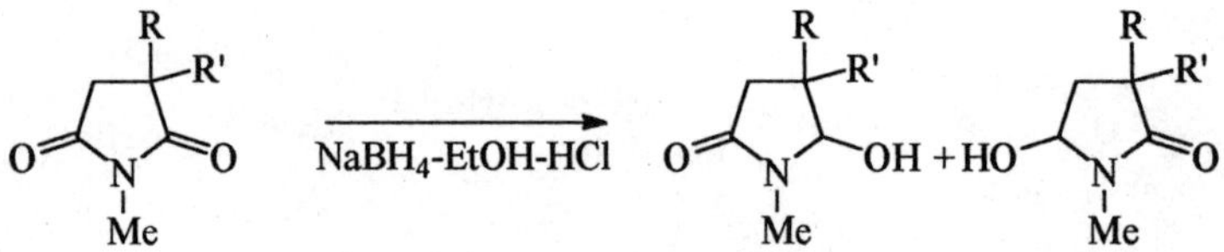

R = R' = Me	79	21
R = R' = Ph	>95	<5
R = Me, R' = H	59	41

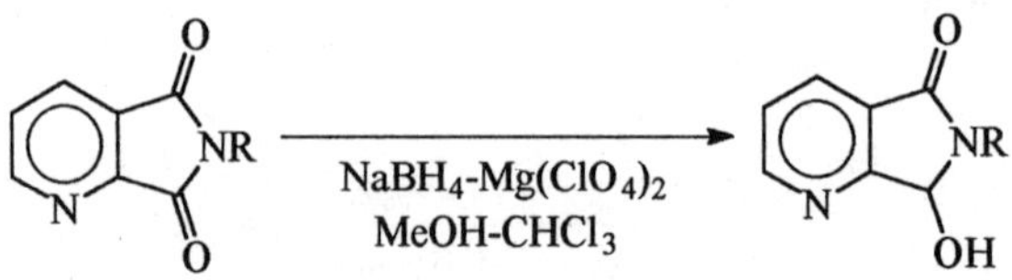

3.201

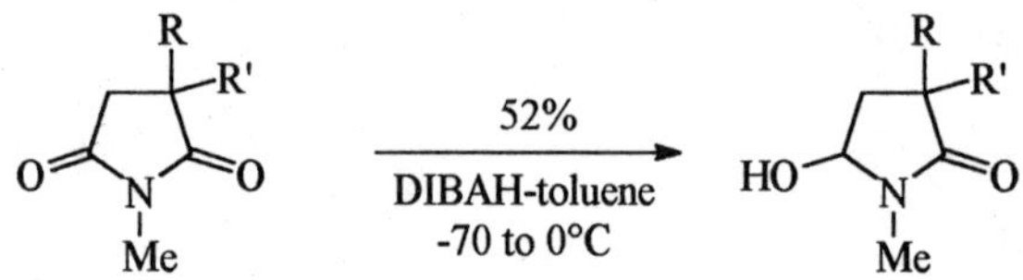

Figure 3.77

$BF_3{\cdot}Et_2O$ in diglyme [MS2]. Thioamides are reduced to amines by $BH_3{\cdot}Me_2S$ [JS4].

3.2.9 α,β-Ethylenic Carbonyl Compounds: α,β-Ethylenic Aldehydes, Ketones, Esters, and Amides: RCH=CHCOY (Y = H, R′, OR′, NR′R″)

Reduction of α,β-ethylenic aldehydes and ketones can lead to three compounds (Figure 3.79):

- Allylic alcohols resulting from attack on the carbonyl;
- Saturated aldehydes and ketones resulting from attack on the double bond;
- Saturated alcohols resulting from the subsequent reduction of intermediate saturated aldehydes and ketones. This is generally observed in the presence of proton donors (most frequently the solvent).

The regioselectivity of these reductions depends on the structure of the starting compound: Aldehydes are more sensitive to the attack at the carbonyl than ketones. Other things being equal, the reduction of the carbonyl group becomes predominant when the double bond is sterically hindered. The regioselectivity also depends on the type of reducing agent and the medium. The more important the electrophilic

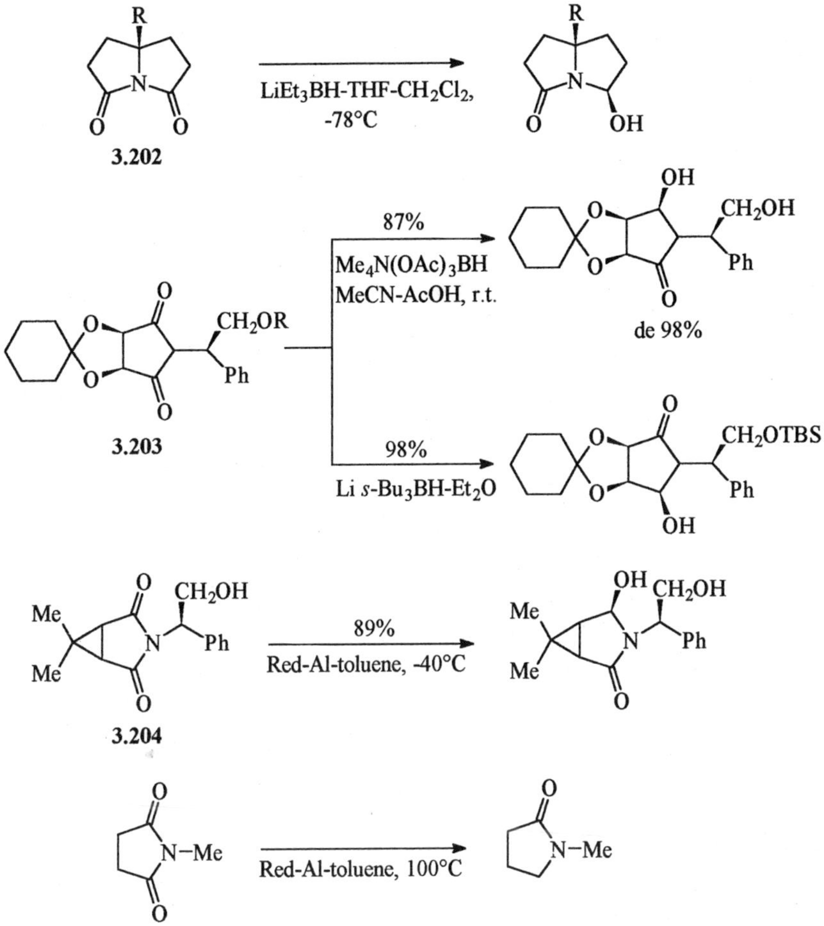

Figure 3.78

assistance by a protic solvent, by the cation associated with the reagent, by the reagent itself, or by an added Lewis acid, the easier the attack on the carbonyl [LL3, LS1, S2]. In contrast, the reduction of the double bond is more prevalent if the reducing agent is bulkier or if it is associated with a cation such as ammonium, not able to induce electrophilic assistance [S2], or with a transition metal such as copper. Aprotic media strongly solvate alkaline cations, and this also favors double-bond reduction. Similar trends are found in the reduction of α,β-ethylenic esters and lactones either to allylic alcohols or to saturated esters and lactones.

Thus the attack on the carbonyl group of α-enones or α,β-ethylenic aldehydes is preferred when one uses LAH in Et_2O [LS1, PR5], AlH_3 in Et_2O [E2, M1], DIBAH in hexane [CG3, PR5, W1], DIBAH–*n*-BuLi ate complex [KA1], $LiAl(OMe)_3H$ in Et_2O [M1, M3], Red-Al in C_6H_6 [M1], $NaBH_4$ in aqueous glycosidic media [DL2],

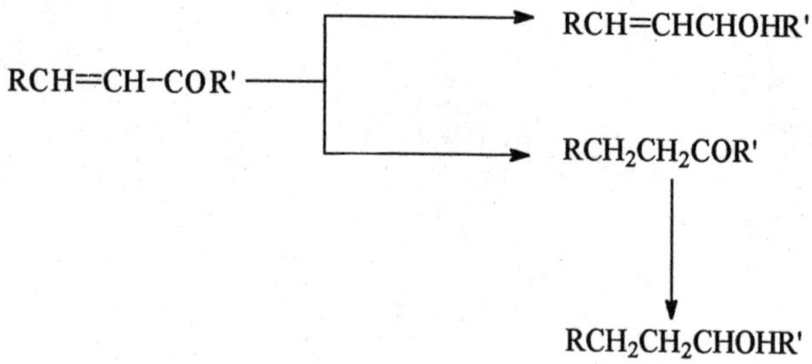

Figure 3.79

borohydride exchange resin in MeOH [SJ3], Li aminoborohydrides [FF2, FS2], diisopropoxytitanium tetrahydroborate [RB3], $Zn(BH_4)_2$ or $NaCNBH_3$–$ZnCl_2$ in Et_2O [IL1, KO1, VM1, YL1], $Zn(BH_4)_2$ on SiO_2 [R3], $BH_3{\cdot}Me_2S$ [HC1], or $BH_3{\cdot}THF$ in the presence of $LiBH_4$ [AH1], $LiBuBH_3$ in Et_2O [KM1], 9-BBN in THF [BK5, KB2, PS1], $Na(AcO)BH_3$ in THF [NB3], and $NaBH_4$–$CeCl_3$ in MeOH [EH1, GL1, KK12, W4]. This latter reduction can also be carried out in CH_2Cl_2–EtOH and is compatible with SePh groups [DD1]. With the last two reagents, ester, nitrile, carbamate, and NO_2 groups are unchanged. α,β-Unsaturated aldehydes can be selectively reduced to primary allylic alcohols by $NaBH_4$–MeOH–CH_2Cl_2 at −78°C, leaving α-enones unchanged [WR1]. Moreover, benzalacetone PhCH=$CHCOCH_3$ is overreduced by AlH_3 to l-phenyl-l-butene PhCH=$CHCH_2CH_3$ [E2]. In the presence of nickelocene, LAH reduces the C=C bond of α-enones [CC2]. $NaCNBH_3$–$BF_3{\cdot}Et_2O$ in THF promotes the overreduction of α-enones to alkenes [SV1]. $BH_3{\cdot}THF$ generally attacks the carbonyl group and the double bond [PS1].

In contrast, in the presence of copper salts, $LiAl(OMe)_3H$, Red-Al in THF [CL3, M1, M3, SS1], or borohydride exchange resin [YS1], various complexes of CuH with organolithiums [BM4, MB1], the complex $(Ph_3PCuH)_6$ [MB2, MS6], DIBAH in THF–HMPA sometimes in the presence of MeCu [TH1], LAH in the presence of cuprates [AL1], n-Bu_4NBH_4 in THF [IL2], and Li and K(s-Bu)$_3$BH in THF [CR1, G1, KH3, OM2] or KPh_3BH [KP2] favor the reduction of the double bond of α-enones (Figure 3.80). When the carbonyl group is precoordinated to a bulky Lewis acid such as aluminum tris(2,6-diphenylphenoxide), 1,4-reduction of α,β-unsaturated aldehydes and of α-enones can be performed with n-BuLi–DIBAH complex at −78°C [SY6]. Catecholborane promotes the conjugate reduction of α-enones, which can lie under the *s-cis* conformation, even when sterically hindered [EF2]. The 1,4-reduction by $LiAl(OMe)_2H_2$–CuBr in the presence of $BH_3{\cdot}Et_2O$ is compatible with the N-COO-t-Bu group [CL3]. The mechanism of the reduction is a conjugate addition of hydride; the enolate formed can then be trapped by an electrophile [CN1, CS1, G1, KS1, MB2, OM2] (Figure 3.80).

The reduction of α,β-unsaturated aldehydes to saturated aldehydes can be carried out by $(Ph_3PCuH)_6{\cdot}Me_3SiCl$ in C_6H_6. When the reaction is run in wet THF without Me_3SiCl, saturated alcohols are formed [BS3].

The conjugate addition of Li and K(s-Bu)$_3$BH in THF to α-enones is sensitive to

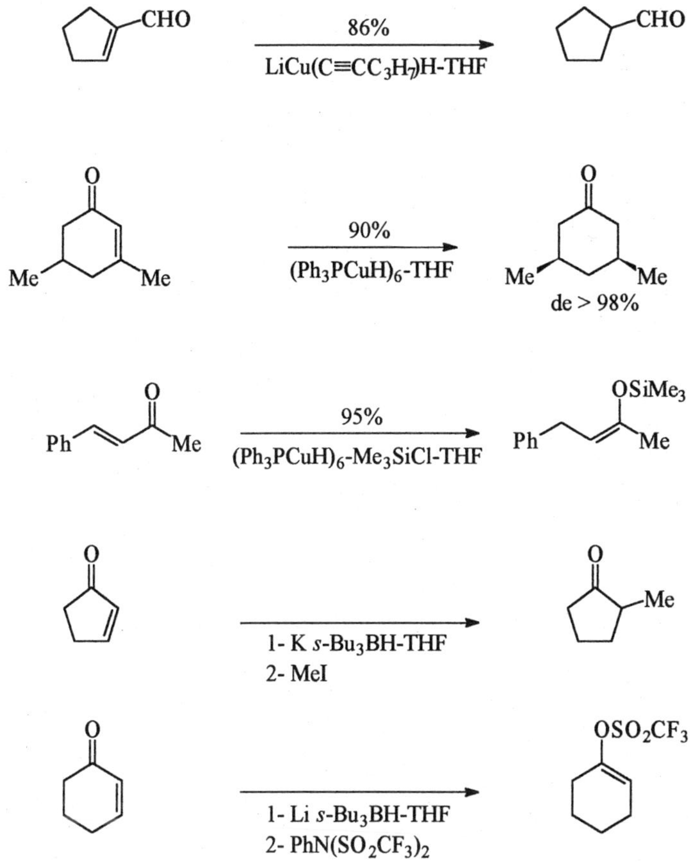

Figure 3.80

steric hindrance of the double bond; 3-methylcyclohexenone gives a mixture of ketone and allylic alcohol [G1]. LTBA in THF [M1] and $LiEt_3BH$ [BK6, CL3, G1], although less bulky, often give rise to mixtures. Similarly, 2α-fluoro-Δ-4-androsten-3,17-dione is reduced by K $(s\text{-Bu})_3BH$ to the 3-α-ol [GM1]. However, in the presence of MAD [aluminum bis(2,6-di-*t*-butyl-4-methyl)phenoxide], Li *n*-Bu $i\text{-Bu}_2AlH$ [NM2], $LiEt_3BH$, or $Li(s\text{-Bu})_3BH$ [CL3] give 1,4-reduction of sterically hindered α-enones.

Cyclopentenones, which are particularly prone to conjugate addition, are reduced by LTBA to the corresponding cyclopentanones [M1], while the 3-substituted cyclohexenones give mainly allylic alcohols [BG1, G1] (see above). The complexes of copper hydrides or DIBAH–MeCu are much less sensitive to steric hindrance [LU1, MB1, MB2, TH1]. Indeed, progesterone **3.205** is selectively reduced to progestanone under these conditions (Figure 3.81). Likewise, bicyclic ketones **3.206** and **3.207** are reduced at the ring junction in a stereoselective fashion (Figure 3.81).

With regard to the reactions with DIBAH–MeCu, trapping of the aluminum

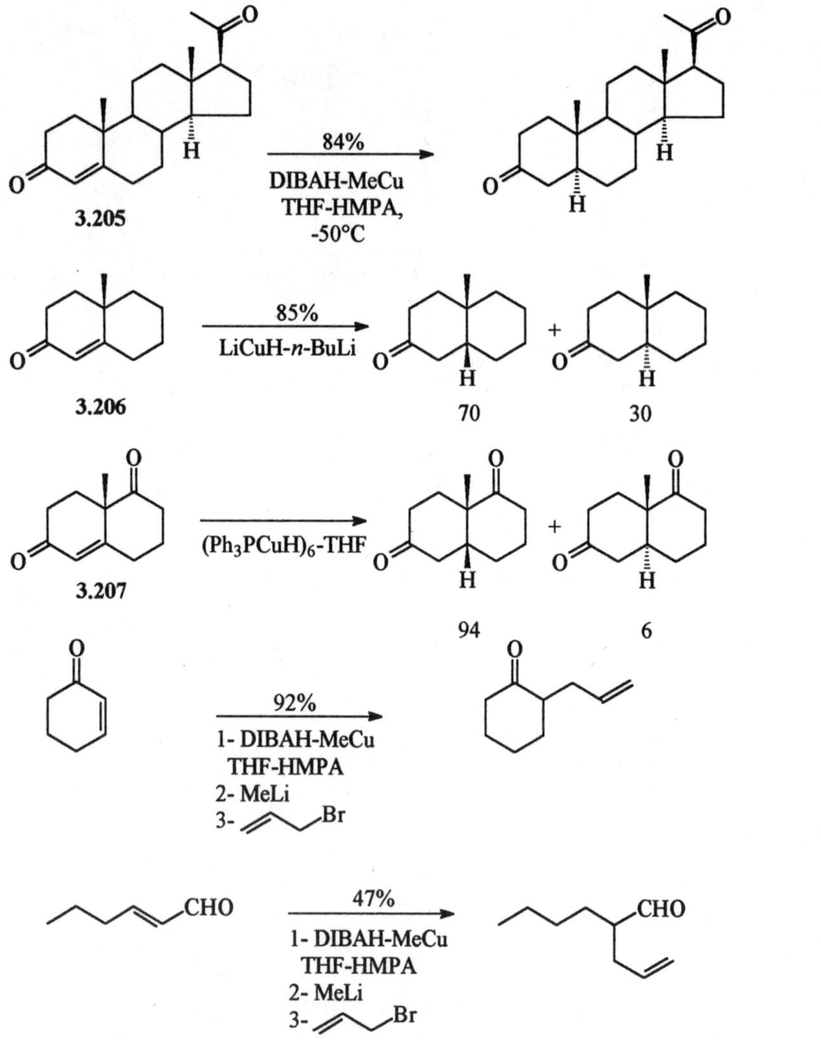

Figure 3.81

enolates obtained with alkyl halides requires their transformation into an -ate complex by reaction with MeLi [TS1] or *t*-BuLi [DK1]. In the latter case, trapping of the Al enolate can be carried out with aldehydes or acyl chlorides; ketones, esters, methyl vinyl ketone or methyl acrylate, MeI, tosylates, and methyl chloroformate do not react. Polyalkylation reactions are thus avoided (Figure 3.81).

The reduction of α-enones by the alkaline borohydrides in alcohols or THF in the presence of a protic solvent most often gives mixtures in proportions that depend on the solvent and on the structure of the substrate [EH1, PS1, VK1]. Aldehydes principally lead to allylic alcohols, as do some linear ketones (Figure 3.82). When

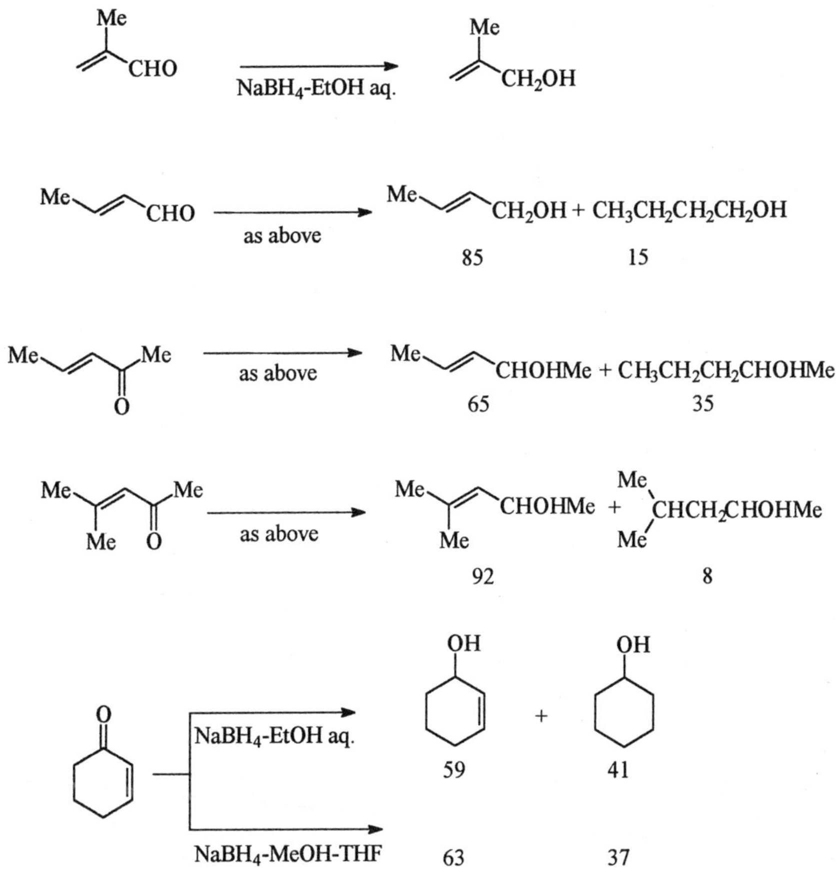

Figure 3.82

starting from cyclic α-enones, one obtains saturated alcohols, the intermediate enolate being protonated to give a saturated ketone, which is then reduced (Figure 3.82). Vinyl ketones are reduced to saturated ketones by $NaBH_4$ on resin in dioxane, but α,β-unsaturated aldehydes give saturated alcohols under the same conditions [NS1]. Sodium borohydride reductions of enones in the presence of micelles, in microemulsions, or in the presence of polyethyleneglycol methyl ether also give mixtures [BK8, FR2, LS7, NS4].

Cyanoborohydrides in the presence of acid in MeOH or HMPA most often lead to allylic alcohols from linear α-enones or α,β-unsaturated aldehydes and to mixtures from cyclic α-enones [HK2], even though some of the steroid ketones could have been reduced to allylic alcohols [VM1].

The relative reactivity of the α-enones and ketones towards different reducing agents has been examined [S2]. When electrophilic assistance is important, the more basic saturated ketone is selectively reduced. This is the case with LTBA in THF [M1], $Zn(BH_4)_2 \cdot 1.5$ DMF in MeCN [HJ1], $NaBH_4$, and $BH_3 \cdot NH_3$ in MeOH [A1,

IL1, TK2]. Provided that steric hindrance does not intervene, the reactivity order with $NaBH_4$ in MeOH–CH_2Cl_2 at −78°C is as follows [WM1, WR1, WR4]:

α-enones < ketones < α,β-unsaturated aldehydes < aldehydes

It is therefore possible selectively to reduce ketones in the presence of α-enones using this reagent–solvent mixture. When an α-enone and a saturated ketone are present in the same molecule, such as in androst-3-enone **3.208** or Wieland–Mischler ketone **3.207**, one obtains the corresponding secondary alcohol [WR1, WR2] (Figure 3.83). A similar selectivity is observed when performing the reduction of **3.207** with $NaBH_4$ in EtOH at −10°C [IT3, TU1]. Curiously, $Zn(BH_4)_2$ in Et_2O does not appear to be very selective toward progesterone **3.205** [IL1]. However, in DME, this reagent reduces saturated ketones, leaving α-enones untouched [SD2]. At −78°C, **3.207** is selectively reduced to **3.209** (Figure 3.83). $NaBH_4$–$CeCl_3$ in MeOH [GL1] reduces the α-enone moiety of progesterone **3.205** to allylic alcohol, but with a low selectivity, while n-Bu_4NBH_4 in THF leads to a mixture of 20-keto-3-ols **3.210** [IL1] (Figure 3.83). The reduction of steroidal diketones **3.205** and **3.208** with Selectrides is poorly selective [WD1].

The presence of a hydroxy group at the α position may direct the double-bond reduction of **3.211** to the face bearing this group [SJ2] (Figure 3.83). Enaminones RCOC(R′)=$CHNMe_2$ are reduced by LAH in Et_2O to β-aminoketones RCOCH(R′)CH_2NMe_2 [SE1].

As previously mentioned (Section 3.2.2), axial attack is more favored for cyclic α-enones than for saturated cyclanones [CG7, N2, WH3]. Therefore, highly stereoselective reductions of enones are expected. Indeed, the reductions of steroids such as **3.212** or **3.213** by various borohydrides [RB3, VM1] or of **3.214** and of testosterone **3.215** by $NaBH_4$–$CeCl_3$ in MeOH [GL1, KA2] give rise to equatorial allylic alcohols (Figure 3.84). The reduction of pulegone **3.216** by aminoborohydrides [FF2, FS2], diisopropoxytitanium tetrahydroborate [RB3], or $NaBH_4$–$CeCl_3$ in MeOH [GL1] is also highly stereoselective (Figure 3.84).

Other related examples are provided in the literature [KY3, NS5, SM8, TH2, WK1, YK6]. The reductions of substituted cyclopentenones, functionalized vinylketones, and carbacyclin precursors by $NaBH_4$–$CeCl_3$ in MeOH have also provided interesting stereoselectivities [BA4, GL1, J1, MH5, SG3]. The asymmetric reduction of α-enones by oxazaborolidines–BH_3 has been described in Section 3.2.3.

In the case of α,β-ethylenic esters, DIBAH in toluene at −70°C and its ate complex with n-BuLi [AH2, KA1, TR3] are the reagents of choice to access allylic alcohols [YG1], the E or Z configuration of the double bond being retained [DD2, MT4] (Figure 3.85). The selective reduction of an ester group may be performed in the presence of a carboxylic acid such as **3.217** [TR3] (Figure 3.85). With $LiEt_3BH$, isolated benzoate esters may be preserved in sugar derivatives, while they are reduced by DIBAH [DD2]. LAH in THF, Et_2O, or C_6H_6 can also induce this reduction when one adds the ester to a cold solution of LAH, but the results are often unsatisfying. When the α,β-unsaturated ester bears an acylamino group, the yield of

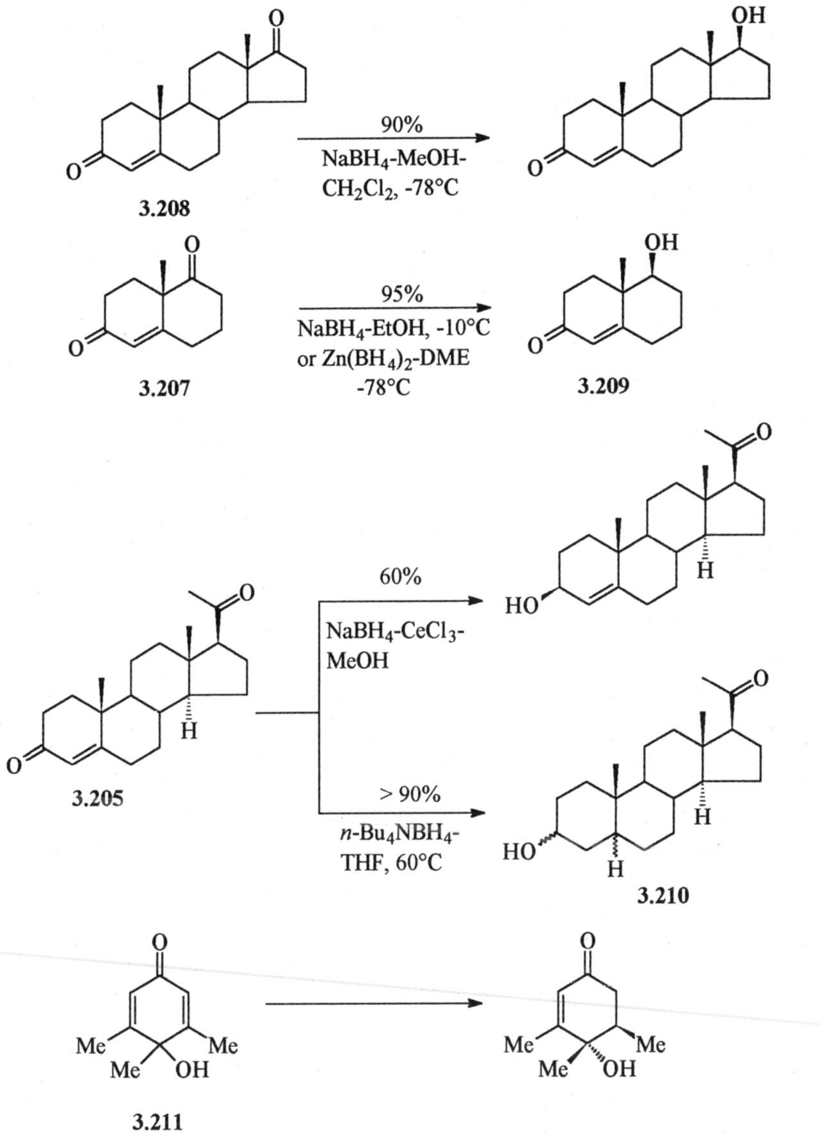

Figure 3.83

the reduction is higher if one begins by adding $BF_3 \cdot Et_2O$ to prevent the complexation of the reagent to the nitrogen site of **3.218** [MH3] (Figure 3.85). A stoichiometric amount of Red-Al in C_6H_6 can give access to allylic alcohols [H3].

Red-Al or $LiAl(OMe)_3H$ in the presence of CuBr in THF–2-butanol leads to saturated esters or lactones [M3, SS1], as does $LiEt_3BH$ in THF–*t*-BuOH [G1]. The role of the alcohol here is to protonate the enolate formed, thus avoiding side condensations. As shown in Figure 3.86, the nonconjugated double bond of **3.219**

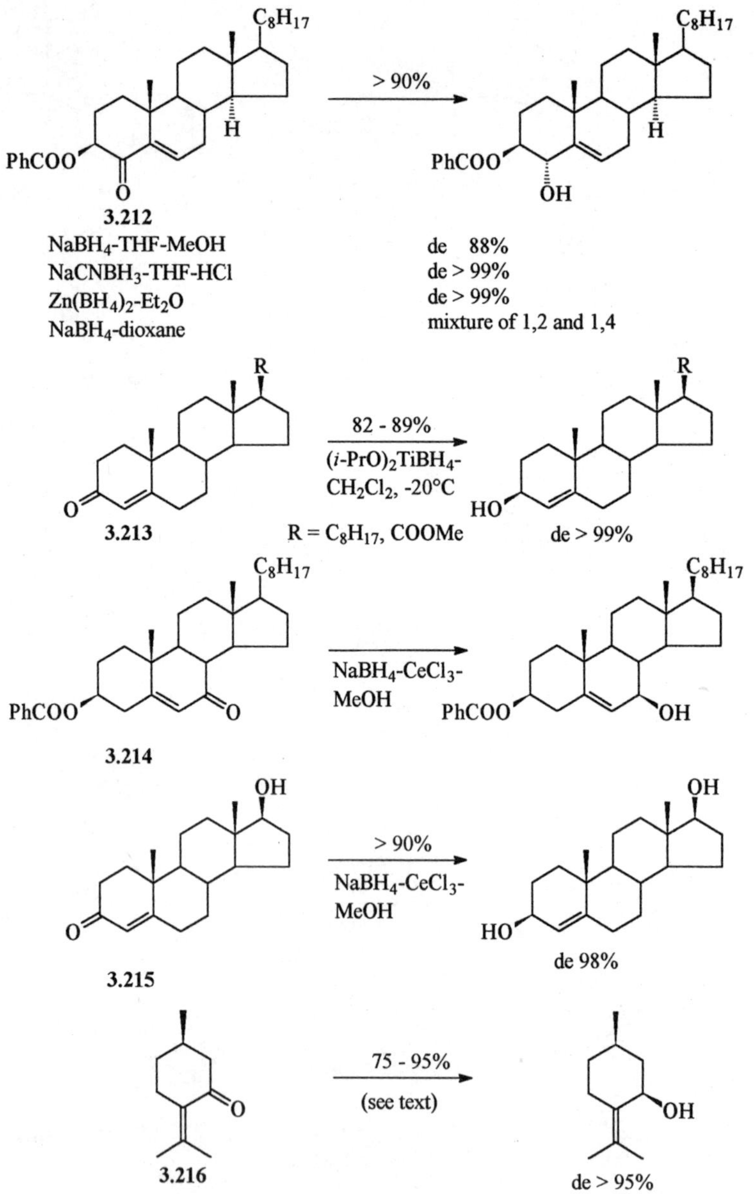

Figure 3.84

remains untouched [BS2]. DIBAH–MeCu also gives access to saturated esters [TH1]. Just as in the case of α-enones, trapping of the enolates formed by reaction with an alkyl halide requires an intermediate ate complex [TS1] (Figure 3.86). Catecholborane also reduces α,β-unsaturated esters into saturated esters, but only under $Rh(PPh_3)_3Cl$ catalysis [EF2] (Figure 3.86).

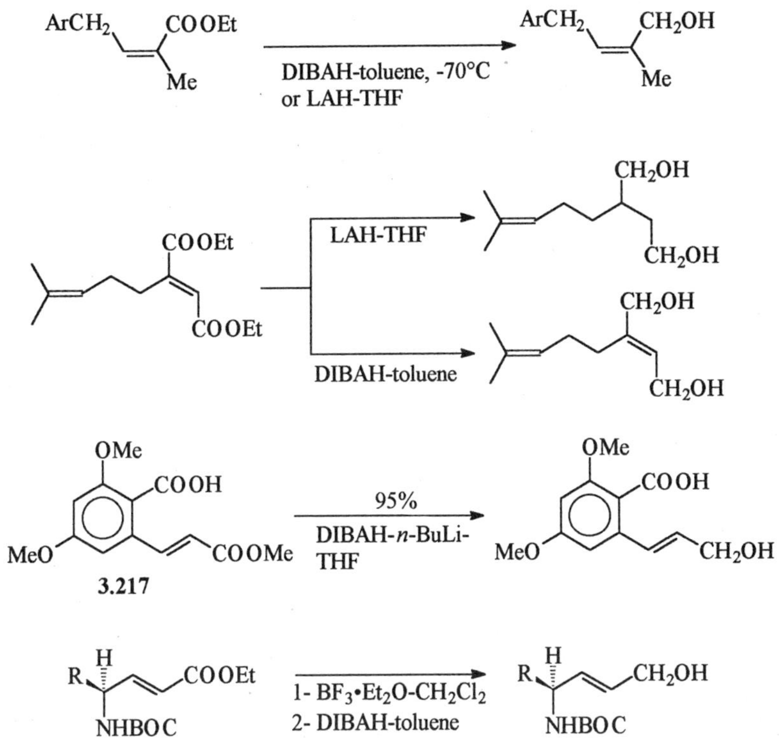

Figure 3.85

In dioxane solution, $NaBH_4$ on a resin reduces α,β-ethylenic esters to saturated esters [NS1], as do $NaBH_4$–Cu_2Cl_2 in MeOH at 0°C [NH1] and $NaBH_4$–$BiCl_3$ in EtOH [RP3]. Disubstituted isolated double bonds remain unchanged [NH1].

Alkaline borohydrides in alcoholic media or in THF–MeOH [SO3] most often give mixtures, while $NaBH_4$–LiCl in THF–EtOH leads to saturated alcohols [JD1]. However, gemdiesters **3.220** or α,β-unsaturated lactone–esters are reduced to saturated esters by $NaBH_4$ or $NaCNBH_3$ in alcoholic media [HR1, PS1, SS3] (Figure 3.86). In some cases, $BH_3{\cdot}Me_2NH$ may be preferred, as shown in the reduction of **3.221** [HS2] (Figure 3.86). $NaCNBH_3$ in an alcoholic medium at pH 3–4 reduces unsaturated gem–ketoesters or the nitrile–esters **3.222** to saturated derivatives without modifying other functional groups [HR1], while $NaBH_4$ reduces the nitrile–esters to alcohols in the same medium [MR4], unless $NaBH_4$ is fixed on a resin [NS1] (Figure 3.86). The reduction can be stereoselective, as shown in the case of **3.223** [BJ3] (Figure 3.86).

AlH_3 reduces α,β-unsaturated amides to allylic amines [BK5] (Figure 3.87). α,β-Unsaturated amides and related compounds such as **3.224** are reduced to saturated amides by Li or K(*s*-Bu)$_3$BH [G1, GL7]. Trapping by alkyl halides has been described in many cases, such as **3.225** [KS1] (Figure 3.87). Conjugate reduction of

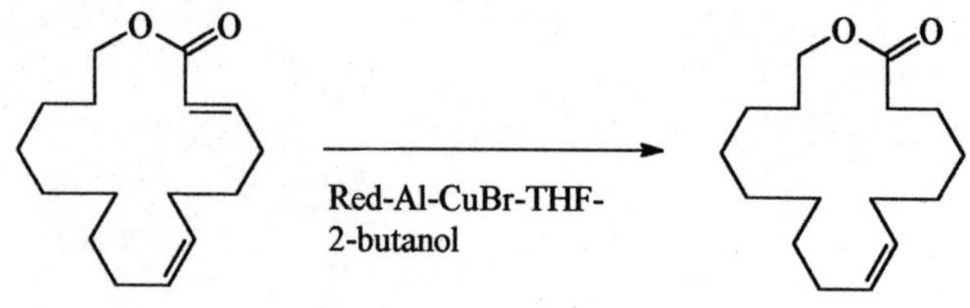

3.219

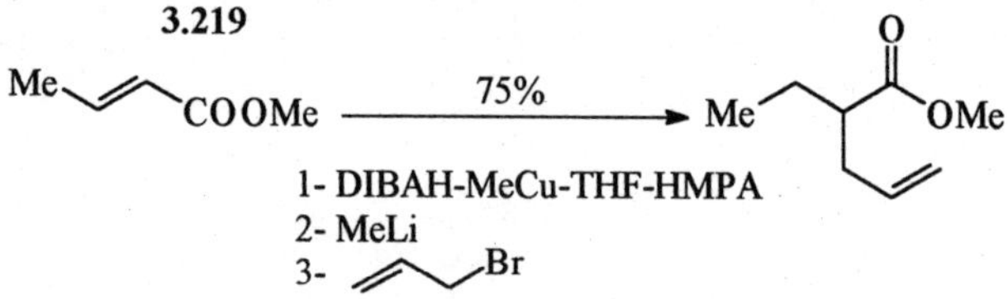

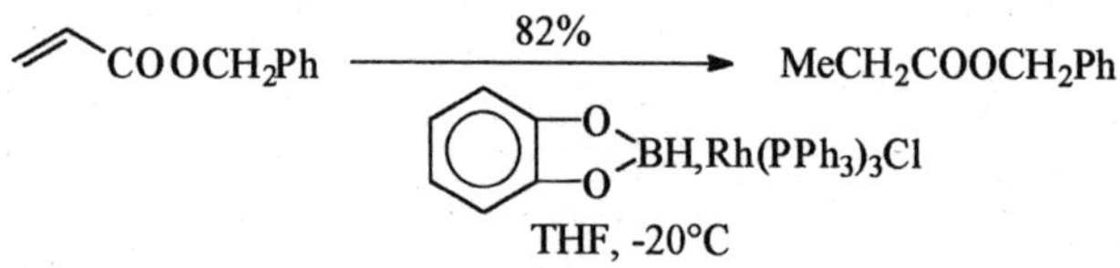

3.220

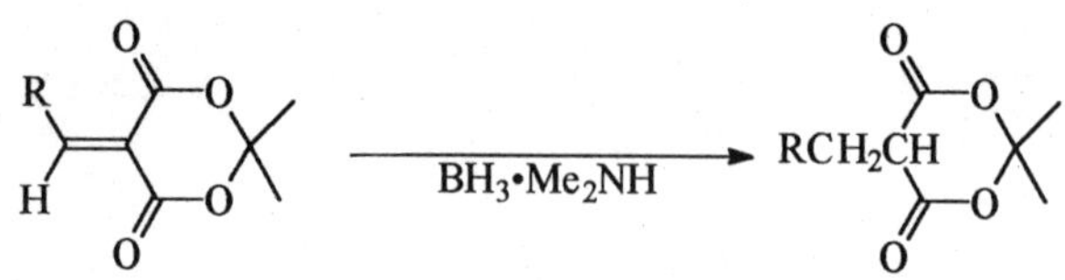

3.221

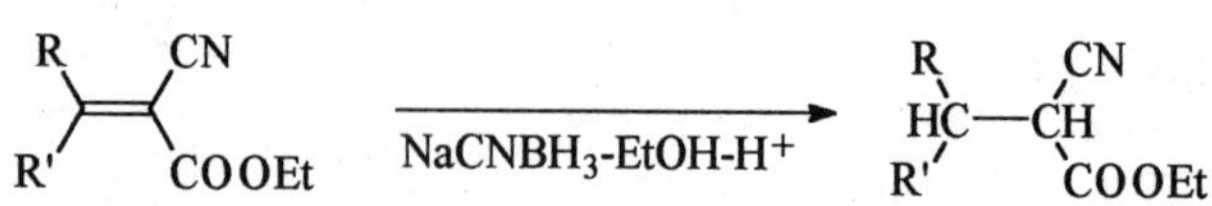

3.222

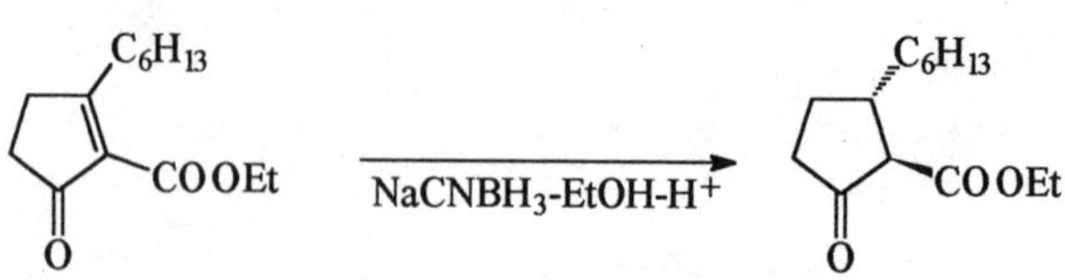

3.223

Figure 3.86

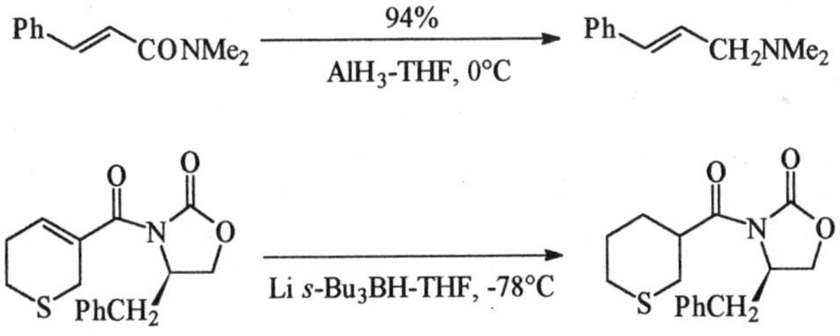

3.224

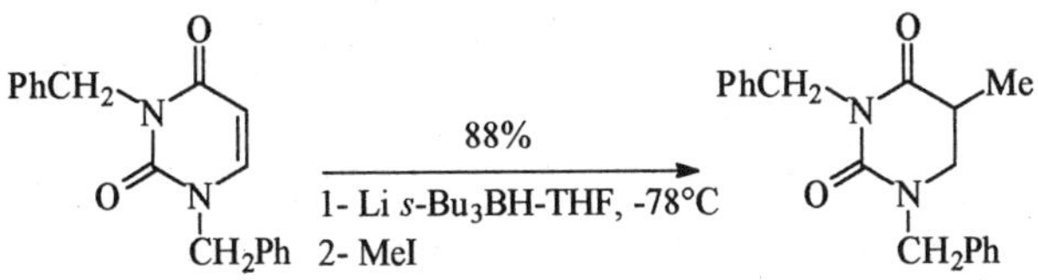

3.225

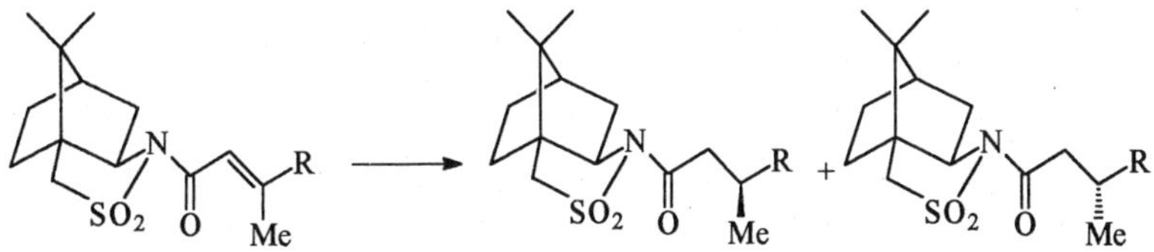

3.226	LAH-CoCl$_2$-THF	92	8
	LAH-CoCl$_2$-Et$_2$O	16	84

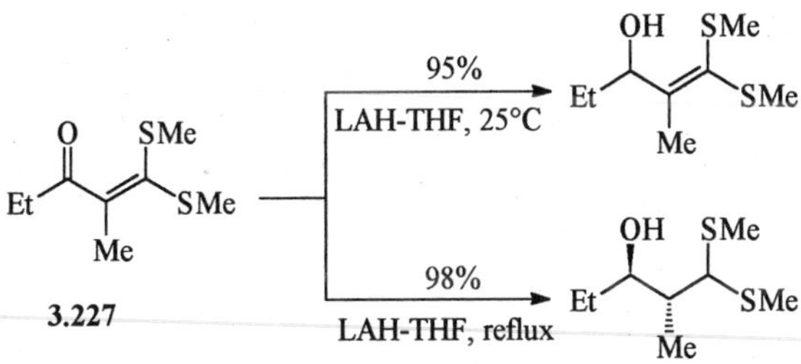

3.227

Figure 3.87

α,β-unsaturated amides or imides can be performed with catecholborane under Rh(PPh$_3$)$_3$Cl catalysis [EF2]. NaCNBH$_3$ also reduces α,β-ethylenic amides that are geminally substituted by another electron-withdrawing group to saturated compounds [HR1]. α,β-Unsaturated lactams can be reduced to saturated cyclic amines by LAH or alkoxyaluminohydrides in ether media, but the results are often disappointing [H3].

The asymmetric reduction of prochiral β,β-disubstituted α,β-unsaturated esters and amides can be performed with NaBH$_4$ in the presence of catalytic amounts of CoCl$_2$ and a semicorrin at r.t. in protic solvents [MP4, P3]. Chiral α,β-unsaturated sultams **3.226** are reduced by LAH–$CoCl_2$ in suspension in THF or Et_2O to saturated derivatives [OM3]; the stereoselectivity is inverted depending on the nature of the solvent (Figure 3.87). Li(*s*-Bu)$_3$BH also promotes such conjugate reduction [OP2].

The transformation of the α-oxoketene dithioketals **3.227** is especially interesting. Reduction by LAH in THF at room temperature leads to allylic alcohols, while under reflux, the hydroalumination of the double bond also takes place. This reaction can be stereoselective [GB2, RC1] (Figure 3.87). $NaBH_4$ in MeOH gives allylic alcohols, while DIBAH leads to the corresponding saturated ketones [RC1].

3.3 CARBON–NITROGEN DOUBLE BONDS: >C=N–

3.3.1 Imines and Iminium Salts: >C=NR, >C=$\overset{+}{N}$RR′

Imines and iminium salts are easily reduced to amines by LAH in THF or Et_2O, Red-Al in C_6H_6 at room temperature [H3, M3, PS1], alkaline borohydrides in alcoholic medium or in AcOH [GN1], or else in the presence of Co or $NiCl_2$ in THF–MeOH [PD2], $Zn(BH_4)_2$ in Et_2O [KY2], $NaBH_4$–$ZrCl_4$ in THF [IS1], alkyl borohydrides in THF [WG1], BH_3·THF [L2], $(CF_3COO)_2$BH·THF [NM1], or amine–boranes in acid media in CH_2Cl_2 [PS1]. In the case of N-triphenylmethylimines, $NaBH_4$ in AcOH must be used because LAH induces unwanted bond cleavage [PR6]. Nevertheless, the common reductions are those run with the cyanoborohydrides at pH 6–8 [L1]. Indeed, under these conditions, ketones and aldehydes are reduced much more slowly. It is then possible to carry out "one-flask" reductive amination of carbonyl compounds by reaction of a primary or secondary amine in the presence of cyanoborohydrides in aqueous MeCN or in MeOH at controlled pH [KO1, L1, PS1]. $NaBH_4$ in the presence of H_2SO_4 [GC2, VG1], $NaBH_4$, and cyanoborohydrides in AcOH or CF_3COOH [GN1], $NaCNBH_3$ in trimethylorthoformate [SB5], preformed Na(AcO)$_3$BH in THF, MeCN, or better $ClCH_2CH_2Cl$ [AC5, AM1, RJ2, YH3] are also valuable reagents. With the last reagent, aromatic and α,β-unsaturated ketones react slowly [AC5]. When using $NaBH_4$ in AcOH, one can still observe the side reaction of alkylation (Section 3.2.8) [GJ2, GN1]. In the presence of Ti(O-*i*-Pr)$_4$, reductive amination either with $NaBH_4$ [AC5, B4, B5, BC10] or with $NaCNBH_3$ [MP1] takes place under mild conditions in diglyme or in EtOH. Hindered amines can also be used provided that the imine is pregenerated in CH_2Cl_2 eventually in the presence of $TiCl_4$ and EtN-*i*-Pr$_2$; reduction is then performed with $NaCNBH_3$ in MeOH [BH5, RK3]. This reaction can also be carried out with a phase-transfer catalyst [HM1, HN3, YK5] or in the presence of pyridine–borane eventually on solid phase [BC9, KA5, KS8, M7, PR2]. Na(AcO)$_3$BH is the most efficient reagent to perform reductive aminations with weakly basic amines even from aldehydes [AC5]. Primary amines and NH_3 give imines, which are then reduced [H3] (Figure 3.88). In acid media, secondary amines are converted to iminium salts, which also undergo reduction (Figure 3.88).

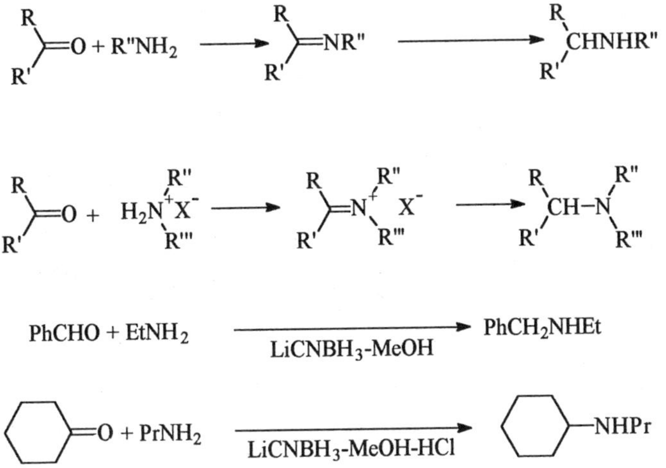

Figure 3.88

The following examples show the compatibility of the reaction with the presence of various functional groups [B4, L1] (Figure 3.89). Nitroimidazoles remain unchanged [YH3]. No epimerization takes place when chiral amino esters are used [SS9], but in the presence of water, some esters can be hydrolyzed [L1]. The methodology involving $Ti(O\text{-}i\text{-}Pr)_4$ and $NaCNBH_3$ [MP1] leaves acid-sensitive groups such as acetals, carbamates, ureas, esters, and amides unchanged. Similarly, α-ketoesters suffer reductive amination with $Na(AcO)_3BH$ in $ClCH_2CH_2Cl$, leading to N-substituted α-amino esters [AC5]. It is possible to obtain amino acids from ketoacids by using $LiCNBH_3$ in MeOH under a careful control of the reaction pH, although only moderate yields result [BB1].

The application in the N-methylation of alkaloids has been published [SH3]. The N-methylation of amines by paraformaldehyde–$NaBH_4$ in CF_3COOH in the presence or absence of THF or by CH_2O–$NaCNBH_3$ in AcOH has also been recommended [GN3]. However, this method is limited because it is not possible to make monomethylated amines from primary amines; the transformation of the intermediate secondary amine to tertiary amine is very rapid [AC5]. The best way to prepare monomethylated amines is then via the carbamates (Section 3.2.8).

Reductive amination can also be accomplished in an intramolecular fashion [BM8, VO1], as shown in Figure 3.90. One predominant stereoisomer **3.228** is formed when generating six-membered rings [AO1]. The stereoselectivity of the formation of pyrrolidines is lower [BM8]. Finally, one can couple the reductive amination and alkylation with an acid (Section 3.2.8) by raising the temperature [GN1] (Figure 3.90). α-Amino esters can easily be obtained from N-silyl-iminoesters **3.229** with $NaCNBH_3$ in MeOH, $NaBH_4$ in MeOH, or $Me_2NH{\cdot}BH_3$ in MeOH [MT5], while LAH converts them to amino alcohols [MT5] (Figure 3.90). Reductive amination can be carried out from acetals [M8].

The stereoselectivity of the reduction of the six-membered cyclic imines has been

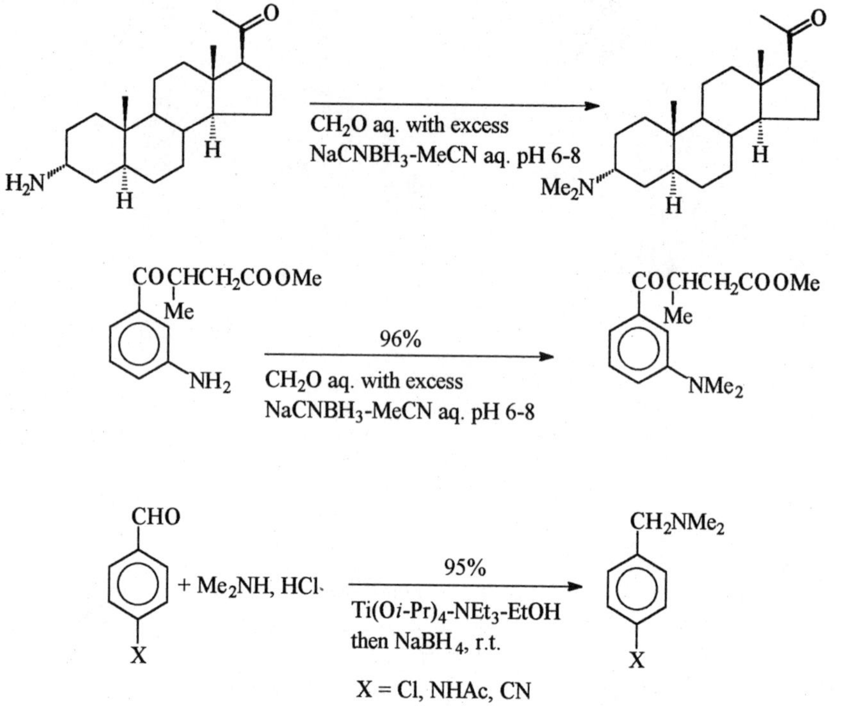

Figure 3.89

examined [HS1, M3, PD2, WG1, ZH1]. The results are comparable to those obtained with the cyclic ketones, as shown in Figure 3.91. One nevertheless observes less axial attack by slightly hindered reagents than with the corresponding ketones [HS1]. A suggested variant involving N-diphenylphosphinylimines such as **3.230** allows the preparation of axial primary amines with an excellent stereoselectivity [HA2, HR2, ZH1] by the action of Li(s-Bu)$_3$BH followed by treatment in an acid medium (Figure 3.91). These observations were extended to other substituted cyclohexyl, cyclopentyl, and bicyclic derivatives. On the other hand, reduction of **3.230** with t-BuNH$_2$·BH$_3$ in MeOH gives predominantly the equatorial amine [HA2] (Figure 3.91). Other reducing agents are less stereoselective. The stereoselective reductive amination of substituted cyclohexanones and of tropinone has been realized using in situ generated acyloxyborohydrides [AC5, ML3]. The stereoselective reduction of bicyclic N-silyliminocyclopentenes with Red-Al has been recently published [HB1].

Asymmetric reductive amination can be carried out on chiral ketones able to form an intermediate imine such as **3.231** or **3.232** [BW1, MN1, PP4] (Figure 3.92). According to the structure of the substrates, $NaBH_4$, $NaCNBH_3$, or $Me_4N(AcO)_3BH$ are the reducing agents that give the best yields or selectivities.

Another way to obtain achiral or chiral β-aminoalcohol derivatives stereoselectively is to use $NaBH_4$ or $Zn(BH_4)_2$ to reduce α-hydroxyimines **3.233** or α-trimethyl-

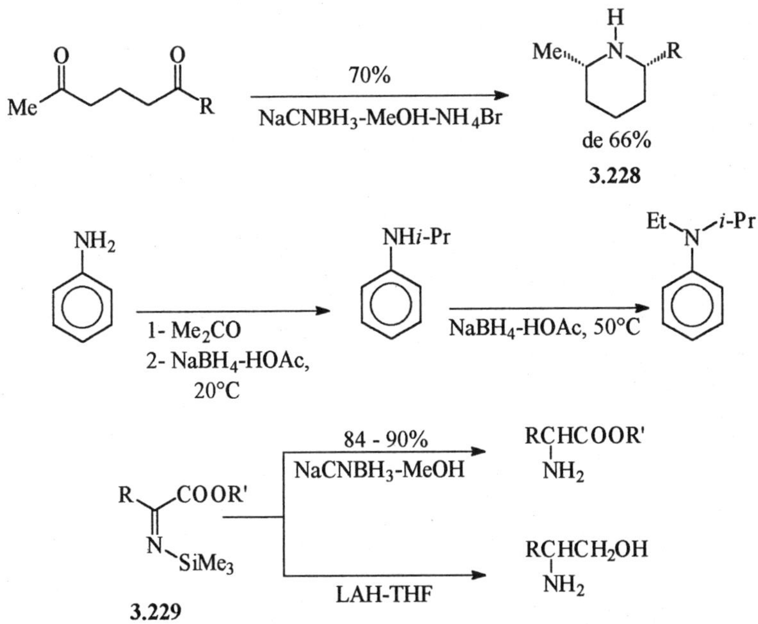

Figure 3.90

silyloxy-N-magnesioimines **3.234**. The latter intermediates are formed by a Grignard reaction with the corresponding protected cyanohydrin [JJ1, KJ1, UA1, ZH1] (Figure 3.92). $NaBH_4$ gives poorer stereoselectivity except when using $OSiMe_2$-*t*-Bu ethers, which are subsequently cleaved by HF to avoid racemization; *syn* isomers are predominant in this case [BD1]. γ-Amino alcohols have also been obtained in a stereoselective fashion by DIBAH reduction of iminoalcoholates **3.235** [TV1] (Figure 3.92). Chiral β-iminosulfoxides also suffer stereoselective reductions by DIBAH–$ZnBr_2$ or by Li $(s\text{-Bu})_3BH$, each reagent giving predominantly one isomer or the other [GL8].

Chiral substituents may also be introduced on the nitrogen of the imine [HM6, ZH1]. The reductions of imines of (*S*)-1-phenethylamine either with $NaBH_4$–$CoCl_2$ or with $LiEt_2NBH_3$ [FB1] do not give high stereoselectivities, except with imines of *t*-BuCOMe. The reduction of a chiral sulfilimine **3.236** by 9-BBN in THF gives a single diastereoisomer, precursor of (*R*)-alanine ethyl ester [HL3] (Figure 3.93). LAH or DIBAH is more or less stereoselective [HM6].

Reduction of prochiral imines with chiral reducing reagents has also been examined [NN1, ZH1]. Binal reduces phosphinylimines with a high enantioselectivity, but in a low chemical yield [HA3]. Itsuno's reagent (borane plus amino alcohol **3.69** in THF) allows the enantioselective reduction of N-phenylarylimines **3.237** with a good selectivity [CC9] (Figure 3.93). CBS reagents are sometimes less efficient [CC9, SY7]. The asymmetric reduction of dialkylarylimines, iminoesters, or -sulfamides under similar conditions [CC9, CG8, SY5] also gives moderate or poor

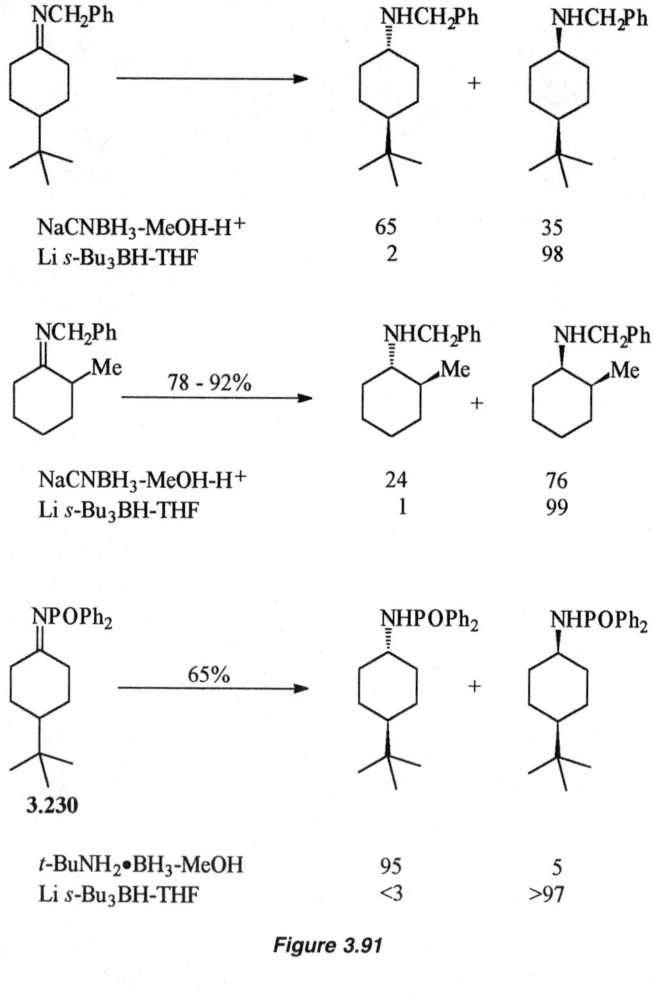

Figure 3.91

enantioselectivities (up to 70%). Oxazaphospholidine–borane reagents are also poorly enantioselective (up to 63% ee) [BB9]. A peculiar reaction leading to nitroalkenes has been observed when reacting 2-nitroimines **3.238** with $NaBH_4$–$CeCl_3$ [DS3] (Figure 3.93).

Iminium salts can also be formed by cleavage of the C—CN bond of aminonitriles either in alcoholic media or in the presence of a cation having sufficient Lewis acid properties. The intermediates are then transformed in situ into amines (Figure 3.94). Aminonitriles are thus reduced to amines by LAH in Et_2O [CT1], AlH_3 in Et_2O [E2], $NaBH_4$ in alcoholic medium [GR1, MR2, YR1] or in diglyme [YA1], $Zn(BH_4)_2$ in ether medium; sometimes the presence of $AgBF_4$ or $Hg(OTf)_2$ is required [FD1, GR1, GR2, S7] (Figure 3.94). The reduction is inhibited if the carbon is too sterically hindered [BM2]. This reductive decyanation can be stereo-

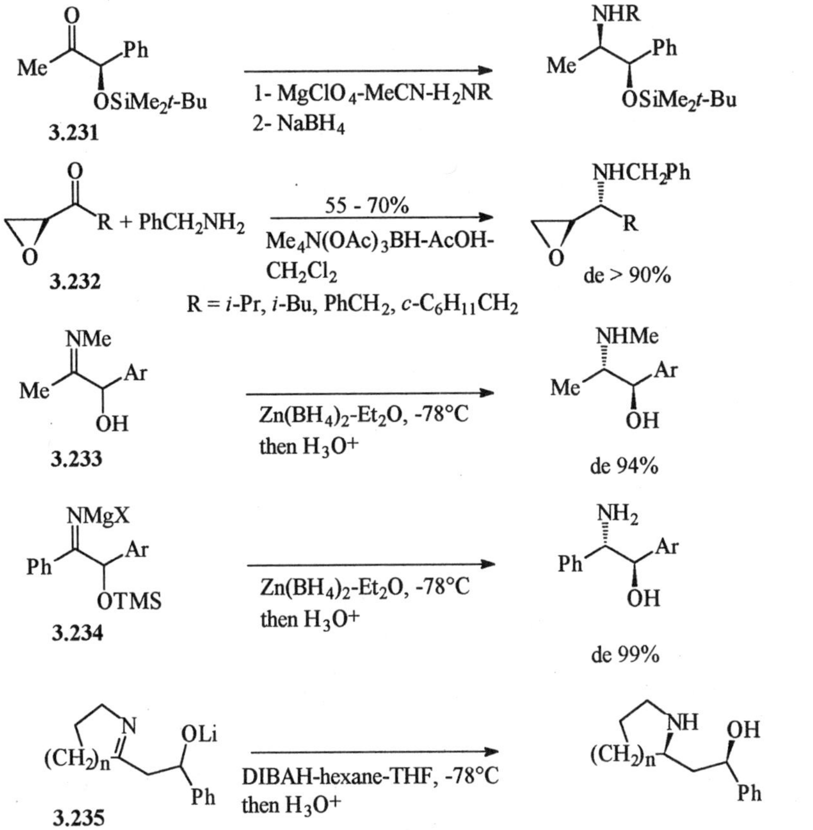

Figure 3.92

selective in cyclic systems, as shown with **3.239** [BR1] (Figure 3.94). In noncyclic molecules, the stereoselectivity is even lower [MR2].

Another way to produce iminium salts is to reduce aminals with LAH, AlH_3, DIBAH, $NaCNBH_3$ in AcOH [GR1, MR2, WR3], or $NaBH_4$ in EtOH (Figure 3.95). Cyclic aminals **3.240** are converted to amino alcohols [GR1, MR2] (Figure 3.95). However, six-membered tetrahydro-1,3-oxazines **3.241** are not reduced by $NaBH_4$. LAH in Et_2O converts aminals such as **3.241** by C—O bond cleavage to γ-amino alcohols, and their corresponding methiodides are converted by C—N bond cleavage to alkyl N-methyl-3-aminopropylethers [AA1] (Figure 3.95). Stereoselective reductions can be observed as with **3.242** [MQ1] (Figure 3.95). The reduction of the five-membered analogues in bicyclic derivatives **3.243** (n = 1 or 2) can be performed either with AlH_3 (n = 1) at low temperature or with Red-Al in refluxing THF (n = 2). The lactam carbonyl is simultaneously reduced in both cases [BM7, MM9] (Figure 3.95). This method provides the route to nonracemic 2-alkylpyrrolidines or piperidines.

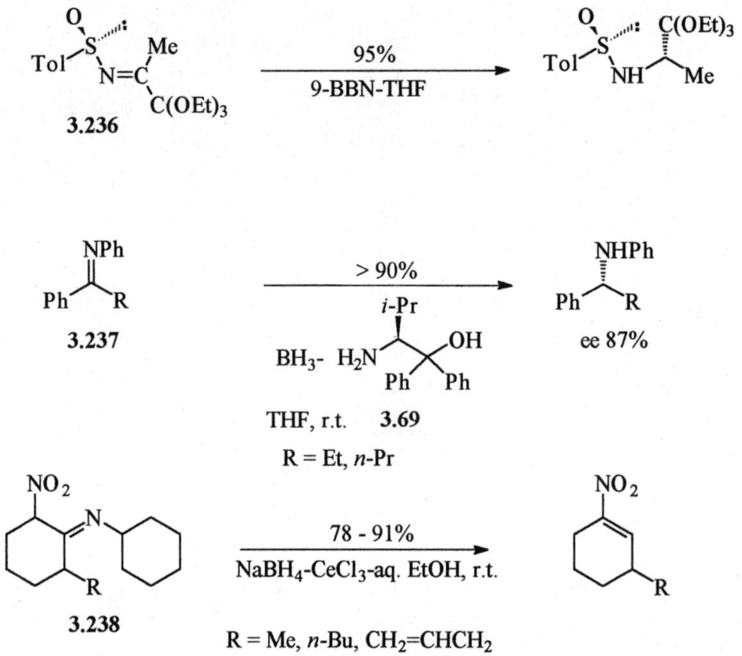

Figure 3.93

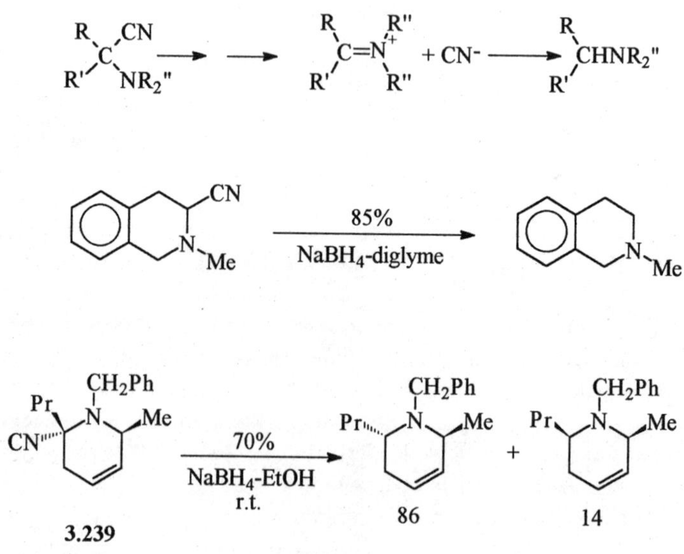

Figure 3.94

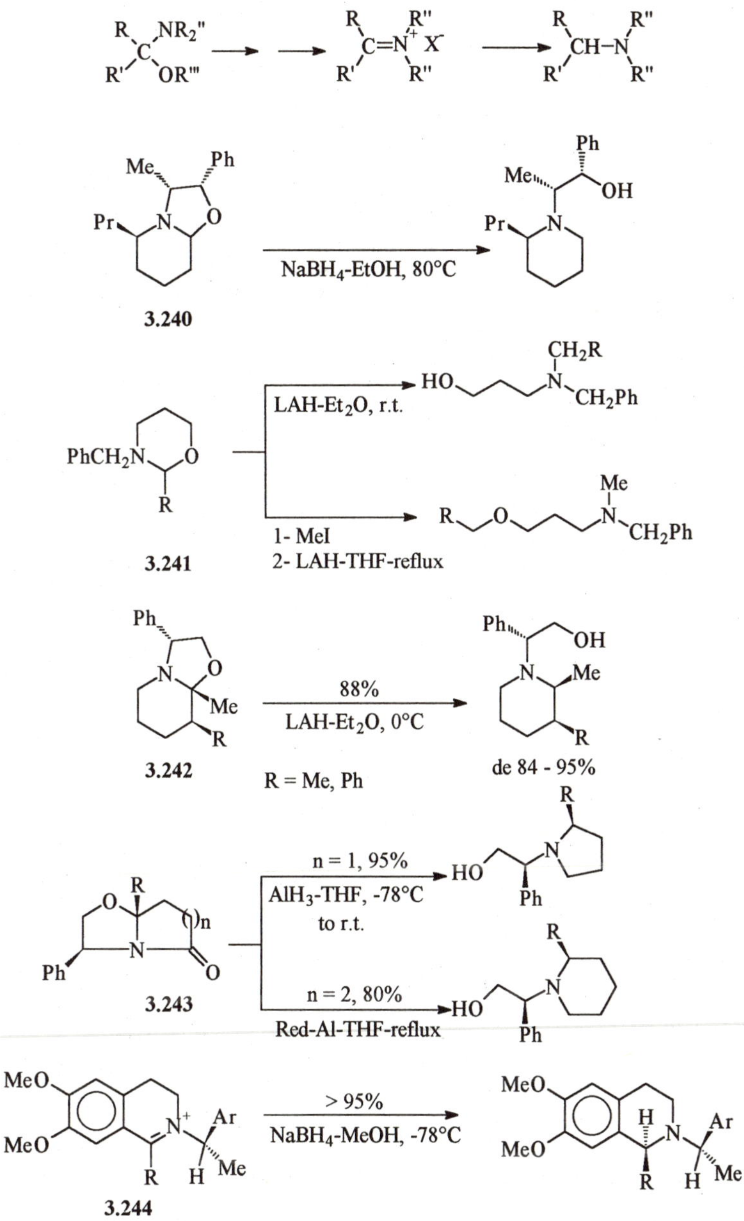

Figure 3.95

Similarly, chiral iminium salts such as **3.244** can be reduced in a diastereoselective fashion by $NaBH_4$ in MeOH at $-78°C$ [PK1] (Figure 3.95). The highest stereoselectivity is observed when the aryl group is 2,6-dichlorophenyl. The bifunctional derivatives can undergo two successive reductions [MR2].

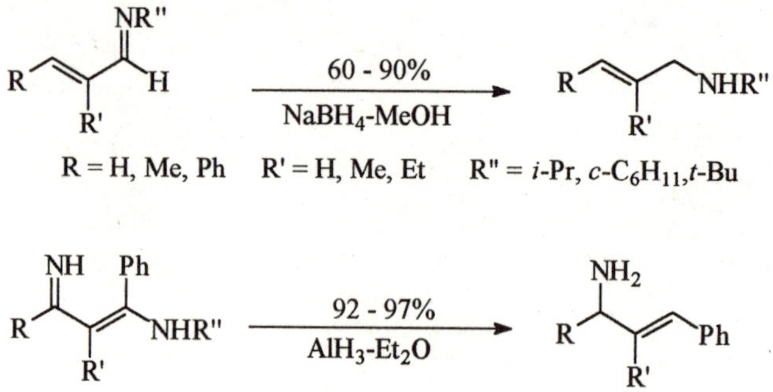

3.245

Figure 3.96

α,β-Unsaturated imines are converted into the corresponding unsaturated secondary amines by $NaBH_4$ in MeOH or EtOH [DS2] (Figure 3.96). Allylic amines can be obtained from 4-aminoazadienes **3.245** and AlH_3 or DIBAH [BA2] (Figure 3.96). If R′ is allyl or benzyl, saturated imines are predominantly formed, but sequential treatment of these azadienes by DIBAH and $NaBH_4$ in MeOH leads to allylic amines.

3.3.2 Enamines: $>C{=}C<NR_2$

The reduction of enamines occurs after their protonation and the tautomerism of the protonated enamine to an iminium salt. Substrates will be accordingly reduced only in the presence of sufficiently strong acids or in protic media. LAH in THF does not reduce enamines. Enamines are transformed into saturated amines by AlH_3 in Et_2O [H3] and by $NaBH_4$ in alcoholic media [BB1], in THF–AcOH [GN1], in the presence of CF_3COOH [GN1] or, better yet, by $Zn(CNBH_3)_2$ in MeOH [KO1] and $NaCNBH_3$ in THF–MeOH [BB1] (Figure 3.97). Since cyanoborohydrides do not reduce esters, β-enaminoesters can be transformed into β-aminoesters with $NaCNBH_3$ in THF–MeOH [BB1]. The stereoselectivity of the reduction of cyclic enamines such as **3.246** and **3.247** has been examined. Reaction with $NaCNBH_3$ in AcOH or $BH_3{\cdot}NH_3$ in AcOH provides predominantly the axial saturated amines [HS1] (Figure 3.97). In heterocyclic systems such as alkaloids, highly stereoselective reductions can be observed; however, stereoselectivities are highly dependent on both substituents and conformation [WF1]. Substituted β-enaminoesters suffer diastereoselective C=N reduction with $Na(AcO)_3BH$ in AcOH–MeCN [CP5].

3.3.3 Nitrogen Heterocycles

Indoles Indoles can be viewed as cyclic enamines. Therefore, they are reduced to indolines in acid medium by $BH_3{\cdot}THF$ in CF_3COOH [LO1, MM1], pyridine–

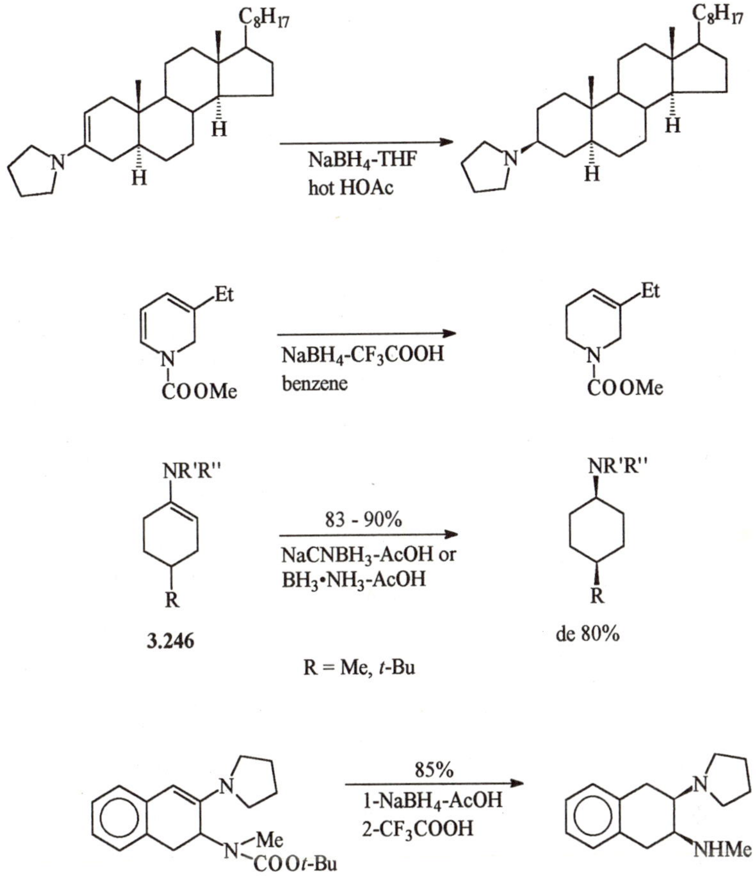

3.247

Figure 3.97

borane in CF_3COOH [K3], and $NaCNBH_3$ in CF_3COOH [GN1, GN2, KL2]. The use of $NaBH_4$ in AcOH is unsuitable because there may be concurrent N-acylations [GN1, GN2, GJ1]. However, if the indole carries a COMe or COEt substituent, it is reduced by $NaCNBH_3$ in CF_3COOH to a CH_2Me or CH_2Et group [KL2]. It is interesting to note that $NaBH_4$ in CF_3COOH does not lead to indolines with NSO_2Ph derivatives [KL1], while $NaCNBH_3$ in CF_3COOH does [KL2]. These reductions are compatible with ester and nitrile substituents. The reduction by $NaBH_4$ in CF_3COOH is compatible with halides and ester groups, and it can be stereoselective, as shown in the reduction of **3.248** [GN1] (Figure 3.98). $Zn(BH_4)_2$ reduces 2,3-dimethylindole into the *trans*-2,3-indoline [DG2].

The reduction of indoles by pyridine–borane in CF_3COOH [K3] or $BH_3{\cdot}THF$ in CF_3COOH [MM1] is compatible with amide, nitrile, or ester groups. It is interesting to emphasize that LAH in ether media reduces these groups without affecting the

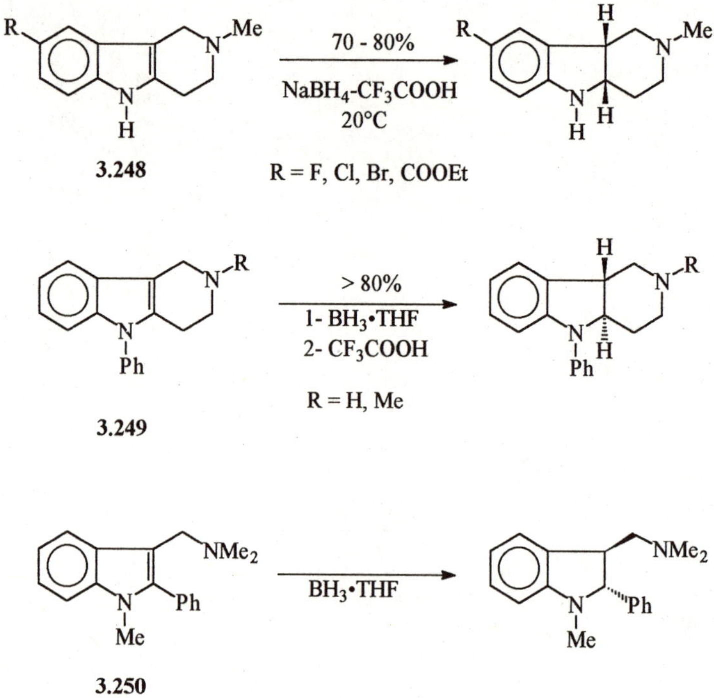

Figure 3.98

indole heterocyclic double bond. The formation of a compound with a *trans* ring junction can be realized starting from an indole ring fused to a nitrogen heterocycle **3.249** by preforming the corresponding amine–borane. This leads to the *trans*-indoline in an intramolecular fashion [EG1] (Figure 3.98). A similar stereoselectivity is observed in the reduction of 2-phenyl-3-dimethylaminomethyl-N-methylindole **3.250** by BH_3 in THF [DG2] (Figure 3.98).

Heterocyclic Imines and Iminium Salts Heterocyclic imines are reduced under the same conditions as linear ones. This reduction is compatible with the same functional groups, as shown in the case of **3.251** [GN1] (Figure 3.99). Moreover, if the acid used is chiral, one can observe an asymmetric induction [GN1] (Figure 3.99). In the reduction of compound **3.252**, the presence of the secondary amine in the six-membered ring prevents the subsequent reduction of the indole, which is not protonated under these conditions (Figure 3.99). 2,6-Dialkylpiperideines or 2,5-dialkylpyrrolines **3.253** can be stereoselectively reduced to *cis*- or *trans*-disubstituted piperidines or pyrrolidines by using either DIBAH or LAH–Me_3Al [BC8, MM5], the other reagents being less stereoselective (Figure 3.99). The reduction of 2-phenyl-3-alkyl-Δ-l-pyrrolines **3.254** by $NaCNBH_3$–AcOH at low temperature is highly diastereoselective towards the *cis*-2,3-disubstituted pyrrolidine [PB2]. At higher temperature, the stereoselectivity decreases, except when R = *i*-Pr (Figure

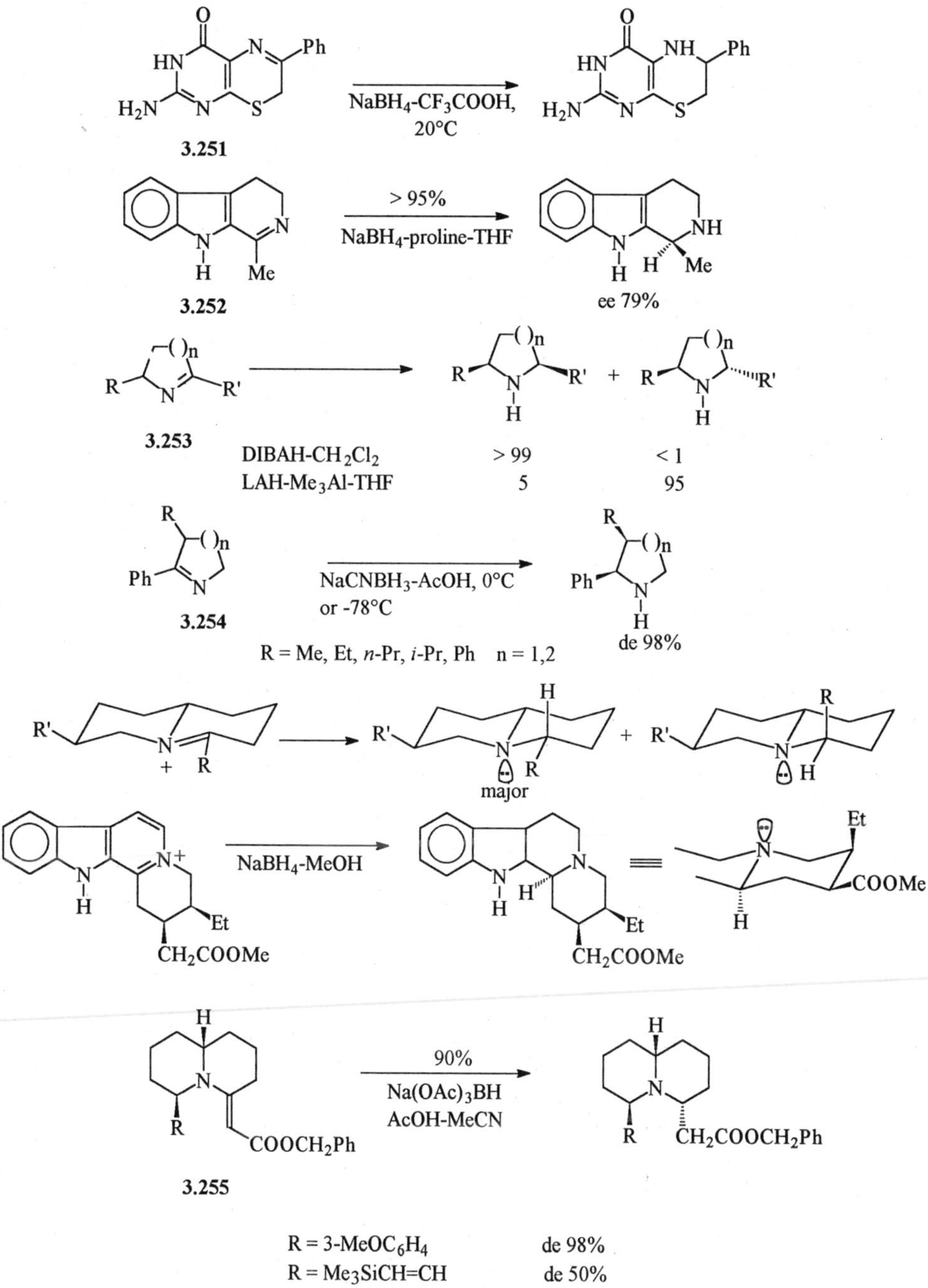
3.251
NaBH4-CF3COOH, 20°C
3.252
> 95%
NaBH4-proline-THF
ee 79%
3.253
DIBAH-CH2Cl2 > 99 < 1
LAH-Me3Al-THF 5 95
3.254
NaCNBH3-AcOH, 0°C or -78°C
R = Me, Et, n-Pr, i-Pr, Ph n = 1,2
de 98%
major
NaBH4-MeOH
3.255
90%
Na(OAc)3BH AcOH-MeCN
R = 3-MeOC6H4 de 98%
R = Me3SiCH=CH de 50%

Figure 3.99

3.99). Cyclic N-oxides (nitrones) are reduced with $NaCNBH_3$ in MeOH to the corresponding oximes [OB3].

The reduction of the bicyclic iminium salts having the nitrogen at the ring junction can be very stereoselective. The hydride enters preferentially on the axial face antiperiplanar to the developing lone pair on nitrogen [D2, HL4, M3, NS3] (Figure 3.99). LTBA or $Na(AcO)_3BH$ can be more stereoselective than $NaBH_4$ or $NaCNBH_3$ [HL4, M3], and the nature of the substituents in the 6-position of **3.255** has a strong influence on the stereoselectivity of the reaction [HH5, HL4] (Figure 3.99). When the nitrogen atom is not at the ring junction, the reduction is often less stereoselective [BB4, SM5].

Isoxazolidines, Isoxazolines, Oxazolines, and Oxazines Isoxazolidines, easily obtained by cycloaddition of nitrones to olefins, are reduced to 1,3-amino alcohols by reaction with LAH in ether media. The synthesis of racemic sedridine from **3.256** is illustrative [TA2] (Figure 3.100).

Quaternary ammonium salts derived from isoxazolidines are also reduced by LAH. Depending on the nature of the substituents, hydride attack takes place α to the oxygen, as previously from **3.257**, or α to the nitrogen from **3.258**. One can then obtain either a 1,3-amino alcohol [TA2] or a substituted hydroxylamine [LS3] (Figure 3.100).

Isoxazolines **3.259** are obtained by cycloaddition of nitrile–oxides to olefins. Their reduction by LAH in Et_2O leads to 1,3-amino alcohols, the *syn* isomer generally being largely predominant [JS2, WP1]. This constitutes an interesting synthetic method (Figure 3.100). When R is a latent carboxyl group such as *p*-anisyl or α-furyl, α-hydroxy amino acids can be obtained in a highly stereoselective fashion [JG1]. Such methodology can also be applied to the synthesis of amino sugars [JG1, JM2]. However, $NaCNBH_3$ in the presence of HCl reduces only the C=N bond and converts the isoxazolines to the corresponding isoxazolidines [JB1], which can be further reduced by LAH–$NiCl_2$ [GO2]. In some cases, $NaBH_4$–$NiCl_2$ in MeOH gives better results than LAH [AC4]. 5,6-Dihydro-1,2-oxazines **3.260** can be reduced by $NaCNBH_3$ in AcOH into the corresponding tetrahydro-1,2-oxazines with a good stereoselectivity [ZA1] (Figure 3.100). Other ethers give similar results. However, if the EtO group is replaced by Me_3SiO, this latter group is removed under these conditions.

The reduction of aryloxazolines **3.261** to 1,2-amino alcohols is carried out by DIBAH in Et_2O or in hexane at 0°C [MH1], and is compatible with a halogen substituent on the aromatic ring (Figure 3.101). Quaternarization of chiral oxazolines with MeOTf followed by treatment with $NaBH_4$ generates the corresponding aldehydes [GM4]. 2-Aminosubstituted phenols can also be prepared by reduction of the oxazoline **3.262** by $NaBH_4$ in THF in the presence of AcOH [YL3] (Figure 3.101). This method leaves the ester groups unchanged, while nitriles suffer some reduction to amines. The N-oxides of chiral oxazoline **3.263** also suffer reduction by $NaBH_4$ in MeOH to the saturated hydroxylamine [DB1] (Figure 3.101).

The stereoselective synthesis of 1,3-amino alcohols having three or four chiral

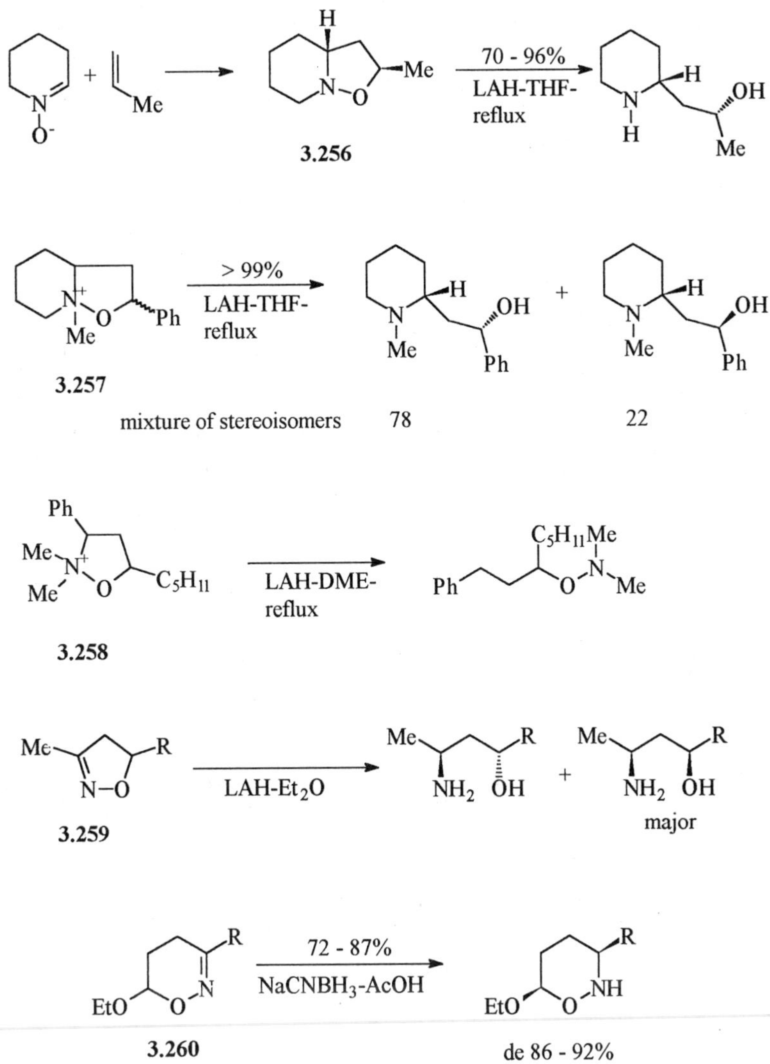

Figure 3.100

centers can be carried out by LAH reduction of 1,3-oxazines **3.264** [BJ4] (Figure 3.101). However, $NaBH_4$ in THF–EtOH can only reduce the C=N bond of **3.265**. This leads to aminals, which are hydrolyzed under acidic conditions to aldehydes [PH1] (Figure 3.101). The reduction of cyclic aminals is described in Section 3.3.1.

Pyridines, Quinolines, and Analogues Pyridines are not reduced by the alumino- and borohydrides unless they carry electron-withdrawing groups at the 3- and 5-positions. In this case, they are converted into the corresponding 1,4-dihydropyridines by $NaCNBH_3$ in AcOH [GN1]; the use of $NaBH_4$ leads to mixtures

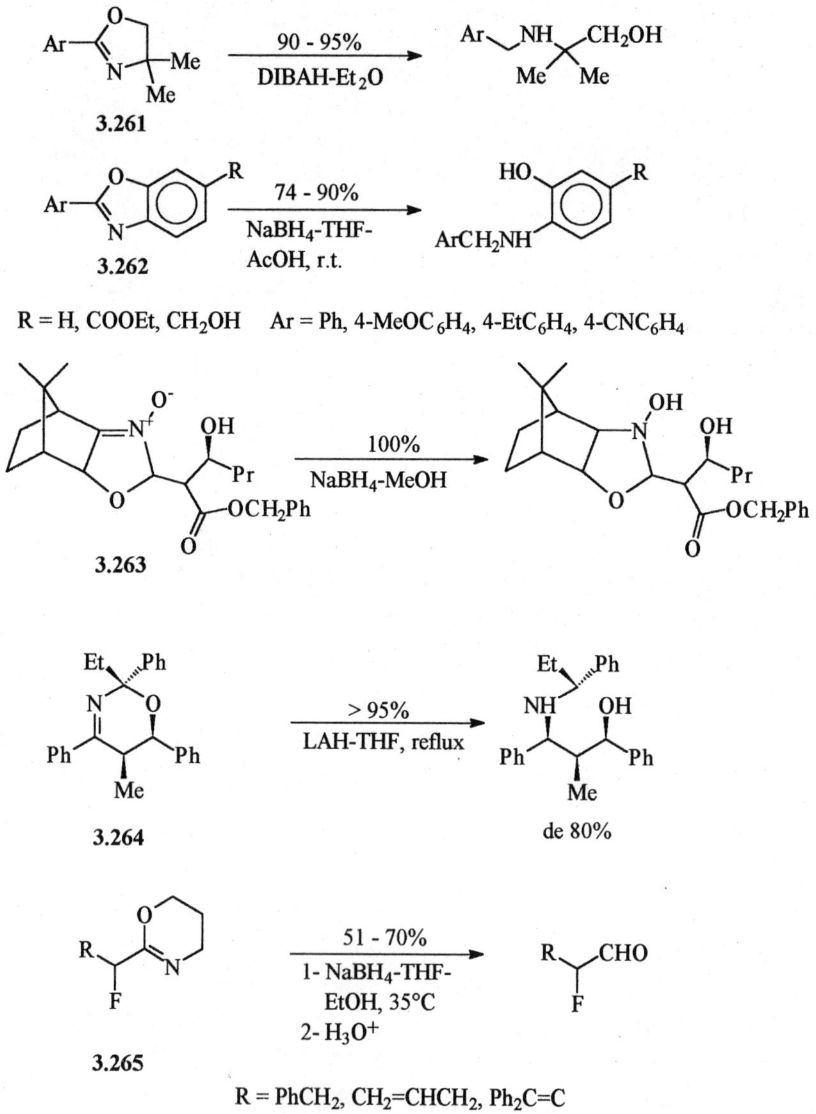

Figure 3.101

(Figure 3.102). LAH in pyridine functions as a reducing agent [LL1]. 1-Methyl-4-phenyl-3,5-dihydro-2-pyridone **3.265a** can be reduced by hydrides. Among them, Li (s-Bu)$_3$BH in THF gives selectively the 3-unsaturated pyridone, while LAH–TiCl$_3$ leads to the 4-unsaturated pyridine [MC4] (Figure 3.102).

Pyridinium quaternary salts **3.266** are, on the other hand, easily reduced by AlH$_3$ or LAH in ether media, Red-Al in C$_6$H$_6$, or alkaline borohydrides in alcoholic media, leading to the 1,2,3,4-tetrahydro-N-alkylpyridines. If diastereoisomers can be generated, the stereoselectivity is usually poor [DG2]. When the pyridinium salt

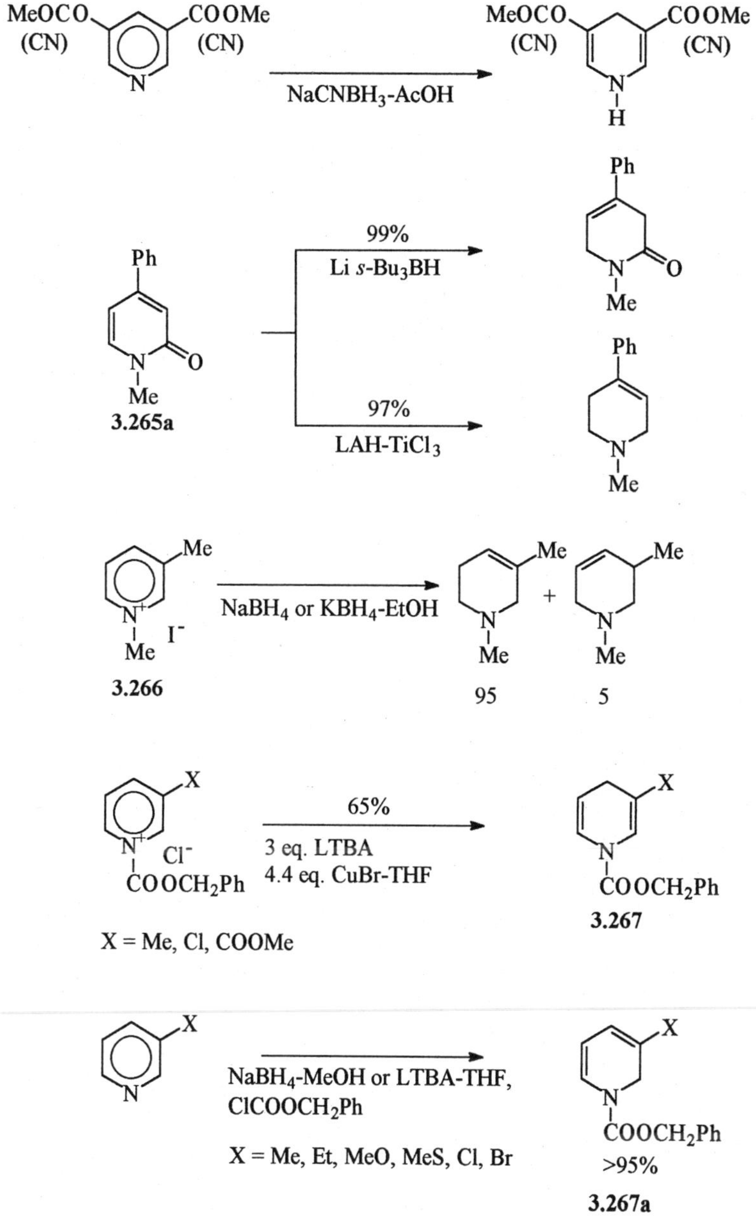

Figure 3.102

bears a substituent at the 3-position, one of the regioisomers is selectively formed (Figure 3.102). Acylpyridinium salts are converted to mixtures by reaction with $NaBH_4$ or $NaCNBH_3$, or with LBTA or Red–Al–CuBr under standard conditions [SS1]. A specific method provides the obtaining of N-acyl-1,4-dihydropyridines **3.267** with a high regioselectivity through the reaction with LTBA–CuBr in THF.

Under these conditions, chlorides and esters remain intact [CA1] (Figure 3.102). It is possible to reduce regioselectively some 3-alkylpyridines with $NaBH_4$ in MeOH or LTBA in THF into N-carbamoyl-1,2-dihydropyridine **3.267a** by performing the addition of chloroformate esters at −78°C to a mixture of the pyridine and the reducing reagent. The regioselectivity is high provided that X is neither too bulky nor an electron-withdrawing group [SH8] (Figure 3.102).

Quinolines and isoquinolines are more easily reduced than pyridines. Aluminohydrides or $BH_3 \cdot THF$ in CF_3COOH [MM1] leave them intact, but $NaBH_4$ or $NaCNBH_3$ in AcOH [GN1], $NaCNBH_3$–$BF_3 \cdot Et_2O$ in refluxing MeOH [SR2], pyridine–borane in AcOH [H3] and $NaBH_4$–$NiCl_2$ in MeOH [GO2] reduce them to tetrahydroquinolines or isoquinolines (Figure 3.103). With $NaBH_4$ in a hot organic acid medium, one can carry out a subsequent N-alkylation as from **3.268** (Section 3.2.8) [GN1] (Figure 3.103). However, $NaBH_4$ in CF_3COOH leads to mixtures. In the presence of a ketone, it is possible to form an N-alkylated amine **3.269** by reduction followed by reductive amination (Section 3.3.1) [GN1] (Figure 3.103).

The treatment of the nitroquinolines **3.270** by $NaBH_4$ in AcOH at 5°C leads to the reduced compounds **3.271**, the NO_2 functional group being retained [GN1]. On heating, the corresponding N-ethylamine **3.272** (Section 3.2.8) [GN1] is obtained (Figure 3.103). Quinoxalines and quinazolines **3.273** or acridine show the same kind of reactivity: $NaBH_4$ in AcOH or CF_3COOH in the cold leads to cyclic secondary diamines, while in hot AcOH, the corresponding bis-N-ethylamines (Section 3.2.8) [GN1] are obtained (Figure 3.103).

3.3.4 Oximes and Hydrazones: $>C{=}NOH$, $>C{=}NNR_2$

Oximes are reduced to amines by LAH or SAH in THF or in Et_2O [CB5, H3]. LAH–N-methylpyrrolidine or AlH_3–Et_3N complexes also reduce oximes to amines [CB7, FS1]. Oximes are inert in the presence of LTBA or $NaBH_4$ (unless $NiCl_2$ in MeOH is added to the latter) [GO2] (Figure 3.104), ion-exchange borohydride in the presence of $Ni(OAc)_2$ [BK10], $TiCl_4$ in DME [KT2, ZH1], or $ZrCl_4$ in THF [IS1]. They also are inert towards diisopropoxytitanium tetrahydroborate [RC2]. When using $NaBH_4$–$NiCl_2$ in MeOH, α,β-ethylenic oximes **3.274** are reduced to saturated amines. In the presence of MoO_3, the double bond is preserved [GO2] (Figure 3.104). The reduction by DIBAH induces some rearrangements [SM4], while, in some cases, reductions by LAH in Et_2O or Red-Al in C_6H_6 can give mixtures of primary and secondary amines, or even aziridines [GW1, M3, PP1, ZH1] (Figure 3.104). The stereoselectivity of the reduction of substituted cyclohexyloximes is poor [ZH1]. However, some chiral oximes have been reduced with good stereoselectivity using $NaCNBH_3$–$TiCl_3$ [ZH1].

The reduction of α-alkoxyoximes **3.275** by LAH or AlH_3 is also poorly stereoselective [IY2] (Figure 3.105).

Oximes are reduced to the corresponding hydroxylamines by $BH_3 \cdot THF$, BH_3 in CF_3COOH, amine–boranes [KK11], $NaBH_4$, or $NaCNBH_3$ in AcOH in the cold. On warming, $NaBH_4$ in organic acids leads to N-alkylhydroxylamines [GN1, MM1]

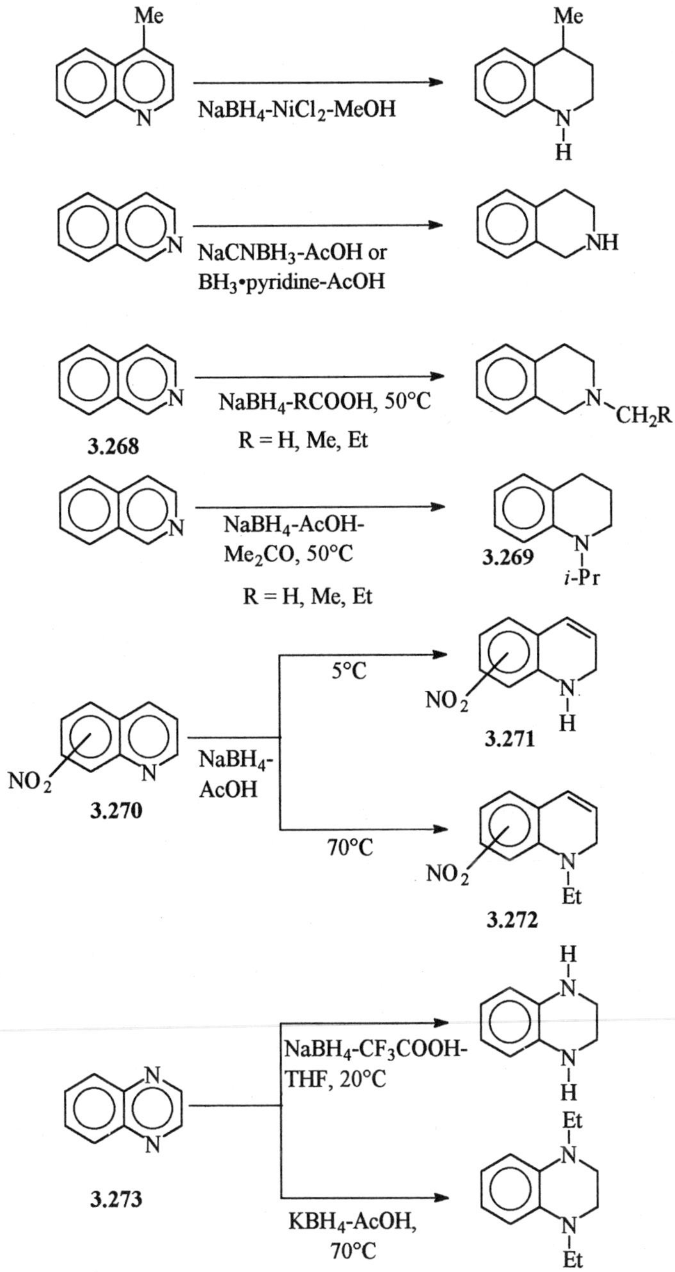
Me
N
NaBH4-NiCl2-MeOH
Me
N
H
N
NaCNBH3-AcOH or
BH3•pyridine-AcOH
NH
N
NaBH4-RCOOH, 50°C
R = H, Me, Et
N
CH2R
3.268
N
NaBH4-AcOH-
Me2CO, 50°C
R = H, Me, Et
3.269
N
i-Pr
NO2
N
3.270
NaBH4-
AcOH
5°C
NO2
N
H
3.271
70°C
NO2
N
Et
3.272
N
N
3.273
NaBH4-CF3COOH-
THF, 20°C
H
N
N
H
KBH4-AcOH,
70°C
Et
N
N
Et

Figure 3.103

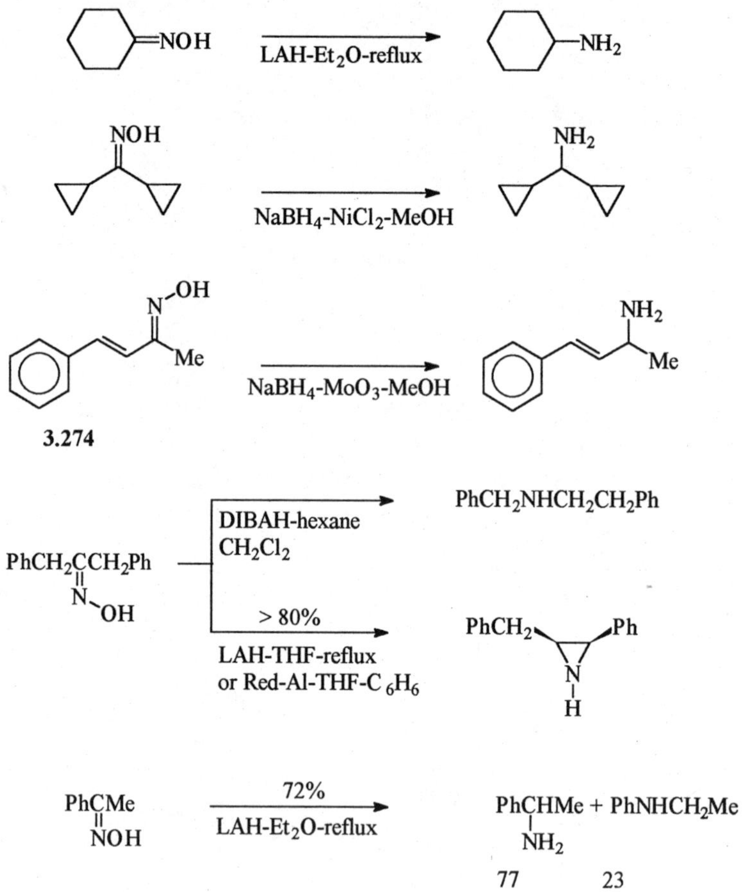

Figure 3.104

(Figure 3.105). Heating in the presence of CF_3COOH provides primary amines [GN1].

Ketoxime ethers are reduced under similar conditions to ketoximes. A high stereoselectivity can be obtained when reducing *syn* β-hydroxyoxime ether **3.276** by LAH in THF, while reduction of *anti* isomer **3.277** requires the presence of MeONa [LY1, NY1, ZH1] (Figure 3.106). Reduction of **3.278** with tetramethylammonium triacetoxyborohydride in AcOH–MeCN also gives interesting stereoselectivities towards *syn* α-hydroxy-N-benzyloxyamines, precursors of the *syn* 1,2-amino alcohols [WO1] (Figure 3.106). The reduction of chiral N-methoxy-α-sulfinyl ketoxime **3.279** by Li(s-Bu)$_3$BH is also highly stereoselective [MT7] (Figure 3.106). Other reducing agents give lower yields.

Asymmetric reduction of ketoxime O-alkylethers to chiral primary amines can be carried out with a high enantiomeric excess by borane or better by $NaBH_4$–$ZrCl_4$ in THF in the presence of a chiral amino alcohol [IS2, WM2, ZH1].

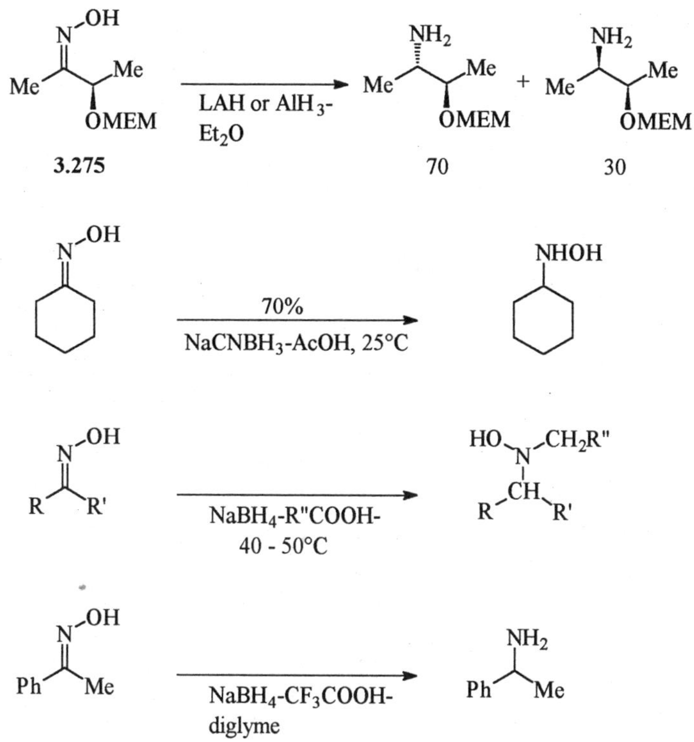

Figure 3.105

Oxime esters **3.280** can be reduced to acyloxyamines by $NaCNBH_3$ in AcOH [GN1, SJ1] (Figure 3.106). On the other hand, $BH_3{\cdot}THF$ or $NaBH_4–I_2$ converts oxime ethers to amines and to the corresponding alcohols [BC11, H3]. $NaCNBH_3–TiCl_4$ in aqueous MeOH converts oximes into amines; this reduction is compatible with ketones, esters, acetals, and isolated double bonds [LK2].

Moreover, α-oximinoesters can be reduced to α-amino acid esters by $NaCNBH_3–TiCl_4$ in aqueous MeOH in buffered conditions. Tartaric acid is the best buffer, although no asymmetric induction is observed [HT3].

Hydrazones are reduced to hydrazines by LAH in ether, but some exceptions are met [EK1], or by $BH_3{\cdot}THF$. Whereas dialkylhydrazones are resistant to reduction with $NaBH_4$, α-nitrohydrazones such as **3.281** are reduced extremely rapidly in EtOH [DS3] (Figure 3.107).

The most interesting reduction is that of tosylhydrazones, due to the presence of the leaving group, which, in a basic medium, converts the tosylhydrazine that is formed into saturated hydrocarbons and nitrogen (Figure 3.107). This reduction can be accomplished with $NaBH_4$ in EtOH [PS1], $BH_3–PhCOOH$, $BH_3–CF_3COOH$ in THF [KB4, MM1] catecholborane [KB7], but, above all, by $NaCNBH_3$ in DMF in the presence of acid or $NaBH_4$ in organic acid media [HN1, L1, MY1]. $NaCNBH_3–ZnCl_2$ in refluxing MeOH has also been used. Under these conditions, epimerization

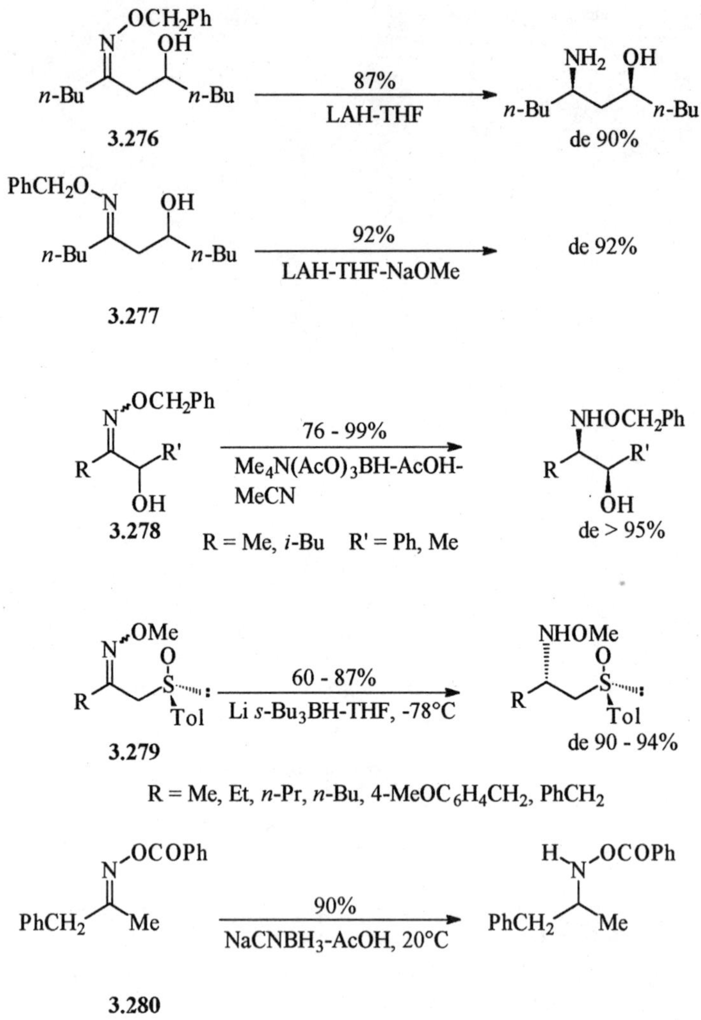

Figure 3.106

of the carbon α to the tosylhydrazone moiety is avoided [SH5]. The reaction can be run in "one-flask" fashion starting from the ketone. This is a modification of the Wolff–Kishner reaction that is compatible with ester and nitrile functional groups, as shown in Figure 3.107 [IT1, L1]. Indeed, while reduction of tosylhydrazone **3.282** with $NaBH_4$ in refluxing THF or MeOH promotes the transformation of the ester groups into the corresponding alcohols, the use of $NaBH_4$ in AcOH leaves them unchanged. When starting from tosylhydrazones derived from arylketones, the transformation to hydrocarbons requires warming in the presence of base. The limitation of the method is the migration of the double bond during the reduction of tosylhydrazones of α-enones [L1] (Figure 3.107).

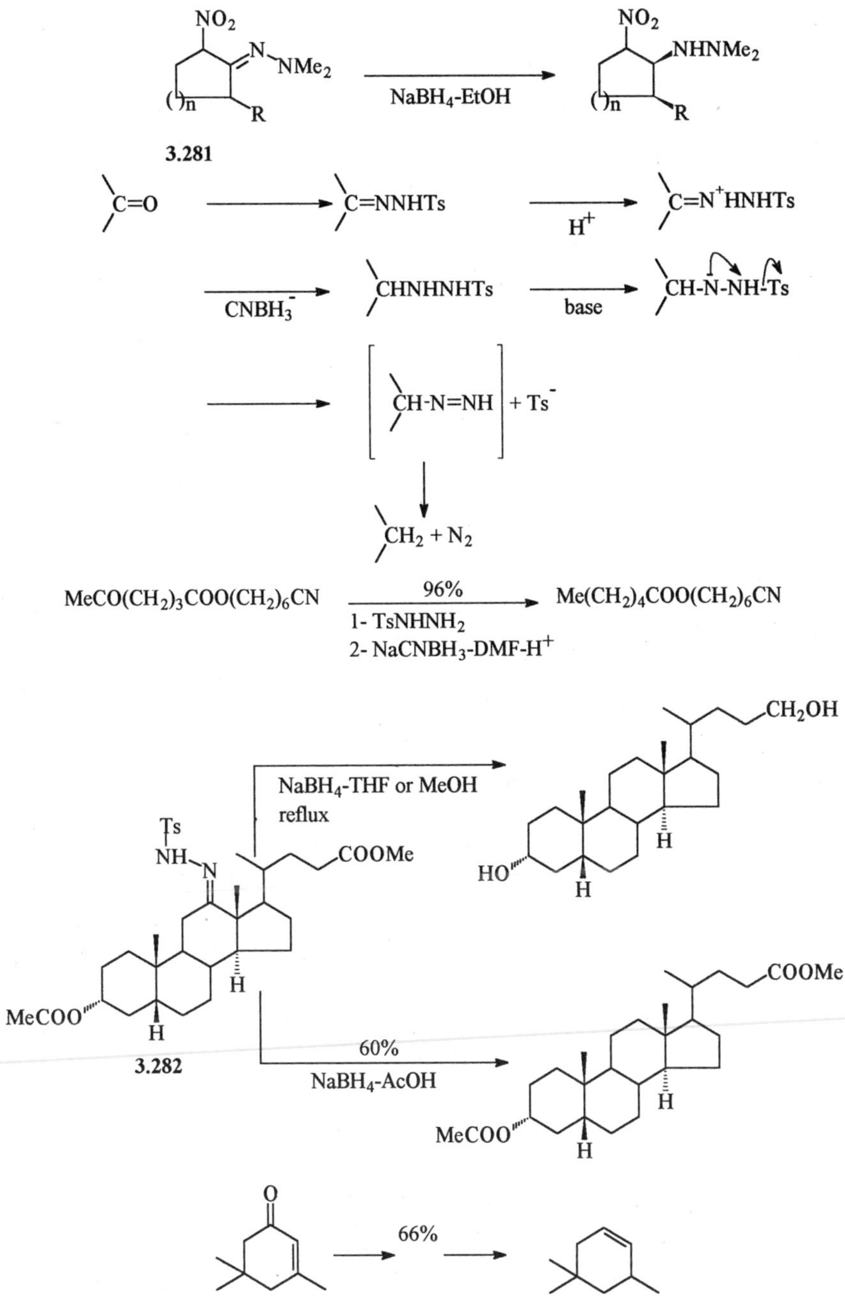
NO2
NMe2
NaBH4-EtOH
NHNMe2
R
3.281
C=O
C=NNHTs
H+
C=N+HNHTs
CNBH3-
CHNHNHTs
base
CH-N-NH-Ts
CH-N=NH
+ Ts-
CH2 + N2
MeCO(CH2)3COO(CH2)6CN
96%
1- TsNHNH2
2- NaCNBH3-DMF-H+
Me(CH2)4COO(CH2)6CN
CH2OH
NaBH4-THF or MeOH
reflux
Ts
NH
N
COOMe
HO
H
MeCOO
3.282
60%
NaBH4-AcOH
COOMe
O
66%

Figure 3.107

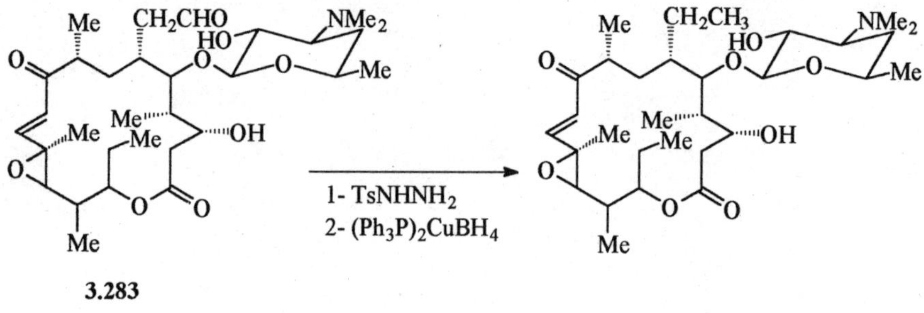

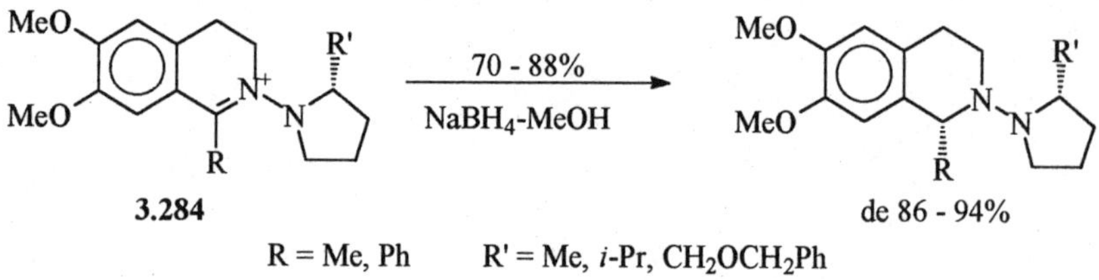

Figure 3.108

$(Ph_3P)_2CuBH_4$ in hot $CHCl_3$ reduces tosylhydrazones of aldehydes or aliphatic or alicyclic ketones to hydrocarbons [FH2], but leaves the tosylhydrazones of aromatic or α,β-unsaturated ketones and aldehydes unperturbed. This method has been applied to the selective reduction of the tosylhydrazone of a multifunctional aldehyde **3.283** [GL3], carrying an α-enone, an epoxide, and a lactone, all of which remain unchanged (Figure 3.108). Pyridine-borane reduces tosylhydrazones to tosylhydrazines, even in the aromatic series [KK8]. In the presence of base and with heating, these tosylhydrazines can be converted to the corresponding hydrocarbons.

Hydrazonium ions can be reduced by various reducing agents. The reduction of chiral substrates **3.284** is highly stereoselective, $NaBH_4$ in MeOH giving the best chemical yields [SA3]. The asymmetric reduction of a benzodiazepine with borane in the presence of (*R*)- or (*S*)-diphenylleucinol in CH_2Cl_2 gives a good enantioselectivity (86%) [LP2].

Chapter 4

Reduction of Triple Bonds

4.1 CARBON–CARBON TRIPLE BONDS: $-C{\equiv}C-$

Carbon–carbon triple bonds undergo facile hydroboration [PS1, HH1] and hydroalumination [HH1, HH3, W1] at room temperature. Therefore, boranes and DIBAH are not used, except for special cases, in the selective reduction of other functional groups. Alkynes can undergo selective monohydroboration by using relatively bulky boranes such as Sia_2BH or $(c\text{-}C_6H_{11})_2BH$. The stereospecific cleavage of the C—B bond of the resulting *cis*-alkenylboranes by an organic acid or by MeOH in the presence of catalytic quantities of organic acid [BM3] is a frequently used method for the synthesis of terminal or 1,2-disubstituted *Z*-alkenes. There are numerous applications of these transformations in the synthesis of pheromones (Figure 4.1).

Reaction with LAH in THF or hot diglyme converts disubstituted alkynes to *trans*-alkenes [DM1, H3, HH1] (Figure 4.1). However, at room temperature, isolated triple bonds are not attacked by LAH, LTBA, AlH_3, or Red-Al [HH1] or by the alkaline borohydrides, except in the presence of Lewis acids, which induce the formation of diborane. In the presence of transition metal salts ($PdCl_2$, CoX_2, NiX_2), $NaBH_4$ in alcoholic media reduces alkynes to alkenes or to saturated hydrocarbons, depending on the conditions [GO2, SK3, W4].

Propargylic amines **4.1** and alcohols **4.2**, **4.3**, and **4.4** are reduced to allylic amines and alcohols by LAH in ethers or by Red-Al in ether or toluene [HH1, HH3, JD2, M1]. With LAH, the reduction is stereoselective only if the solvent is basic enough (THF or DME) [DJ1, DM1, KJ2, M1]. *E*-Isomers are obtained, and a mechanism involving an intramolecular regioselective transfer of hydride is supported by trapping of the adduct by D_2O (Figure 4.2). Although some authors have pointed out that the reaction leads to allenes if R = H [HH3], propargylic alcohol

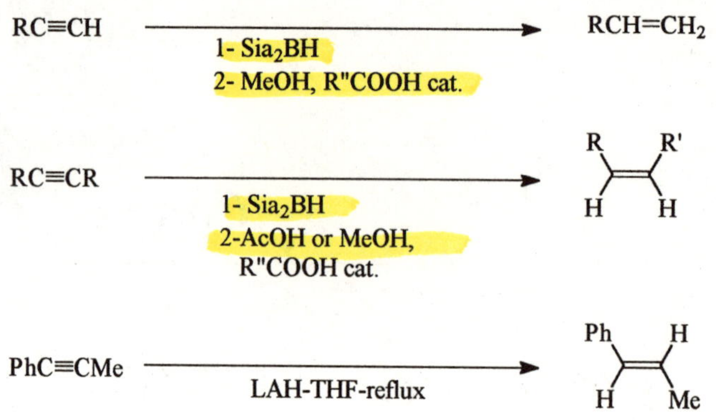

Figure 4.1

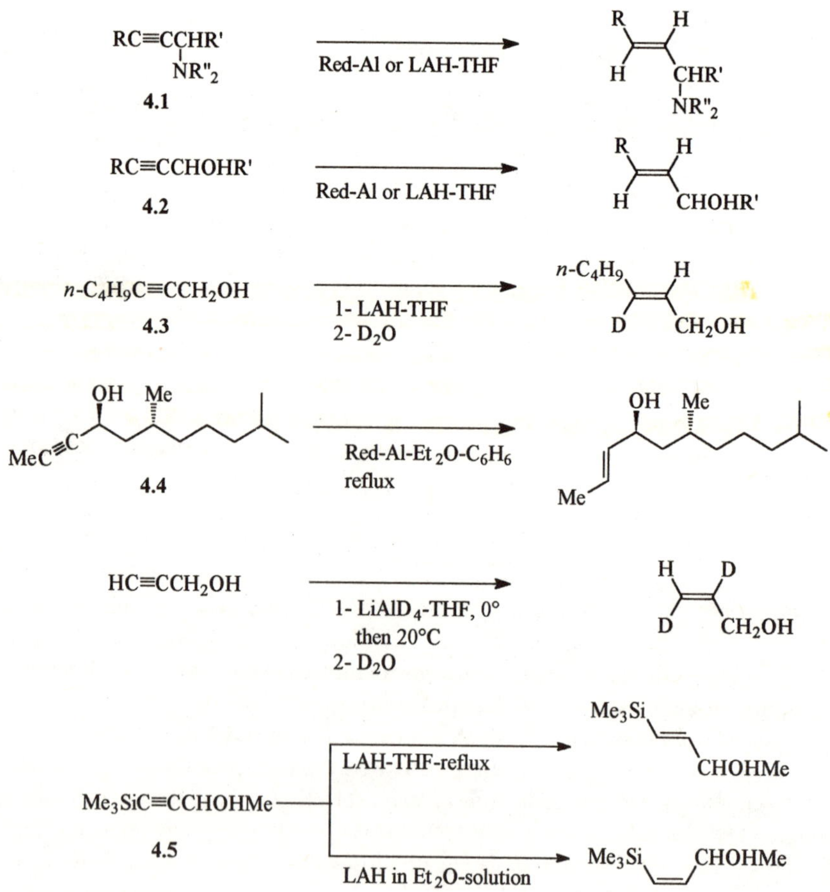

Figure 4.2

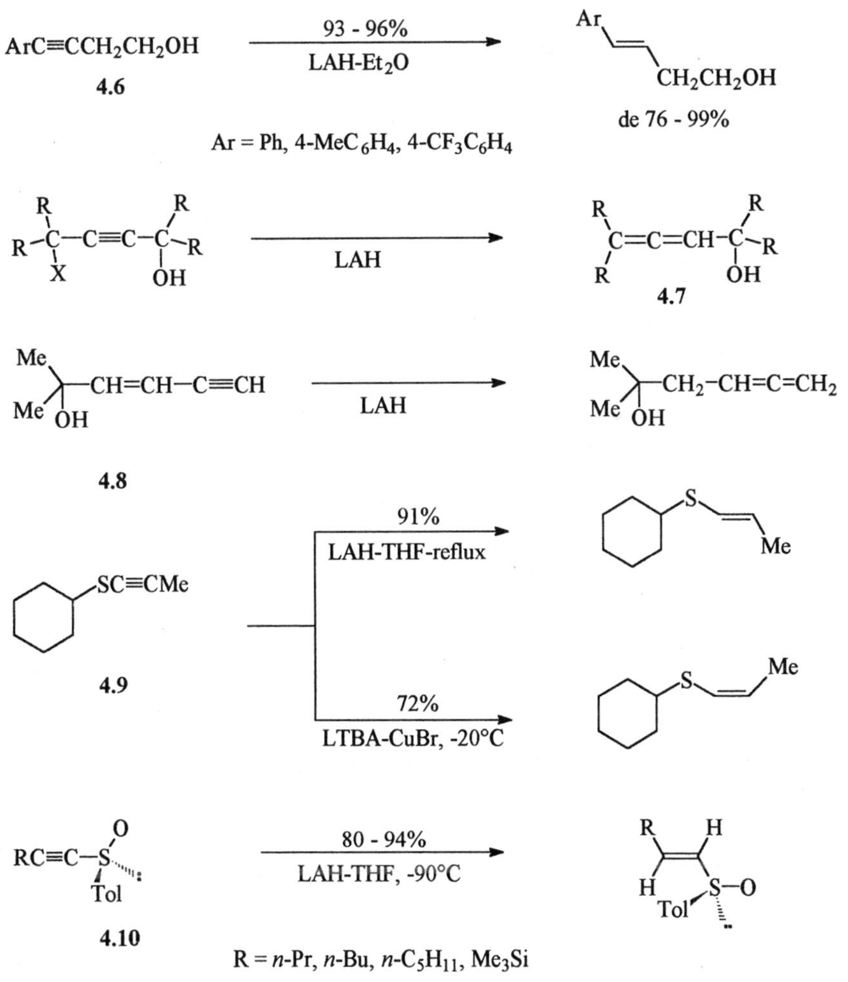

Figure 4.3

has been selectively reduced to the *trans*-dideuterated alcohol by precisely mastering the experimental conditions [BB2] (Figure 4.2). Starting from silyl derivatives **4.5**, it is possible to obtain selectively the *E*- or *Z*-allylic alcohol [MH2] (Figure 4.2). A suspension of LAH in Et_2O gives a mixture of *E*- and *Z*-isomers. The reduction of $Me_3SiC{\equiv}CCH_2OH$ follows the same trends, but is less stereoselective [KJ2] unless it is performed with Red-Al in ether–toluene [OC1].

The reduction of 4-aryl-3-butyn-1-ols **4.6** is stereoselective towards the *E*-α,β-unsaturated alcohol only in Et_2O. In THF, a mixture of *E*- and *Z*-isomers is formed [KJ2] (Figure 4.3). The presence of a leaving group α′ to the triple bond induces the formation of α-allenic alcohols **4.7** [HH3] (Figure 4.3). An allenic alcohol is also formed from conjugated allylic alcohol **4.8** (Figure 4.3).

1-Alkynylsulfides such as **4.9** are reduced to 1-alkenylsulfides. Depending on the reagent, *E*- or *Z*-isomers are formed: $Li(MeO)_3AlH$ or LAH in THF lead to *E*-isomers, while $Li(MeO)_3AlH$–CuBr gives *Z*-isomers [M3, NK2] (Figure 4.3). Chiral α-acetylenic sulfoxides **4.10** are converted to *E*-α,β-unsaturated analogues by DIBAH or, better, by LAH in THF at low temperature [KK10] (Figure 4.3).

4.2 α,β-ACETYLENIC KETONES AND ESTERS: RC≡CCOY (Y = R′, OR′)

The alkoxyaluminohydrides or LAH modified by amines or aminoalcohols [GH1, M1], DIBAH, and cyanoborohydrides in acid media [HK2] reduce the α-ynones to propargylic alcohols. The use of chiral ligands can give high asymmetric induction [GH1, M1, MS5] (Section 3.2.3). Asymmetric reduction of α-ynones to optically active propargylic alcohols can also be carried out via acetals formed from chiral 1,3-diols (Section 2.4.3). The reduction of **4.11** shows that LTBA selectively reduces the ketone and leaves the triple bond untouched, in spite of the presence of the alcoholic functional group [SA1] (Figure 4.4).

The selective reduction of the α,β-acetylenic ketones such as **4.12** to α,β-ethylenic ketones is accomplished by reaction with DIBAH in THF-HMPA [TY2]. In the presence of a catalytic quantity of MeCu, the reduction is faster and the stereoselectivity is modified (Figure 4.4). Nevertheless, the stereoselectivity is never very high [TY2]. The reagent does not reduce the isolated triple bond of **4.13** (Figure 4.4).

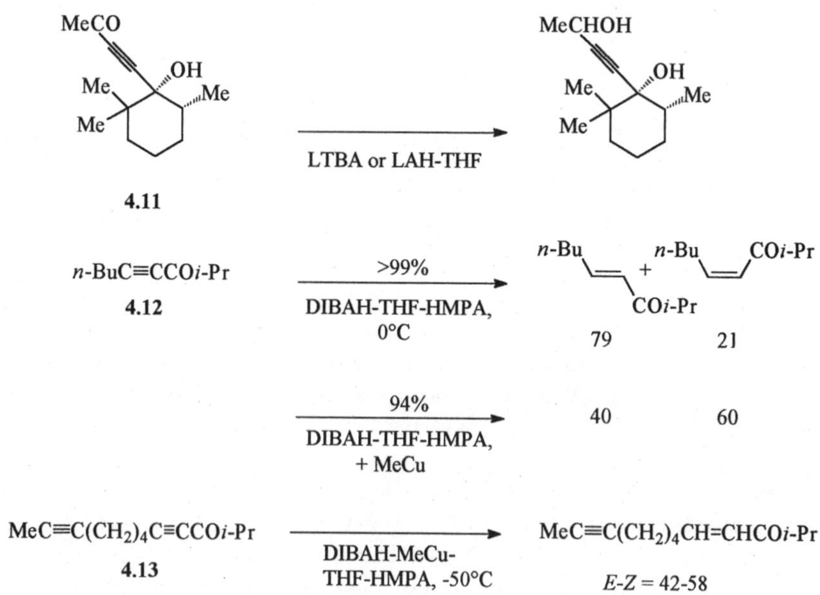

Figure 4.4

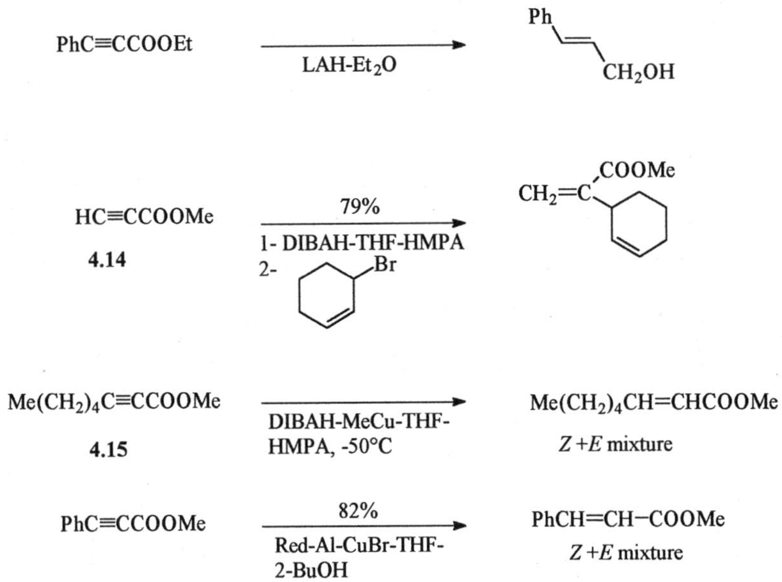

Figure 4.5

α,β-Acetylenic esters are reduced to *E*-allylic alcohols by LAH in Et_2O [DM1] (Figure 4.5). Reduction of these esters by DIBAH in THF–HMPA takes place only if the alkyne contains HC≡C residue such as **4.14**; the adduct thus formed may then be trapped by an allylic bromide [TY1] (Figure 4.5). Substituted α,β-acetylenic esters **4.15** are reduced only in the presence of MeCu (Figure 4.5). They are also reduced to α,β-ethylenic esters by Red-Al–CuBr in the presence of 2-butanol. In all these cases, the reaction leads to a mixture of *Z*- and *E*-α,β-unsaturated esters [SS1] (Figure 4.5).

4.3 CARBON–NITROGEN TRIPLE BONDS: NITRILES RC≡N

The reduction of nitriles occurs in two stages:

- Formation of an imine, which can be hydrolyzed to an aldehyde;
- Double reduction to an amine.

Depending on the reagents and experimental conditions, either process may be observed [C5, HH3] (Figure 4.6).

Nitriles are not reduced by LAH on SiO_2 [KH2], alkaline borohydrides in alcohol media or in ethers at room temperature [BK5, PS1], $(CF_3COO)_2BH$ [MM1], cyanoborohydrides [GN1, L1], or diisopropoxytitanium tetrahydroborate [RC2]. Nevertheless, the 2-cyano- or 4-cyanopyridines are reduced to amines by $NaBH_4$ in EtOH under reflux [HH3, KK2].

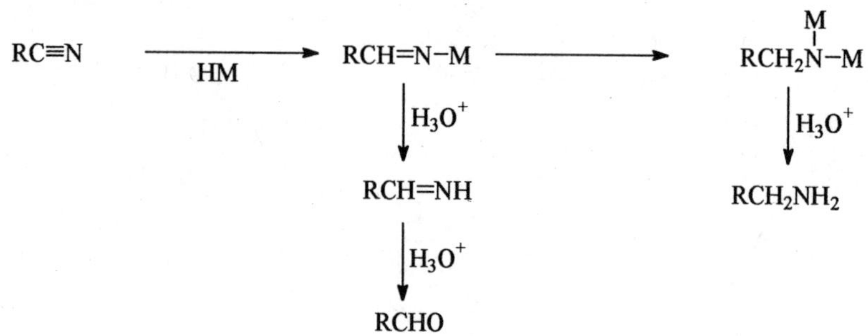

Figure 4.6

A few reducing agents lead to the imines and, after hydrolysis, to the aldehydes. Such is the case of trialkoxyaluminohydrides. Among these, $Li(EtO)_3AlH$ in THF proves to be the best, while LTBA is unreactive [BK5, C5, H3, M3] (Figure 4.7). The intermediate $RCH{=}NAl(OEt)_2$ can be trapped by Me_3SiCl, and this leads to N-trimethylsilylimines [AC2]. However, the initial configuration of nitriles that are substituted at the α-position is not always retained. For example, starting with a pure *Z*-nitrile **4.16**, a mixture of stereoisomeric aldehydes is produced [PS2] (Figure 4.7). Catecholalane, generated from catechol and AlH_3, allows the reduction of aromatic, α,β-unsaturated and aliphatic nitriles into the corresponding aldehydes in THF at r.t. [CC10].

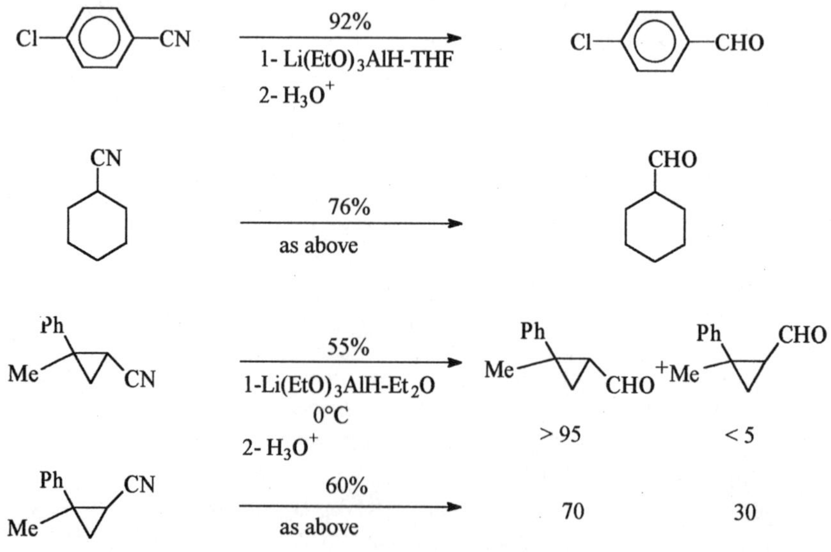

4.16

Figure 4.7

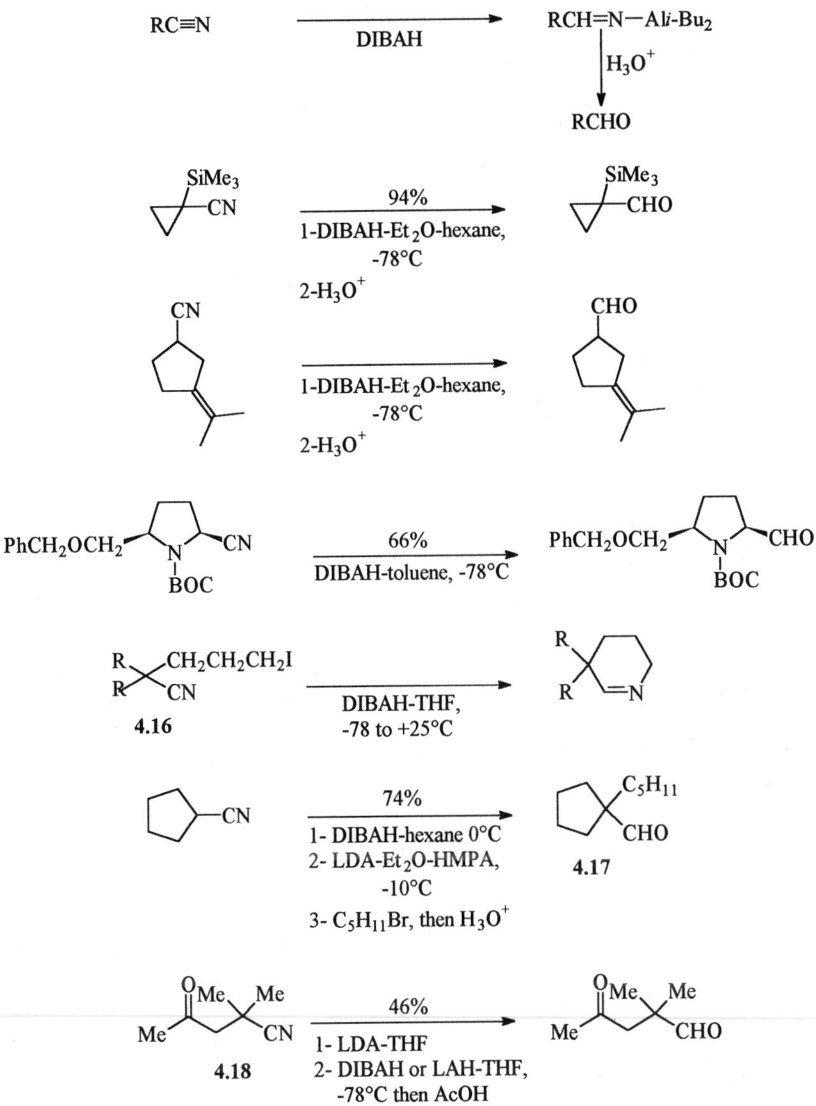

Figure 4.8

If employed in the cold and in stoichiometric quantities, DIBAH leads to the iminoaluminate, which is hydrolyzed to aldehyde [K2, W1, YG1] (Figure 4.8). Toluene or hexane are usually the best solvents for this transformation. The reaction proceeds with aliphatic, aromatic, α,β-unsaturated (see Section 4.4), or cyclopropane nitriles [HH3, WY1] and is compatible with N-BOC groups [KN3] (Figure 4.8). The iminoaluminate thus formed can undergo an intramolecular alkylation, as

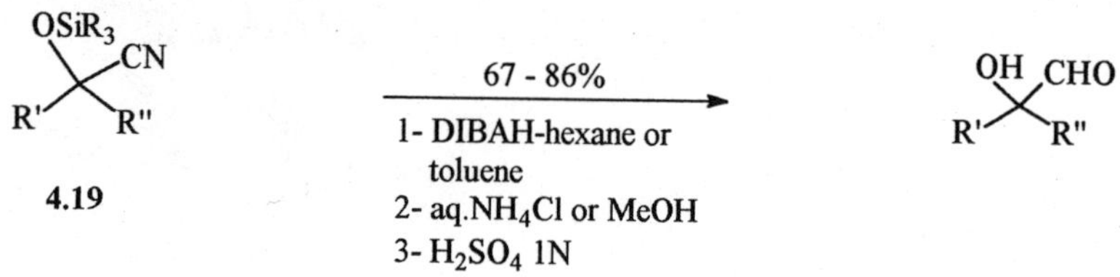

4.19

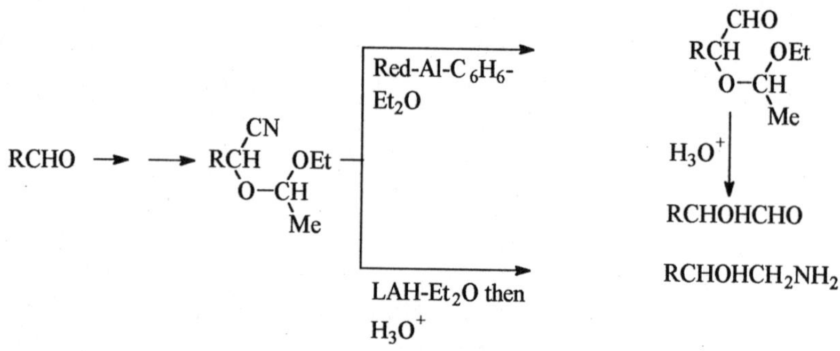

Figure 4.9

shown with compounds **4.16** [OB1] (Figure 4.8). The deprotonation at the α-position of the iminoaluminate by reaction with LDA, followed by alkylation, allows access to the branched aldehyde **4.17** [GT1] (Figure 4.8). As in the case of ketoesters, selective reduction of the ketonitriles **4.18** in which the CN is located on a tertiary carbon can be carried out provided that the ketone enolate is preformed [KF2] (Figure 4.8). α-Trimethylsilyloxynitriles **4.19** (R = Me), easily obtained from ketones, are transformed into the α-hydroxyaldehydes by DIBAH at 0°C followed by subsequent hydrolysis of the imine and the silylether [HY1]. When starting from aldehydes, *t*-butyldimethylsilyloxynitriles **4.19** (R_3 = *t*-BuMe_2, R′ = H) suffer the same transformation at low temperature without racemization when chiral [HY1] (Figure 4.9).

Starting from aldehydes, which are converted into cyanohydrins whose hydroxyl group is protected as an acetal, one can also obtain α-hydroxyaldehydes by reaction with Red-Al or α-hydroxyamines via reduction by LAH [SB1] (Figure 4.9).

The use of $NaEt_2AlH_2$ in the presence of a Lewis acid for converting aliphatic nitriles to aldehydes has also been described [YK2]. Sodium triaminoaluminohydrides or diethylpiperidinoaluminohydride reacts with aromatic nitriles in THF at r.t., allowing their transformation into the corresponding aldehydes [CJ1, YA2].

The reduction of nitriles to amines can be carried out by LAH or SAH in an ether medium, LAH–N-methylpyrrolidine complex, AlH_3 in Et_2O or $AlH_3{\cdot}Et_3N$ [BK5, CB5, CB7, E2, FS1, H3, L2, M3, MC1, MP2, PS1]. $Li(MeO)_3AlH$ or Red-Al at 80°C reduces aromatic nitriles, while aliphatic nitriles remain untouched [M3]. This transformation can also be accomplished with $BH_3{\cdot}THF$ or aminoboranes under

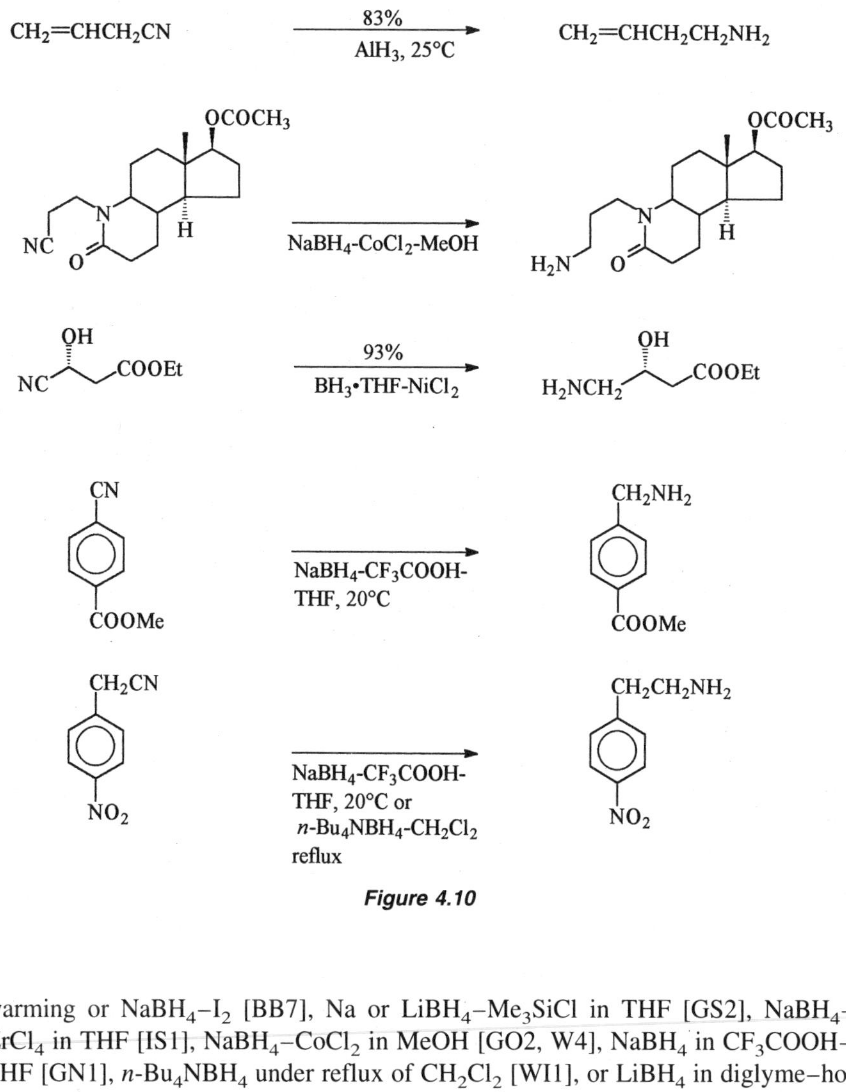

Figure 4.10

warming or $NaBH_4$–I_2 [BB7], Na or $LiBH_4$–Me_3SiCl in THF [GS2], $NaBH_4$–$ZrCl_4$ in THF [IS1], $NaBH_4$–$CoCl_2$ in MeOH [GO2, W4], $NaBH_4$ in CF_3COOH–THF [GN1], n-Bu_4NBH_4 under reflux of CH_2Cl_2 [WI1], or $LiBH_4$ in diglyme–hot MeOH [SO3], $LiBH_4$–$(MeO)_3B$ in Et_2O at 25°C [BN3]. With the last reagent, sulfones, sulfoxides, NO_2 groups, and pyridine rings remain unchanged. Therefore, it is possible to perform a number of selective reductions [GN1, GO2] (Figure 4.10) with $NaBH_4$–$CoCl_2$ [PF1] or $BH_3{\cdot}THF$ in the presence of $NiCl_2$ [LM2].

Trialkylborohydrides also reduce nitriles to amines [BK5]. An exception is Li (s-Bu)$_3$BH, which leaves aliphatic and aromatic nitriles intact unless the latter are para-substituted by an electron-donating group [SM1]. An aldehyde is then obtained (Figure 4.11). Whereas thexylborane and 9-BBN react slowly with nitriles [PS1], thexylchloroborane reduces aliphatic nitriles into corresponding amines [BN5]. Surprisingly, *gem*-dicyanoepoxides **4.20** are reduced to α-cyano α-epoxymethyl amines

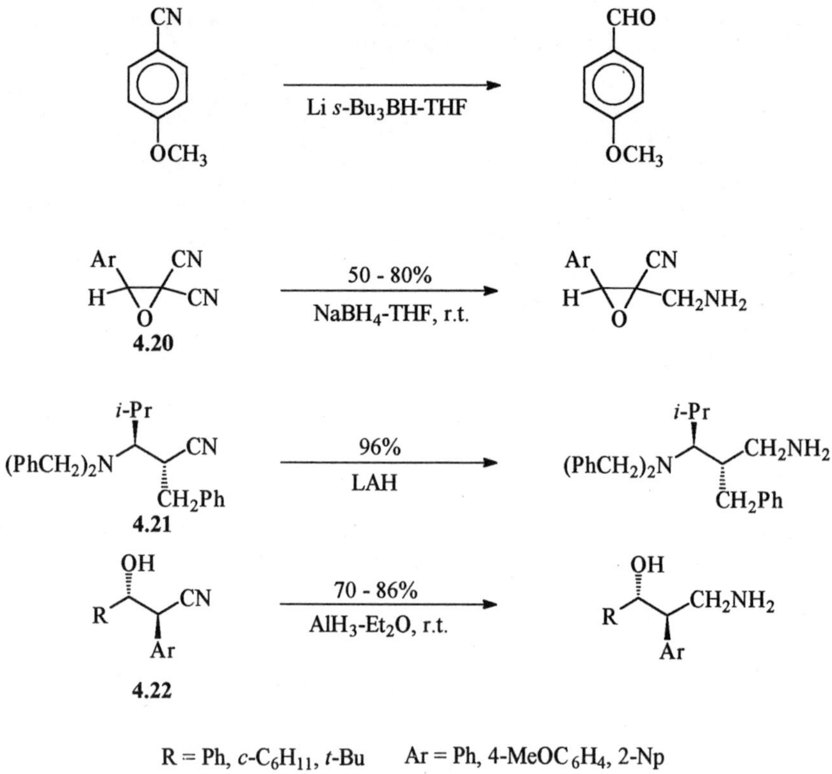

Figure 4.11

by $NaBH_4$ in aqueous THF [MR4] (Figure 4.11). The reductions of chiral nitriles such as **4.21** or **4.22** by LAH or AlH_3 take place without epimerization [CL11, RK2] (Figure 4.11).

A C—CN bond cleavage has been observed in the reduction of 5-cyano-5-isopropylsulfonylnorborn-2-ene with LAH in THF, likely via a SET process in the propagation chain [MS10].

4.4 α,β-UNSATURATED NITRILES: RCH=CH—CN

α,β-Ethylenic nitriles are reduced to α,β-unsaturated aldehydes by DIBAH in toluene at low temperature [K2, TK3] (Figure 4.12). The presence of an acetal group that can coordinate to DIBAH decreases the extent of the reduction. But the addition of a Lewis acid such as Et_2AlCl solves this problem, and the formation of the aldehyde from compound **4.23** is carried out in a satisfying yield [TT2] (Figure 4.12).

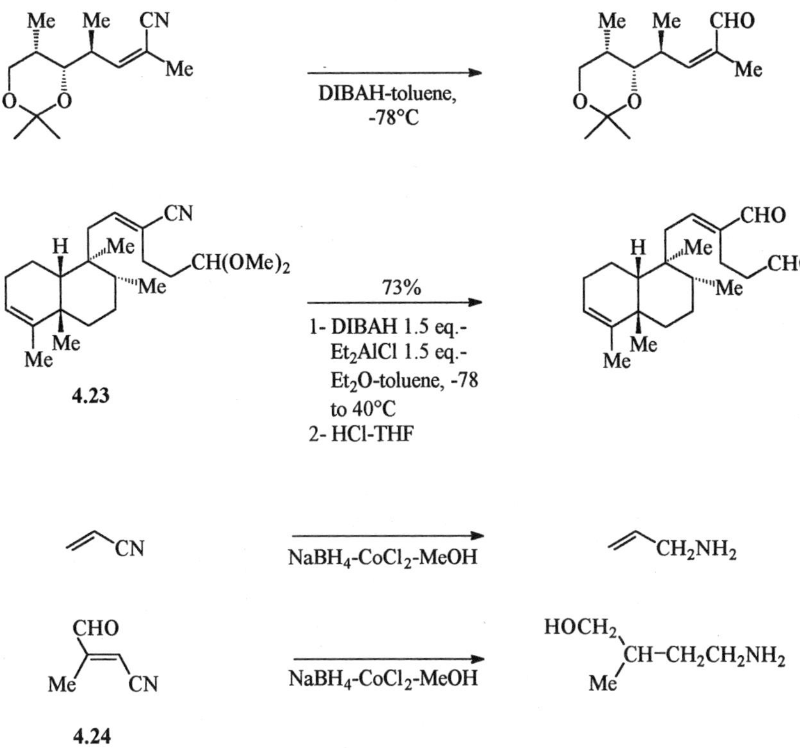

Figure 4.12

$NaBH_4$–$CoCl_2$ in MeOH also reduces α,β-unsaturated nitriles without affecting the double bond, but one obtains an α,β-ethylenic amine [GO2] (Figure 4.12). However, if the double bond is conjugated either to another double bond or to an aldehyde such as **4.24**, the totally reduced product is obtained [GO2] (Figure 4.12).

LAH reduces α,β-unsaturated nitriles to saturated amines [H3]. The formation of an alcoholate in a suitable position, followed by an intramolecular hydride transfer, allows the stereoselective reduction of an α,β-unsaturated nitrile to the saturated nitrile by reaction with LAH [LM1, LV1] or even with $LiBH_4$ in THF under reflux [LV1] (Figure 4.13).

Reduction of α,β-unsaturated nitriles to saturated nitriles is generally carried out with Semmelhack's system [M3, SS1]. This involves reduction with Red-Al in the presence of CuBr in THF–hexane–2-BuOH, as shown in Figure 4.13 [OP1]. Alkaline borohydrides in alcoholic media do not reduce the α,β-ethylenic nitriles. However, $NaBH_4$ in MeOH–pyridine under reflux [RB2] and $NaBH_4$ in EtOH do work in some cases [KS6, UC1]. On the other hand, if the double bond is activated by another electron-withdrawing group, reduction to the saturated compound takes place [MC2, MR4]. If the experimental conditions are appropriate, the functional groups remain unchanged (Figure 4.13).

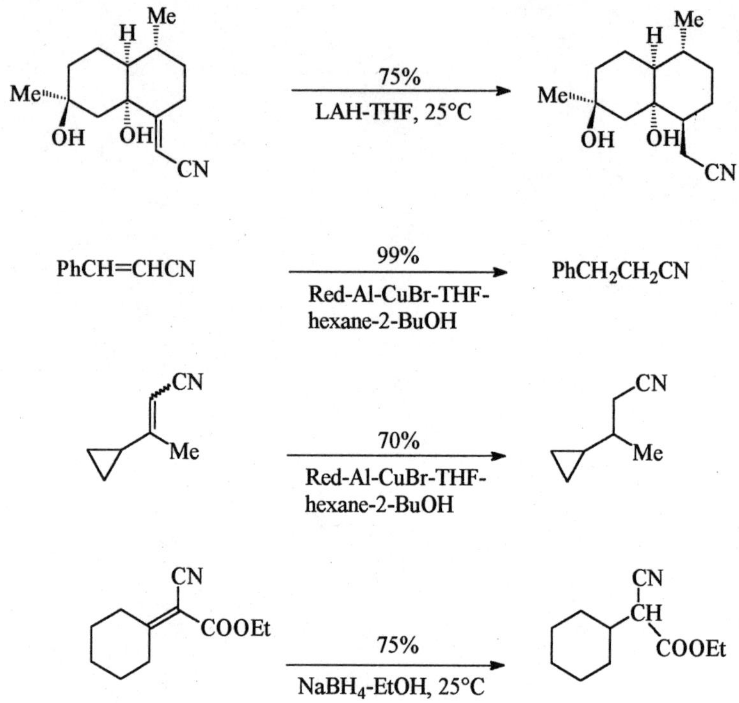

Figure 4.13

The reduction of α,β-acetylenic nitriles to *E*-α,β-ethylenic nitriles by 0.5 equivalents of LAH in Et_2O under reflux [VK4] or $NaBH_4$ in EtOH [KS6] has been described.

Chapter 5

Other Derivatives

5.1 NITRO AND NITROSO DERIVATIVES: RNO_2, RNO

Nitro and nitroso compounds are among the most difficult to reduce using alumino- and borohydrides.

LAH and SAH in an ether medium and $Li(MeO)_3AlH$ reduce aromatic nitro and nitroso derivatives to azo compounds, ArN=NAr, as does Red-Al [CB5, CB7, FS1, H3, M3]. When an excess of SAH is used, PhNHNHPh is obtained [CB5]. On the other hand, nitro derivatives are untouched by AlH_3 in Et_2O [BK5, E2] or by LAH on SiO_2 [KH2]. Borohydrides and boranes leave nitro groups unchanged under the usual conditions (ether or alcohol solvent) [L2, PS1, NM3, R3] or in acid media [MM1]. However, 4-nitroimidazoles, nitropyrazoles, and nitropyridines are reduced by $NaBH_4$–NaOMe in MeOH [SW1]. The reduction of aliphatic nitro derivatives by LAH or SAH gives amines [CB5, H3, M3], although some side reactions can be observed when the reaction is performed with tertiary nitro compounds [WN1]. $LiBH_4$ in diglyme–MeOH under reflux allows the reduction of aliphatic and aromatic nitro derivatives to the corresponding amines [SO3] (Figure 5.1). In the presence of transition metal derivatives ($SnCl_2$, $Cu(acac)_2$, $CuBr{\cdot}SMe_2$, $CuSO_4$, $(Ph_3P)_4Ni$, $NiCl_2$, $Ni(OAc)_2$, $CoCl_2$, $BiCl_3$), reduction of aromatic nitro derivatives to amines by $NaBH_4$ takes place in ether, dioxane, or alcohols [DG1, GO2, PV1, RP3, W4, YC2, YL5]. Under these conditions, halogen or acid groups remain unchanged. In the presence of $SnCl_2$, $Cu(acac)_2$, or $Ni(OAc)_2$, reduction can also be compatible with ketone, ester, amide, and nitrile groups [C4, W4, YC2] (Figure 5.1). Similarly, reduction of aromatic nitro compounds to amines by KBH_4–Cu_2Cl_2 in MeOH is compatible with bromide and ester substituents, but iodides are reduced [HZ1]. The $NaBH_4$–$CuSO_4$ system in alcoholic medium also reduces aliphatic nitro compounds, but more slowly than aromatic ones [YL5]. Ketones are reduced faster

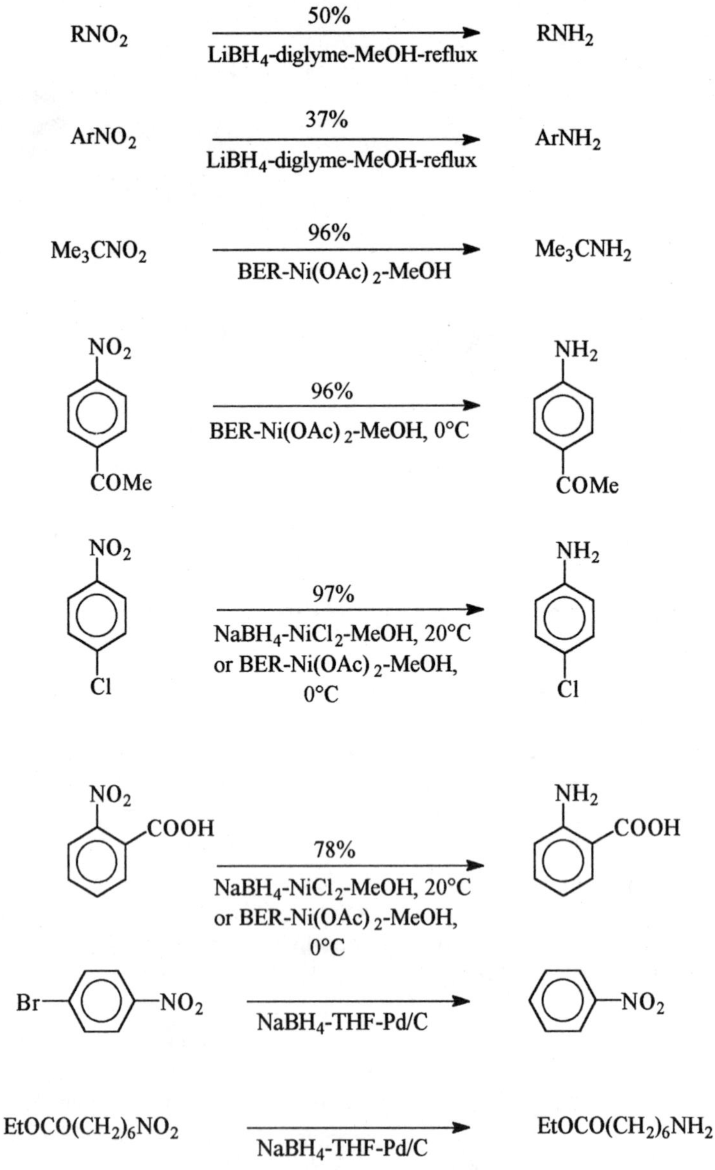

Figure 5.1

than nitro groups, but esters and nitriles can remain untouched [YL5]. The borohydride exchange resin (BER)-$Ni(OAc)_2$ reagent in MeOH rapidly reduces aliphatic nitro compounds at r.t. [YC2] (Figure 5.1).

Primary and secondary aliphatic or aromatic nitro derivatives can also be reduced to amines by $NaBH_4$ in THF in the presence of palladium on charcoal [PB1]. The reduction is compatible with ester and nitrile groups and also with chlorides, but aryl bromides are reduced under these conditions (Figure 5.1).

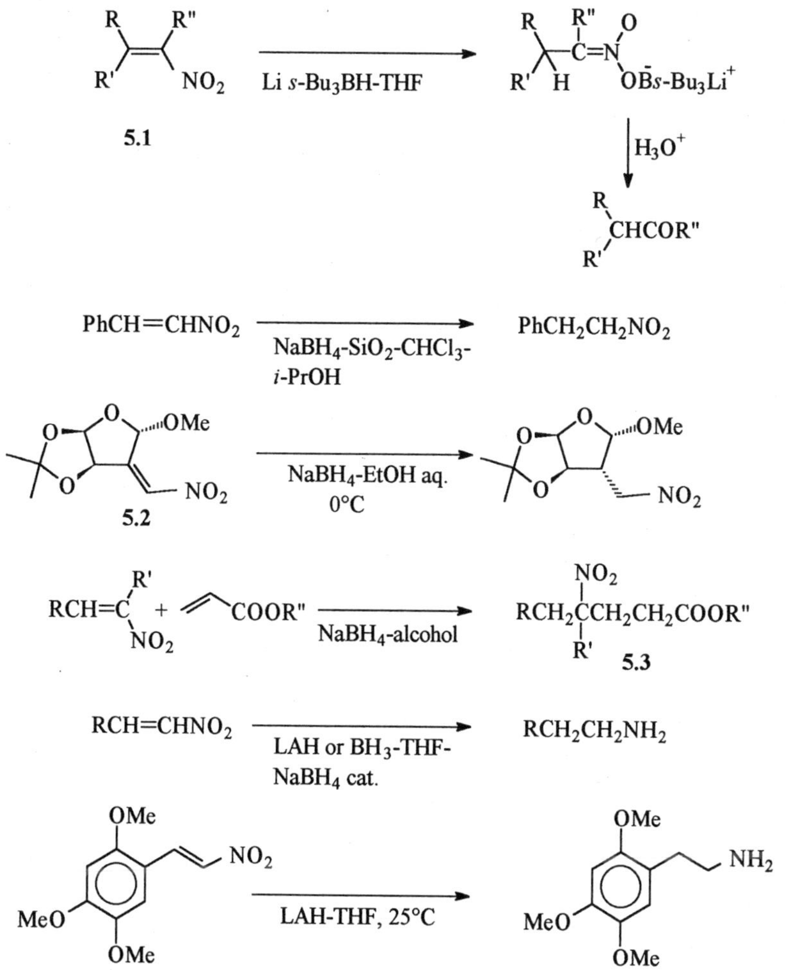

Figure 5.2

α,β-Ethylenic nitro derivatives **5.1** are converted to saturated nitro compounds by reaction with $NaBH_4$, $NaCNBH_3$, borohydride–ion-exchange-resin reagent in alcohol media or in THF–MeOH [GW3, VK3], or with Li(*s*-Bu)$_3$BH or $LiEt_3BH$ in THF [MV1]. Treatment in an acid medium allows access to the corresponding ketones (Figure 5.2). In the case of nitrostyrenes, polymeric products are also obtained, but the formation of these byproducts is avoided if the reaction is carried out on SiO_2 in $CHCl_3$-*i*-PrOH [SB2]. The reduction can be stereoselective as shown in the case of compound **5.2** [NS2] (Figure 5.2).

It is possible to trap in situ the carbanion formed during the reduction of α,β-ethylenic nitro derivatives by an acrylate; γ-nitroesters **5.3** are thus obtained (Figure 5.2). Reduction of the same compounds by LAH or $BH_3{\cdot}THF$ in the presence of catalytic amounts of $NaBH_4$ leads, on the other hand, to saturated primary amines or to saturated hydroxylamines [MV2, MV4, VK2] (Figure 5.2). Similarly, LAH in THF [KS4] or $NaBH_4$–Me_3SiCl in THF [GS2] effects the reduction of an α,β-ethylenic aromatic nitro derivative to saturated amine (Figure 5.2). However, reduction of conjugated nitroalkenes with $Zn(BH_4)_2$ in DME can lead either to saturated nitro derivatives or to saturated oximes depending upon the substituents [R3].

5.2 AZIDES: RN_3, ArN_3

The RN_3 and ArN_3 derivatives are reduced to amines by LAH in ether media [S1]. $NaBH_4$ in alcohol or in THF reduces azides with difficulty, except for certain sugars [S1] and for $ArOSO_2N_3$, which gives sulfoxamides $ArOSO_2NH_2$ [HG1]. $(i\text{-PrO})_2TiBH_4$ leaves azides untouched [RC2]. Nevertheless, under phase-transfer conditions, reduction of the aryl azides by $NaBH_4$ takes place at room temperature, and alkyl azides are reduced at 80°C [R1] (Figure 5.3). These reductions can be carried out with borohydride supported on an ion-exchange resin or by $Zn(BH_4)_2$ in DME under sonication [KW2, RS1]. In a similar fashion, arylsulfonylazides are converted into arylsulfonamides [KW2, RS1] (Figure 5.3).

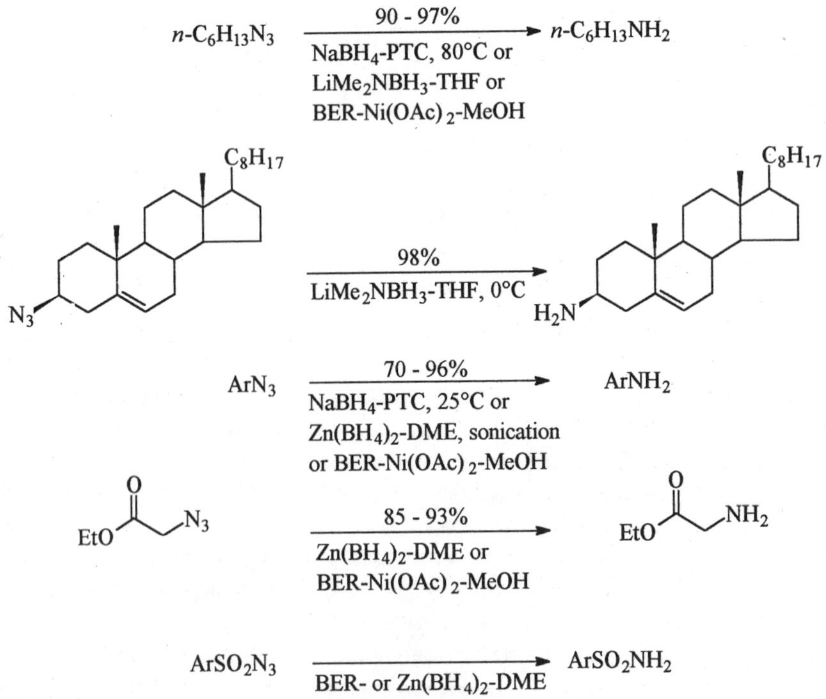

Figure 5.3

$NaBH_4$ in THF under reflux in the presence of MeOH also reduces the primary aliphatic or aromatic azides to amines. The reduction is compatible with Cl and NO_2 substituents on the aromatic ring, which remain unchanged [SY3]. Aliphatic secondary azides are not reduced under these conditions. Aminoborohydrides [AF1, FF2] and borohydride–exchange resin–$Ni(OAc)_2$ [YC3] reduce primary and secondary azides and aromatic azides in high yields. The latter reagent is compatible with chlorides and esters (Figure 5.3). Dichloroborane reduces all azides to amines, leaving bromides, nitro groups, esters, and nitriles untouched [SB3] (Figure 5.3). Aroylazides when substituted by an electron-donating group are transformed by $NaBH_4$–CF_3 COOH into trifluoroethylanilines [KS7].

5.3 ORGANOMETALLICS

Reduction by hydrides of two types of organometallic compounds has received some synthetic applications; so only these cases will be discussed.

5.3.1 Organomercurials: RHgX

Solvomercuration reactions of unsaturated compounds have been the topic of many studies. Organomercurials thus formed, when treated by alkali borohydrides in alcoholic media or better in PTC conditions [BE1] or by LAH in ether, are reduced to saturated functionalized compounds [R2, W5] (Figure 5.4). The proposed mechanism for the reduction involves the formation of an intermediate mercury hydride **5.4**, which undergoes homolytic cleavage and starts a chain reaction (Figure 5.4). Recently, alkylmercury hydrides **5.4** have been prepared from RHgCl and LAH or $NaBH_4$ [BG6]. The mercury hydride formed during reductions can be trapped by α,β-unsaturated compounds [BG6, G2, RL2] in an intra- or intermolecular fashion (Figure 5.4). Alkylmercurials can also be generated from cyclopropanes. These reductions are compatible with $P(O)(OEt)_2$, COOEt, CN, SO_2Ph, and $SiPh_3$ groups [RJ1, RL2]. Vinylmercurials can be reduced to alkenes, but a mixture of *Z*- and *E*-isomers is obtained [RJ1].

Aminomercuration leads to substituted organomercurials **5.5,** which also suffer demercuration with sodium borohydride, preferentially under PTC conditions [BE1, EB4] (Figure 5.5). The mechanism proposed for this reduction in protic solvents is an ionic one, implying the intermediate formation of aziridinium salt [L3]. This method has been applied to the synthesis of cyclic amines from α,β-ethylenic precursors **5.6** [EB4] (Figure 5.5). When the reduction in run in alcohol or water, mixtures of five- and six-membered cyclic amines are obtained from each precursor **5.6** (n = 1 or 2).

5.3.2 Palladium Complexes

Complexes of π-allylpalladium formed by reaction with Pd(0) complexes with allylic derivatives (acetates, ethers, thioethers, and sulfones) can be reduced in a

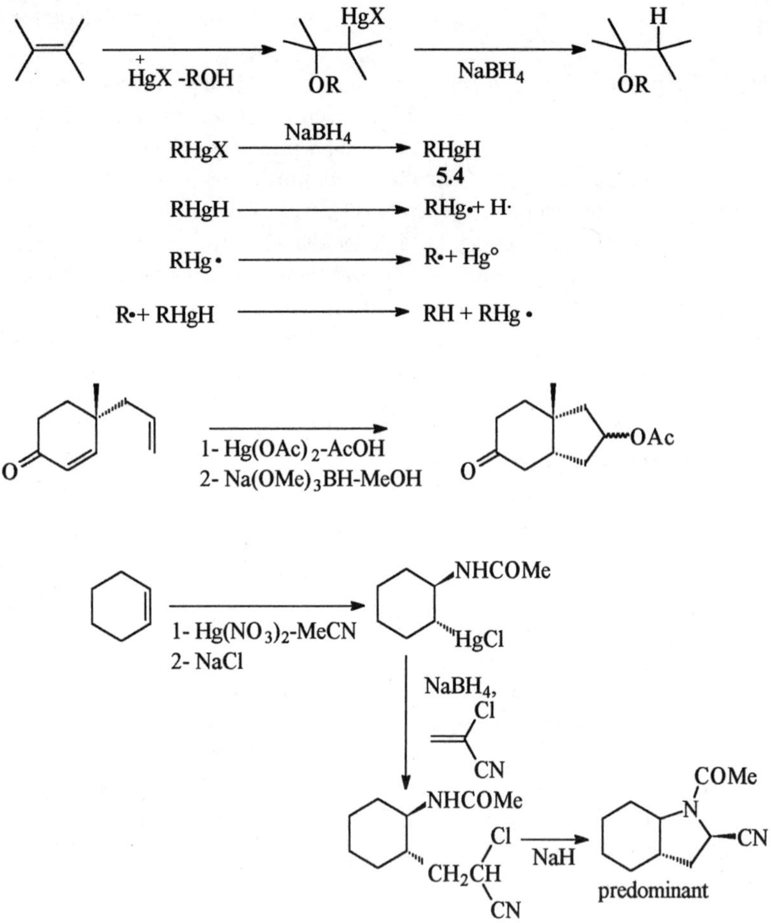

Figure 5.4

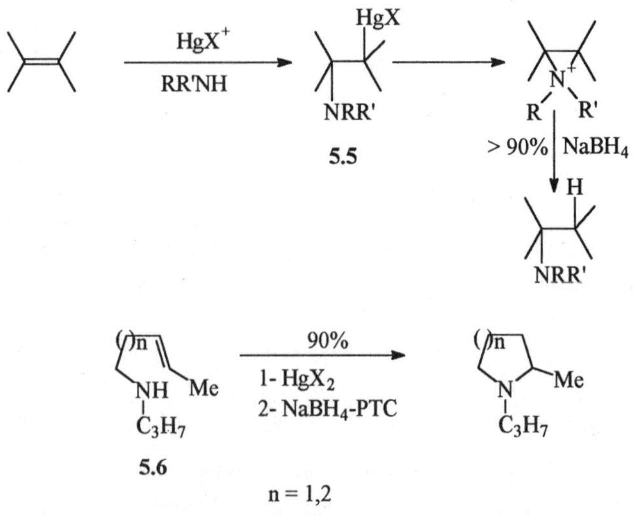

Figure 5.5

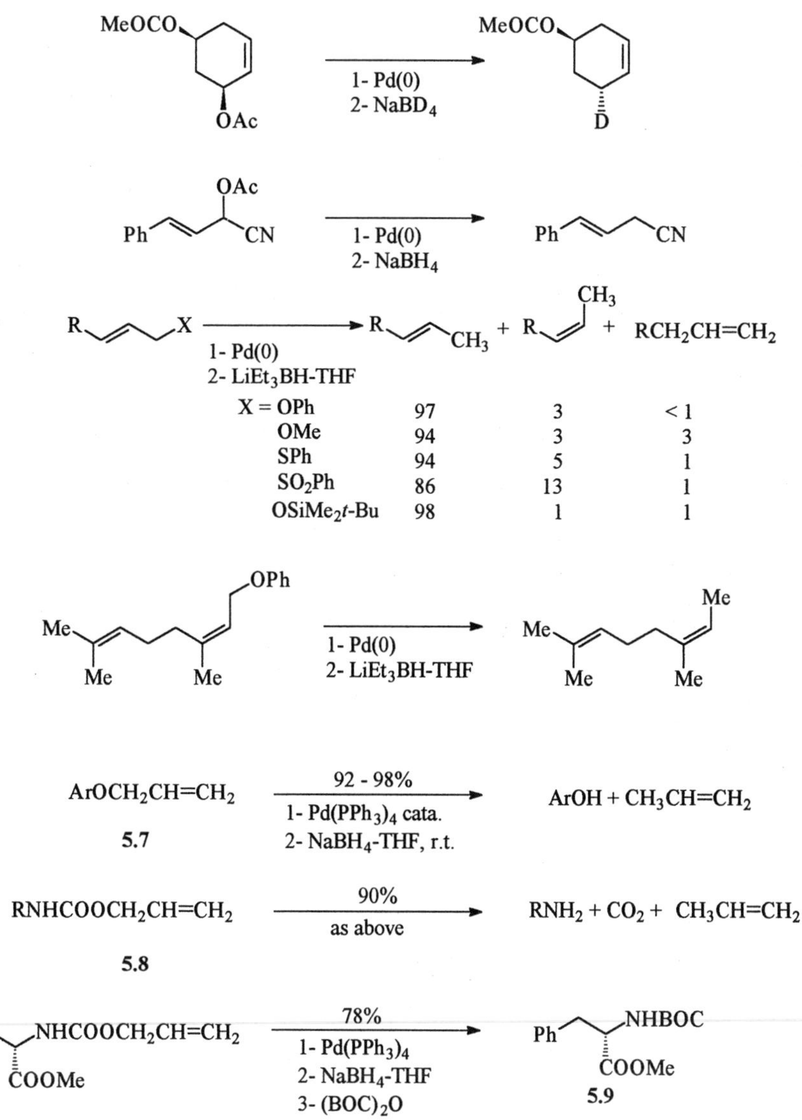

Figure 5.6

regio- and stereoselective fashion by $NaBH_4$ or $LiEt_3BH$ in THF to the corresponding internal alkene derivatives, as shown in Figure 5.6 [HL1, KR1, ME1, TM3]. Reductions in which X is a different group can be less selective (Figure 5.6), as can reactions performed with LAH or $NaCNBH_3$ [HL1, TM3]. The method preserves the isolated double bonds (Figure 5.6). On the other hand, reduction of these π-allyl complexes by ammonium formate gives terminal olefins [TM3]. An interesting application is the Pd(0)-catalyzed deprotection of allyl aryl ethers **5.7** and carba-

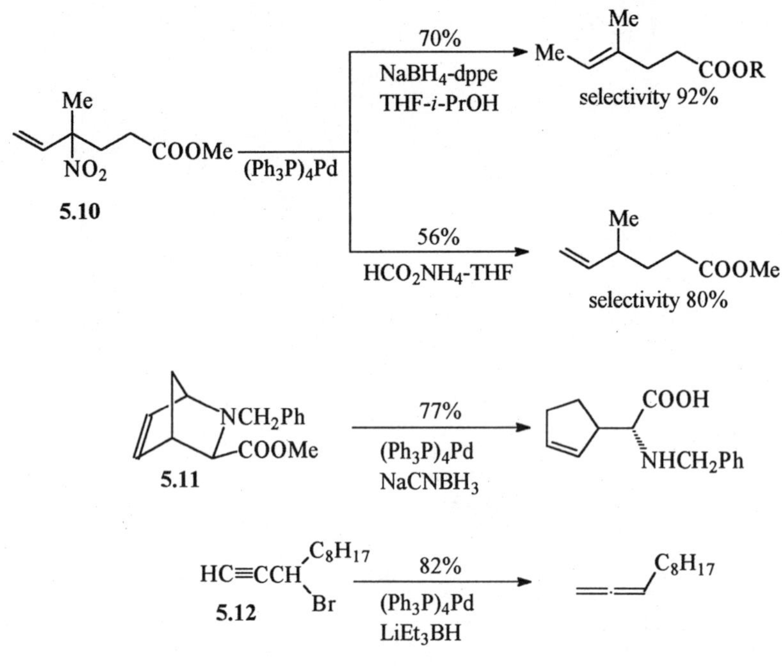

Figure 5.7

mates **5.8** [BB10, BN7], which are, respectively, transformed into propene and phenols or amines (Figure 5.6). This methodology is compatible with NO_2 groups, acids, nitriles, and amides. When performing the reaction in the presence of BOC anhydride, protected amines such as **5.9** are formed (Figure 5.6). Peptide coupling can also be realized under these conditions [BN7].

Allylic nitro derivatives such as **5.10** can also suffer similar reductions via π-allyl complexes [TM3] (Figure 5.7). Tertiary allylic amines such as **5.11** can be transformed into secondary amines via π-allylpalladium complexes that are reduced with $NaCNBH_3$ [TM3] (Figure 5.7).

Terminal propargylic bromides **5.12**, mesylates, and phosphates are also transformed into palladium complexes that are hydrogenolyzed to allenes with a high regioselectivity. $LiEt_3BH$ is the best reagent for this reaction [MT3] (Figure 5.7). However, 1,2-disubstituted alkynes lead to mixtures [MT3].

5.4 SULFIDES, THIOETHERS, SULFOXIDES, SULFONES, AND AMINE-OXIDES: RSR′, RSOR′, OR RSO_2R'

Sulfoxides and sulfones are reduced to sulfides by LAH or AlH_3 in an ether medium [H3, HL5], LAH–$TiCl_4$ in THF [AM5], $NaBH_4$–Me_3SiCl in THF [GS2], and DIBAH [HL5]. Sulfamides are reduced to amines by Red-Al in refluxing toluene [GB9, RG3]. SAH, LAH, and AlH_3 reduce disulfides to thiols [BY1, CB5, CB7, H3,

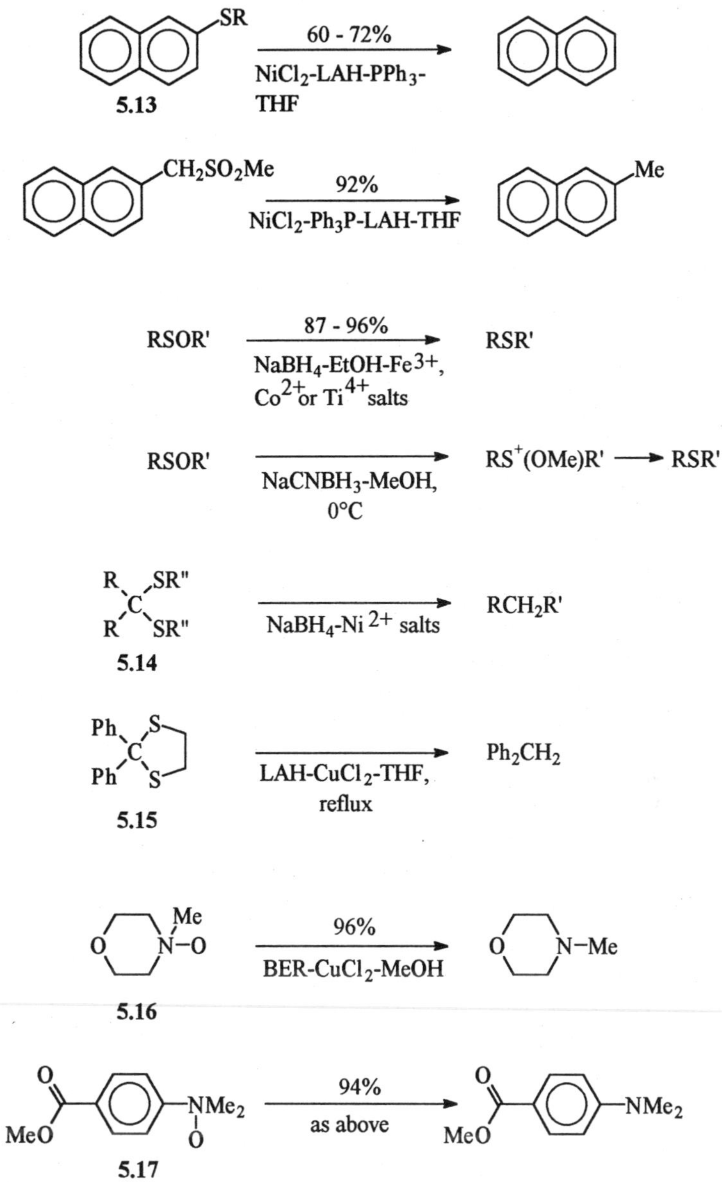

Figure 5.8

M4], as do DIBAH in hot toluene [H3, M3, M4, YG1], nickelocene–LAH, and $NiCl_2$–PPh_3–LAH [CC2]. Diaryl or dialkyldisulfides are converted to thiols by LTBA in THF at room temperature [KA3, M3]. This reduction is faster with diaryl compounds such as **5.13** and is compatible with MeO, Cl, and CN substituents (Figure 5.8). In the presence of copper salts, desulfurization takes place upon heating, and the corresponding hydrocarbons are obtained [GO2] (Figure 5.8). Total reduction also takes place in the presence of $NiCl_2$–PPh_3 [HL2] (Figure 5.8).

Sulfones and sulfoxides are not affected by borohydrides in alcohol media [RJ1] except in the presence of transition metal salts such as $FeCl_3$, $CoCl_2$, or $TiCl_4$ [CH3, GO2, KT2, LZ1, W4] (Figure 5.8). The sulfides obtained are not reduced under these conditions but are desulfurized in the presence of the Ni^{2+} salts [GO2]. Another possibility is to treat the sulfoxides with $NaCNBH_3$ in MeOH. This reduction takes place via the sulfoxonium salt [PS1] (Figure 5.8).

Reductive desulfurization of the dithioketals **5.14** and **5.15** is performed under the same conditions as for thioethers [GO2]: LAH in the presence of copper salts or borohydrides in the presence of nickel salts (Figure 5.8). The deoxygenation of tertiary amine-oxides such as **5.16** and **5.17** can be performed with borohydride exchange resin–copper sulfate in methanol at room temperature or under reflux. This reaction tolerates other functional groups such as carbon–carbon double bonds, chlorides, epoxides, esters, amides, nitriles, sulfoxides, and sulfones [SA4] (Figure 5.8).

5.5 PHOSPHINE OXIDES AND PHOSPHATES: R_3PO AND $ROP(OR')_3$

Phosphine oxides remain untouched by LAH [NW2] or borohydrides [EW1, HJ5]. They are reduced to corresponding phosphines by LAH–$CeCl_3$ in THF [GO2] or in a few cases by LAH in hot THF [TA4] (Figure 5.9). Phosphates are cleaved by LAH in THF to corresponding alcohols [JF1], while enol phosphates **5.18** are converted into carbonyl compounds. However, by using LAH–$CuBr_2$ or DIBAH [IK3], it is

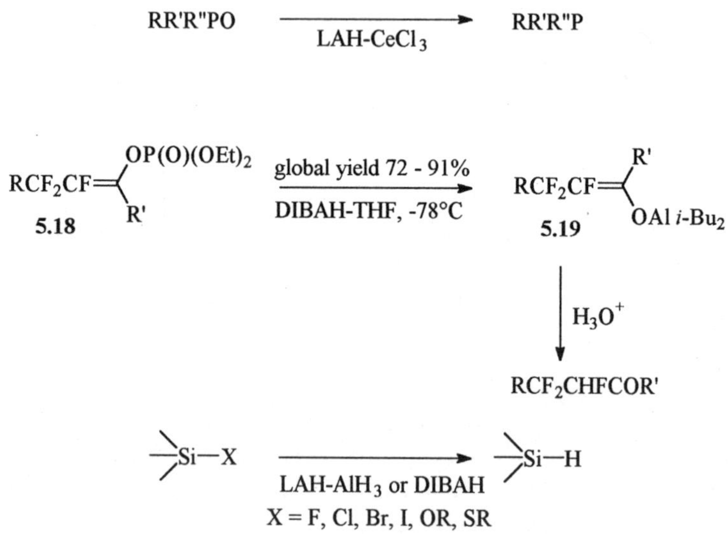

Figure 5.9

possible in some cases to generate the corresponding aluminum enolates **5.19** (Figure 5.9). These enolates can suffer in situ condensation with aldehydes [IK3] (Figure 5.9).

5.6 SILYL DERIVATIVES: R_3SiX

The silicon–halogen, silicon–oxygen, and silicon–sulfur bonds of the halogenosilanes, silyl ethers, and silyl thioethers are cleaved by reaction with LAH, AlH_3, or DIBAH, and the corresponding silyl hydrides are obtained [CG1, CG2]. Ultrasound activation can be applied [LG2] (Figure 5.9). Anionic pentacoordinated silicon compounds are reduced to hydrogenosilanes by LAH or DIBAH [BC6].

A study focusing on the stereochemistry of reduction has been carried out on molecules chiral at silicon. Depending on the Lewis acid character of the reducing agent and the nature of the X group, one observes more or less retention or inversion of configuration at the silicon atom. For a given leaving group, the amount of retention increases as the Lewis acid character of the reducing reagent increases:

$$\text{DIBAH in hexane} > \text{DIBAH in Et}_2\text{O} > \text{DIBAH in THF} > \text{AlH}_3 \text{ in Et}_2\text{O} > \text{LAH in Et}_2\text{O} > \text{LAH in THF} > \text{LAH in THF-[2.1.1]}$$

For a given reducing agent, the degree of retention of configuration becomes higher as the X group becomes harder: RO > F, RS > Cl,Br. A theoretical interpretation of these results has been suggested [CG1, CG2].

5.7 BORON DERIVATIVES

B—Cl and B—Br bonds are converted to B—H bonds by LAH in stoichiometric amounts or by K(i-PrO)$_3$BH [BC3, DK4]. LAH also converts boronates $RB(OR')_2$ into ate complexes $Li^+RBH_3^-$, which are cleaved to give the corresponding alkylboranes by Me_3SiCl [BJ5]. If the R alkyl group is chiral, its configuration is retained, opening a route to asymmetric hydroboration reactions [S3].

Synoptic Tables

Products	Substrates	Section	Reagents
	Alcohols 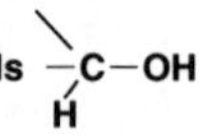		
R$\overset{*}{C}$HOHR′	cyclic 1,2-diol sulfates	§2.2	$NaBH_4$; $NaCNBH_3$
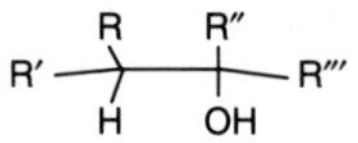	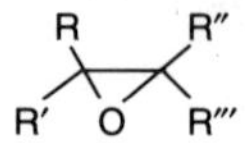	§2.3	LAH; LAH·R_3N; AlH_3; AlH_3·Et_3N; DIBAH $NaCNBH_3$–BF_3; $LiBH_4$–BEt_3; BH_3–BF_3; $LiEt_3BH$; Li 9-BBNH $Zn(BH_4)_2$–SiO_2
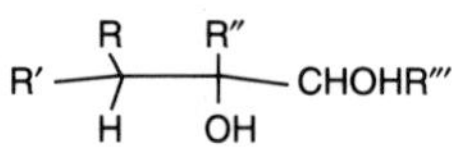 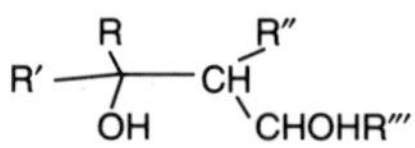	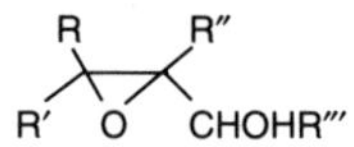	§2.3	Red-Al; LAH; DIBAH $LiBH_4$–Ti salt
ROH	$ROSiMe_3$	§2.4.2	LAH; $LiAl(OR)_{4-n}H_n$; MBH_4; DIBAH
RCH_2OH	RCH_2OTHP	§2.4.3	AlH_3; BH_3·THF; $NaCNBH_3$–BF_3
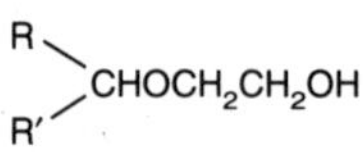	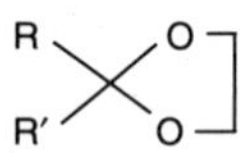	§2.4.3	AlH_3
RCH_2OH	RCHO	§3.2.1 §3.2.2	LAH; $LiAl(OR)_{4-n}H_n$; SAH; AlH_3; Red-Al; MR_3BH; BH_3·R_3N, THF or Me_2S aminoborohydrides; MBH_4; $MCNBH_3$–acid; $M(AcO)_3BH$; $(Ph_3P)_2CuBH_4$–acid
RCH=CHCH$_2$OH	RCH=CHCHO	§3.2.9	$NaBH_4$–alcohols; LAH $MCNBH_4$–acid; AlH_3; DIBAH; Red-Al; BH_3·Me_2S
RCH_2OH	RCOSEt	§3.2.5	*n*-Bu_4NBH_4 under heating
RCH_2OH	RCOOR′ or lactones	§3.2.5	LAH; LAH·R_3N; SAH; AlH_3; DIBAH–*n*-BuLi; $LiBH_4$ under heating; LiR_3BH; $NaBH_4$–MeOH or other additives; BH_3·Me_2S under heating; 9-BBN and ThexBHCl under heating; Li 9-BBNH; $Ca(BH_4)_2$
RCH=CHCH$_2$OH	RCH=CHCOOR′	§3.2.9	DIBAH; LAH–Et_2O; AlH_3; DIBAH–*n*-BuLi

(continued)

Products	Substrates	Section	Reagents
$RCH{=}CHCH_2OH$	$RC{\equiv}CCOOR'$	§4.2	Red-Al-CuBr, LAH–Et_2O; DIBAH–MeCu
RCHOHCHOHR′	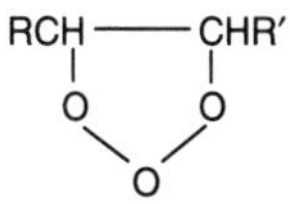	§2.4.4	LAH; $NaBH_4$; $BH_3{\cdot}Me_2S$
RCH_2CH_2OH	$RCH{=}CHCOOR'$	§3.2.9	LAH–THF
RCH_2OH	RCOOH or $(RCO)_2O$	§3.2.6	LAH; SAH; AlH_3; Red-Al (under heating); $LiBH_4$–hot MeOH; $BH_3{\cdot}THF$ or NR_3; $NaBH_4$–$ZnCl_4$ or $TiCl_4$; $(i\text{-}PrO)_2TiBH_4$; in situ generated borane
RCH_2OH	RCOOCOOEt	§3.2.6	$NaBH_4$
$RCH{=}CHCH_2OH$	$RCH{=}CH{-}COOCOOEt$	§3.2.6	$NaBH_4$–SmI_3–THF
RCH_2OH	RCOCl	§3.2.7	LAH; SAH; AlH_3; DIBAH; $(i\text{-}PrO)_2TiBH_4$; MBH_4; 9-BBN; aminoborohydrides; $Zn(BH_4)_2$
RCH_2OH	$RCONR'_2$	§3.2.8	LiR_3BH; 9-BBN; Sia_2BH; $LiBH_4$–hot MeOH; Li pyrrolidinoborohydride
RCHOHR′	RCOR′	§3.2.1 §3.2.2 (stereo-selectivity) §3.2.3 (asymmetric reductions)	LAH; $LiAl(OR)_{4-n}H_n$; AlH_3; Red-Al; MBH_4; $BH_3{\cdot}THF$, R_3N or Me_2S; MR_3BH; $MCNBH_3$–acid; $NaBH_4$–$CeCl_3$–MeOH
$RCH{=}CHCHOHR'$	$RCH{=}CHCOR'$	§3.2.9 §3.2.3 (asymmetric reductions)	LAH–Et_2O; AlH_3; DIBAH; DIBAH–n-BuLi; $LiAl(OMe)_3H$; Red-Al in benzene; $NaCNBH_3$–$ZnCl_2$; $BH_3{\cdot}Me_2S$; 9-BBN; $NaBH_4$–$CeCl_3$–MeOH; $(i\text{-}PrO)_2TiBH_4$; Li aminoborohydrides; $Zn(BH_4)_2$
$RC{\equiv}CCHOHR'$	$RC{\equiv}CCOR'$	§3.2.3 (asymmetric reductions) §4.2	LAH; $LiAl(OR)_{4-n}H_n$; DIBAH; $MCNBH_3$–acid; LAH + ligand

(continued)

Products	Substrates	Section	Reagents
RCHOHCO (OR′; NR'_2) RCHOHCHCO (OR′; NR'_2), with R″ on CH	RCOCO (OR′; NR'_2) RCOCHCO (OR′; NR'_2), with R″ on CH	§3.2.2 §3.2.4 (stereo-selectivity)	BH_3-*t*-$BuNH_2$; DIBAH; $LiAl(Et_3CO)_3H$; M-*s*-Bu_3BH; Red-Al; LAH; $Zn(BH_4)_2$; $NaBH_4$–$CaCl_2$; $LiEt_3BH$; $LiBH_4$–LiBr
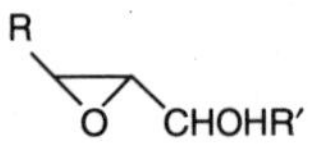	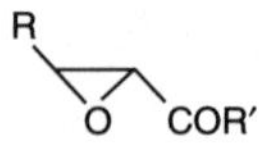	§3.2.2 §3.2.4 (stereo-selectivity)	$Zn(BH_4)_2$; MBH_4; $NaBH_4$–$CeCl_3$ or $CaCl_2$
R′CHCHOHR″, with (HO)RO on CH R′CHCHCHOHR″, with (HO)RO and R‴ on CH	R′CHCOR″, with (HO)RO on CH R′CHCHCOR″, with (HO)RO and R‴ on CH	§3.2.4 (stereo-selectivity)	$Zn(BH_4)_2$; LAH; Red-Al; LTBA; DIBAH; M*s*-Bu_3BH; BR_3 or $MeOBR_2$–Na or $LiBH_4$; catecholborane; $TiCl_4$–Et_4NCNBH_3; $R_4NB(OAc)_3H$
RCHOHCHCHOHR′, with R″ on CH	RCOCHCOR′, with R″ on CH	§3.2.4 (stereo-selectivity)	LTBA; LTBA–$TiCl_4$; $NaBH_4$–$CeCl_3$; $LiEt_3BH$
$RCHOHCH_2SR'$	$RCOCH_2SR'$	§3.2.4 (stereo-selectivity)	DIBAH
$RCHOHCH_2SOTol$	$RCOCH_2SOTol$	§3.2.4 (stereo-selectivity)	DIBAH; DIBAH–$ZnCl_2$; LAH
$RCHOHCH_2POPh_2$	$RCOCH_2POPh_2$	§3.2.4 (stereo-selectivity)	$NaBH_4$; $NaBH_4$–$CeCl_3$
R′CHCHOHR″, with R_2N on CH R′CHCHCHOHR″, with R_2N and R‴ on CH	R′CHCOR″, with R_2N on CH R′CHCHCOR″, with R_2N and R‴ on CH	§3.2.4 (stereo-selectivity)	LAH; LAH–$TiCl_4$; LTBA; DIBAH–$ZnCl_2$; MBH_4
$RCHCH_2OH$, with NH_2 on CH	RCCOOR′, with $=NSiMe_3$ on C	§3.3.1	LAH
$R\overset{*}{C}HCH_2OH$, with NH_2 on CH	$R\overset{*}{C}HCOOH$, with NH_2 on CH	§3.2.6	in situ generated borane
$R\overset{*}{C}HCH_2OH$, with NHEt on CH	$R\overset{*}{C}HCOOH$, with NHCOMe on CH	§3.2.6	in situ generated borane
$R\overset{*}{C}HCH_2OH$, with NHBOC on CH	$R\overset{*}{C}HCOOH$, with NHBOC on CH	§3.2.6	in situ generated borane; (*i*-PrO)$_2TiBH_4$

(continued)

Products	Substrates	Section	Reagents
RCH(NHR′)CHOHAr	RC(=NR′)CHOHAr	§3.3.1 (stereo-selectivity)	$Zn(BH_4)_2$
R′CH(NHR)CH_2CHOHR″	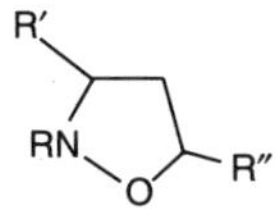	§3.3.3	LAH; LAH–$NiCl_2$
R′CH(NH_2)CH_2CHOHR″	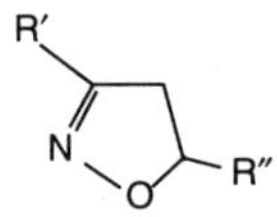	§3.3.3	LAH–$NiCl_2$; $NaBH_4$–$NiCl_2$
	ALDEHYDES RCHO		
RCHO	RCOOPh	§3.2.5	LTBA
RCHO	RCOOEt	§3.2.5	DIBAH; $Na(Et_2N)_3AlH$; piperazinoaluminohydride
RCHO	RCOSEt	§3.2.5	DIBAH
RCH=CHCHO	RCH=CHCOOEt	§3.2.5	LAH–Et_2NH–pentane
RCHO	RCOOH	§3.2.6	piperazinoaluminohydride; ThexBHCl
RCHO	RCOCl	§3.2.7	LTBA; DIBAH; $NaBH_4$–$CdCl_2$–DMF; $(Ph_3P)_2CuBH_4$; $(Ph_3P)_2CuCNBH_3$
RCHO		§3.2.8	LAH; Red-Al
RCHO	RCON(OMe)Me	§3.2.8	LAH; DIBAH
RCH=CHCHO	RCH=CHCON(OMe)Me	§3.2.8	DIBAH
RCHO	$RCONR'_2$	§3.2.8	LTEA; SAH; DIBAH–*n*-BuLi; LiR_3BH
RCH_2CH_2CHO	RCH=CHCHO	§3.2.9	DIBAH–MeCu
RCH_2CH(R′)CHO	RCH=CHCHO	§3.2.9	DIBAH–MeCu–R′X

(continued)

Products	Substrates	Section	Reagents
RCHO	RCN	§4.3	LTEA; DIBAH; Na aminoaluminohydride; $NaEt_2AlH_2$–Lewis acid; catecholborane
RR′R″CCHO	RR′CHCN	§4.3	DIBAH–LDA–R″X
RCH=CHCHO	RCH=CHCN	§4.4	DIBAH
AMINES (Amino Alcohols: see Alcohols) Primary: R-NH$_2$			
RCH_2NH_2	$RCONH_2$	§3.2.8	LAH; AlH_3; Red-Al; BH_3·THF; $NaBH_4$–$TiCl_4$; $LiBH_4$–diglyme–MeOH; in situ generated borane
RR′CHNH$_2$	RR′C=NOH	§3.3.4	SAH; LAH; $NaBH_4$–$NiCl_2$, $TiCl_4$, or $ZrCl_4$; AlH_3·Et_3N; $NaBH_4$–CF_3COOH–diglyme (under heating); $NaCNBH_3$–$TiCl_4$ or $TiCl_3$
RCH(NH$_2$)COOR′	RC(=NOH)COOR′	§3.3.4	$NaCNBH_3$–$TiCl_4$
RCH=CH–CH(NH$_2$)R′	RCH=CH–C(=NOH)R′	§3.3.4	$NaBH_4$–MoO_3
RCH_2NH_2	RCN	§4.3	SAH; LAH; AlH_3; BH_3; $NaBH_4$–$CoCl_2$; $NaBH_4$–acid; in situ generated borane; Red-Al (under heating); $LiBH_4$–diglyme–MeOH; $LiEt_3BH$
RCH=CHCH$_2$NH$_2$	RCH=CHCN	§4.3	$NaBH_4$–$CoCl_2$
$RCH_2CH_2CH_2NH_2$	RCH=CHCN	§4.4	LAH
RCH(NH$_2$)COOR′	RC(=NSiMe$_3$)–COOR′	§3.3.1	$NaCNBH_3$–MeOH; $NaBH_4$–MeOH; Me_2NH·BH_3–MeOH
RCH=CHCH(NH$_2$)R′	RC≡CCH(NH$_2$)R′	§4.1	LAH; Red-Al
RCH(NH$_2$)CH$_2$CHOHR′	RC(=NOCH$_2$Ph)CH$_2$CHOHR′	§3.3.4 (stereoselectivity)	LAH
RCH(NHOCH$_2$Ph)CHOHR′	RC(=NOCH$_2$Ph)CHOHR′	§3.3.4	$Me_4N(AcO)_3BH$

(continued)

Products	Substrates	Section	Reagents
RNH_2	RNO_2	§5.1	$LiBH_4$–diglyme–hot MeOH; MBH_4–transition metal salt or Pd-C
$ArCH_2CH_2NH_2$	ArCH=$CHNO_2$	§5.1	LAH; BH_3; Cl_2BH in situ generated borane
RNH_2	RN_3	§5.2	LAH; $NaBH_4$–THF–hot MeOH; $NaBH_4$–PTC; aminoborohydrides; $Zn(BH_4)_2$
	Secondary and Tertiary: RNHR′ and RNR′R″		
$RCH_2NR'R''$	RCONR′R″	§3.2.8	LAH; AlH_3; Red-Al; DIBAH; $LiBH_4$ (under heating); MBH_4–hot acid; BH_3·THF; $NaBH_4$–$TiCl_4$; in situ generated borane
RCH=$CHCH_2NR'R''$	RCH=CHCONR′R″	§3.2.8 §3.2.9	DIBAH; AlH_3
RCH_2NHR'	RCH=NR′	§3.3.1 (stereo-selectivity)	$NaBH_4$–$ZrCl_4$; LAH MBH_4–$CoCl_2$ or $NiCl_2$ Red-Al; MBH_4–MeOH; MR_3BH; BH_3; $MCNBH_3$; $LiEt_2NBH_3$
RCH=$CHCH_2NHR'$	RCH=CHCH=NR′	§3.3.1	$NaBH_4$–MeOH
$RCH_2CH_2NR'R''$	RCH=CHNR′R″	§3.3.2	AlH_3; MBH_4–acid; $MCNBH_3$–acid; BH_3–acid
$RCH_2NR'R''$	RCHO + HNR′R″ (reductive amination)	§3.3.1	$MCNBH_3$; $NaBH_4$–acid; $Na(AcO)_3BH$; $NaBH_4$ or $NaCNBH_3$ + Ti(O-*i*-Pr)$_4$
RNR'_2	RNH_2	§3.2.8	$NaBH_4$–acid
$RN(R')CH_3$	RN(R′)COOR″	§3.2.8	LAH; $NaBH_4$–acid
$RR'CHNR''_2$	$RR'C(CN)(NR'')_2$	§3.3.1	$NaBH_4$–MeOH; LAH; $Zn(BH_4)_2$; AlH_3
RCH=CHCHR′ with NR''_2 on CHR′	RC≡CCHR′ with NR''_2 on CHR′	§4.1	LAH; Red-Al
RR′R″N	RR′R″N^+Me, X^-	§1.5	LAH; $LiEt_3BH$

(continued)

Products	Substrates	Section	Reagents
	NITROGEN HETEROCYCLES: see §3.3.3 and 3.2.8		
	UNSATURATED DERIVATIVES 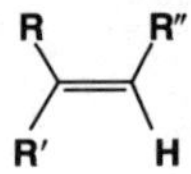 **or RC≡CH or ArH**		
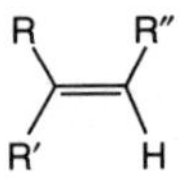	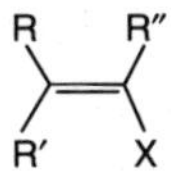	§1.1	LAH; $NaBH_4$–transition metal salt; LAH + $CeCl_3$
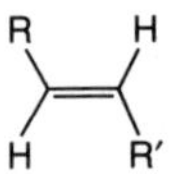	RC≡CR′	§4.1	LAH (under heating)
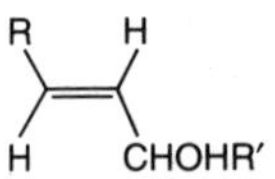	RC≡CCHOHR′	§4.1	LAH, Red-Al
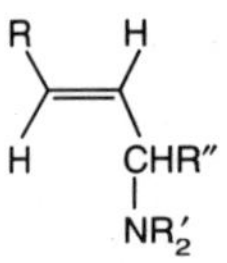	RC≡CCHR″ (with NR'_2 on the CH)	§4.1	LAH; Red-Al
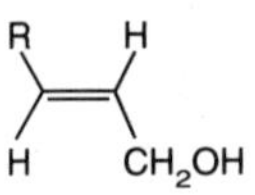	RC≡CCOOEt	§4.2	LAH
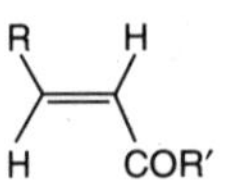	RC≡CCOR′	§4.2	DIBAH–MeCu; DIBAH–HMPA if R = H
CH_2=C(COOR)R′	HC≡CCOOR	§4.2	DIBAH + R′X
RCH=CHCOOR′	RC≡CCOOR′	§4.2	DIBAH–MeCu; Red-Al–CuBr
RCH=CHCN	RC≡CCN	§4.3	LAH; $NaBH_4$–MeOH
RCH=CHS(O)R′	RC≡CS(O)R′	§4.1	DIBAH; LAH–THF
RCH=CHSR′	RC≡CSR′	§4.1	LAH; $Li(MeO)_3AlH$ with or without CuBr
RCH=C=CH_2	HC≡CCH(R)Br	§5.3	Pd(0)–$LiEt_3BH$

(continued)

Products	Substrates	Section	Reagents
RC≡CH	RC≡CX	§2.1	Red-Al; LAH–$CeCl_3$
Ar—H	Ar—X	§2.1	Red-Al; LAH; $NaBH_4$–transition metal salt; $NaBH_4$–DMF-*hv*
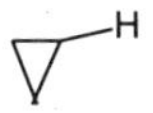		§2.1	$NaBH_4$–DMF (X = Br) Red-Al; LAH
$R_2C{=}C{=}CH_2$	$R_2C(Cl)C{\equiv}CH$	§2.1	LAH
$RR'CHCH{=}CHR''$	$RR'C{=}CHCH(R'')P(O)(OEt)_2$	§2.6	LAH
	SATURATED DERIVATIVES: $-\overset{\vert}{\underset{\vert}{C}}-H$		
RCH_3 or RCH_2R'	RCH_2Cl or $RR'CHCl$	§2.1	DIBAH–*n*-BuLi; LAH; Red-Al; $LiEt_3BH$; $NaCNBH_3$; LAH–$CeCl_3$
RCH_3 or RCH_2R'	RCH_2OSO_2R or $RR'CHOSO_2R$	§2.2	LAH; $LiBH_4$; $LiEt_3BH$ DIBAH; $NaBH_4$–hot DMSO
RCH_3 or RCH_2R'	RCH_2Br or $RR'CHBr$	§2.1	LAH; Red-Al; LTBA; $NaBH_4$–hot DMSO or DME; $LiEt_3BH$; $NaCNBH_3$
RCH_3 or RCH_2R'	RCH_2I or $RR'CHI$	§2.1	*n*-Bu_4NCNBH_3; LAH; LiR_2NBH_3; Red-Al; LTBA; $LiEt_3BH$; $NaCNBH_3$
R_3CH, $ArCH_3$, $RCH{=}CHCH_3$	R_3CX; $ArCH_2X$; $RCH{=}CHCH_2X$	§2.1	*n*-Bu_4NCNBH_3; LAH $NaBH_4$–alcohols; $Zn(BH_4)_2$; $NaBH_4$–$CuCl_2$; $NaCNBH_3$–ZnI_2 or $SnCl_2$; 9-BBN·*n*-BuLi
RC≡CCH3	$RC{\equiv}CCH_2X$	§2.1	Li 9-BBNH
$RCH{=}CHCH_3$	$RCH{=}CHCH_2OSO_2R$	§2.2	LAH; AlH_3
$RCH_2CH(R')CH_3$	$RCH{=}C(R')CH_2OAc$	§2.2	$NaBH_4$–$NiCl_2$
R_3CH; Ar_2CH_2 $ArCH_3$ $RCH{=}CHCH_2R'$	R_3COH; Ar_2CHOH $ArCH_2OH$ $RCH{=}CHCHOHR'$	§2.4.1	$NaBH_4$–CF_3COOH or CF_3SO_3H; AlH_3; $NaBH_4$–$AlCl_3$; $NaCNBH_3$–ZnI_2 or BF_3; $NaBH_4$–ZnI_2
$ArCH_3$	$ArCH_2N^+Me_3$, X^-	§2.5	$NaCNBH_3$

(continued)

Products	Substrates	Section	Reagents
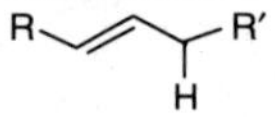	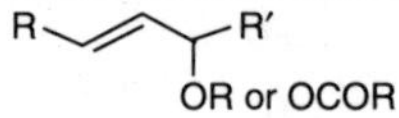	§2.4.2 §5.3.2	$NaBH_4$–$NiCl_2$; Pd(0)–$NaBH_4$ or $LiEt_3BH$
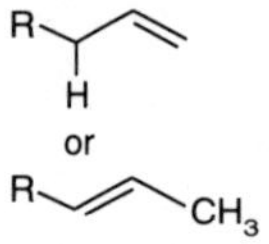	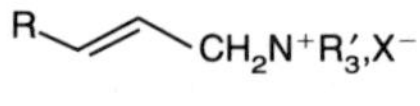	§2.5	LAH; Red-Al; $NaBH_4$–alcohols
$PhCH_3$	$PhCH_2P^+R_3,X^-$	§2.6	LAH
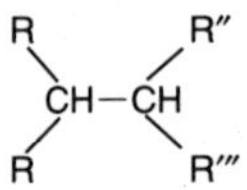	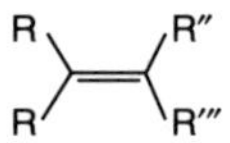	§3.1	$NaBH_4$–$CoCl_2$; $LiEt_3BH$
$ArCH_2CH_2R$	ArCH=CHR	§3.1	$LiEt_3BH$; $NaBH_4$–$BiCl_3$
$ArCH_2R$	ArCOR	§3.2.1	$NaBH_4$–acid; AlH_3; $NaBH_4$–$AlCl_3$; MBH_4–$Ni(OAc)_2$; $NaBH_4$–ZnI_2; $NaBH_4$–$PdCl_2$; $NaCNBH_3$–acid; *t*-$BuNH_2 \cdot BH_3$–$AlCl_3$
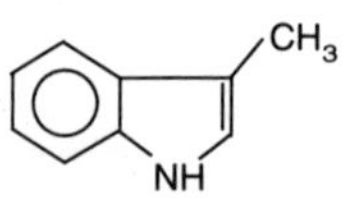	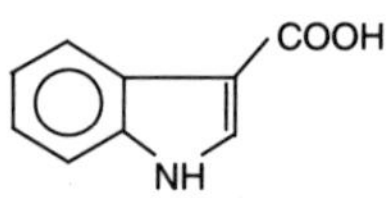	§3.2.6	$BH_3 \cdot THF$
RCH_2R'	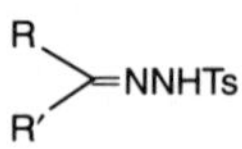	§3.3.4	$NaBH_4$–RCOOH or MeOH; BH_3–RCOOH; $NaCNBH_3$–RCOOH or $ZnCl_2$; $(Ph_3P)_2CuBH_4$ in a few cases; catecholborane
RCH_2CH_2R'	RC≡CR′	§4.1	$NaBH_4$–transition metal salt
RCH_2CH_2COR'	RCH=CHCOR′	§3.2.9	Red-Al–CuBr; DIBAH–MeCu; Li and K *s*-Bu_3BH; LTBA; LAH–nickelocene; R_4NBH_4(resin); $(Ph_3PCuH)_6$
RCH_2CH_2COOR'	RCH=CHCOOR′	§3.2.9	Red-Al–CuBr; DIBAH–MeCu; $LiAl(OMe)_3H$–CuBr; R_4NBH_4 (resin); $NaBH_4$–Cu_2Cl_2
$RCH_2CH_2CONR'_2$	$RCH{=}CHCONR'_2$	§3.2.9	Li and K *s*-Bu_3BH; LTBA; catecholborane-Rh complex
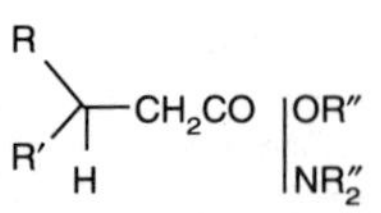	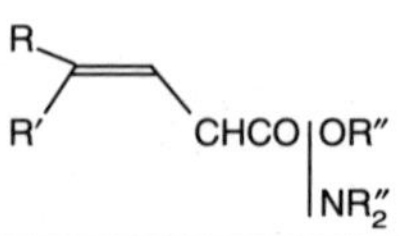	§3.2.9	$NaBH_4$–Co_2Cl_2 semicorrin

(continued)

Products	Substrates	Section	Reagents
RCH_2CH_2CN	$RCH{=}CHCN$	§4.4	LAH in a few cases; Red-Al–CuBr
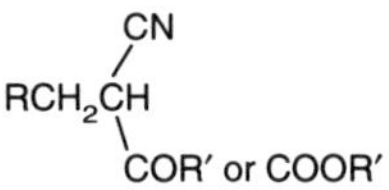	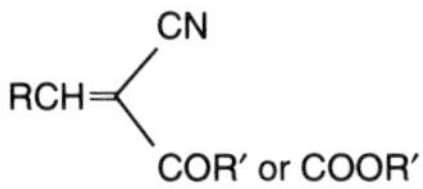	§4.4	$NaBH_4$
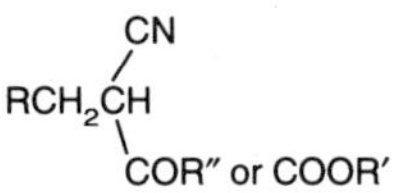	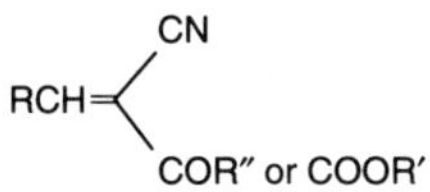	§3.2.9	$NaBH_4$; $NaBH_3CN$
$RCH_2CH_2NO_2$	$RCH{=}CHNO_2$	§5.1	$NaBH_4$–alcohol; $NaCNBH_3$–alcohol; $LiEt_3BH$
RH	RSH or RSSR	§5.4	LAH–Cu^{2+} salts; MBH_4–Ni salts
RH	RHgX	§5.3.1	$NaBH_4$; $NaB(OMe)_3H$ MBH_4–PTC
	ETHERS		
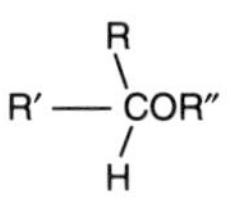	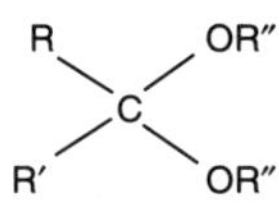	§2.4	LAH–$TiCl_4$; DIBAH under heating; AlH_3; BH_3; $AlBr_2H$; $AlCl_2H$; $ClBH_2{\cdot}Me_2S$; $NaBH_4$ or $NaCNBH_3$–acid or $TiCl_4$; $Zn(BH_4)_2$–Me_3SiCl or $TiCl_4$
	HYDROXYLAMINES: RNHOH		
RCH_2NHOH	$RCH{=}NOH$	§2.3.4	BH_3; $NaBH_4$–RCOOH; $NaCNBH_3$–RCOOH; aminoboranes
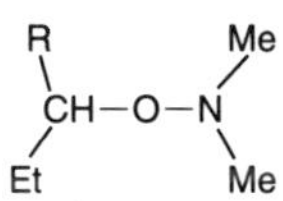	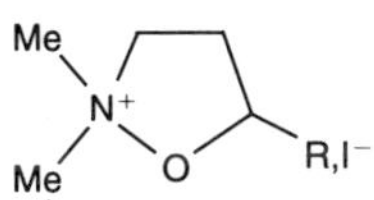	§3.3.3	LAH
	HYDRAZINES		
RCH_2NHNHR'	$RCH{=}NNHR'$	§3.3.4	LAH; BH_3
$RCH_2NHNHTs$	$RCH{=}NNHTs$	§3.3.4	BH_3–pyridine
	IMINES		
$RCH{=}NSiMe_3$	RCN	§4.3	$Li(OEt)_3AlH$, then Me_3SiCl
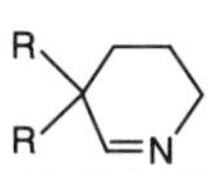	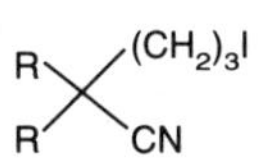	§4.3	DIBAH–THF

(continued)

Products	Substrates	Section	Reagents
	LACTOLS, LACTONES; LACTAMS		
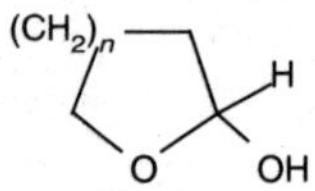	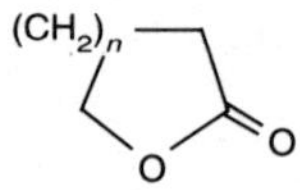	§3.2.5	DIBAH
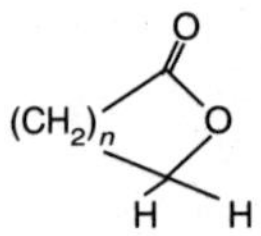	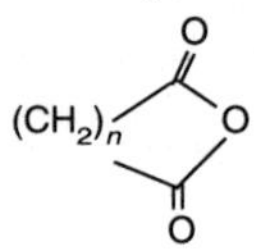	§3.2.6	LAH in calc. amount; $LiBH_4$; $NaBH_4$–THF–MeOH or DMF; MR_3BH; DIBAH–*n*-BuLi
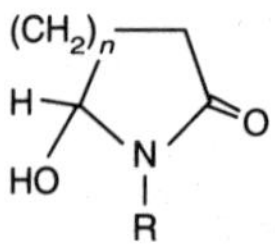	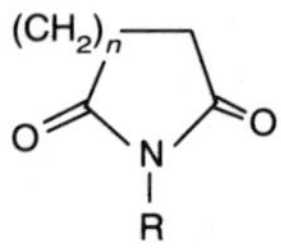	§3.2.8	$NaBH_4$–acid; DIBAH; $NaBH_4$–$CeCl_3$
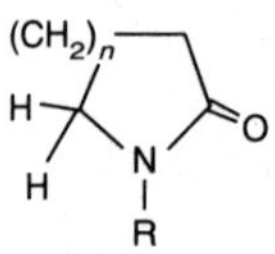	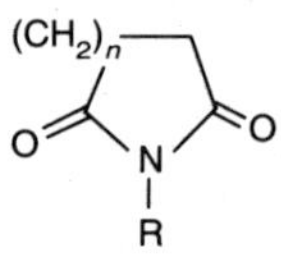	§3.2.8	Red-Al
	PHENOLS		
ArOH	ArOMe	§2.4.2	DIBAH; $LiEt_3BH$
ArOH	$ArOCH_2CH{=}CH_2$	§5.3.2	Pd(0)–$NaBH_4$

REFERENCES

The references are classified according to the initial letter(s) of the surname(s) of the (two) first author(s)

A1 G. C. Andrews, *Tetrahedron Lett.* **21**, 697 (1980); G. C. Andrews and T. C. Crawford, *Tetrahedron Lett.* **21**, 693 (1980).

AA1 A. Alberola, M. A. Alvarez, C. Andres, A. Gonzalez, and R. Pedrosa, *Synthesis*, 153 (1990).

AB1 E. N. Alesso, D. E. Bianchi, L. M. Finkielsztein, B. Lantaño, G. Y. Moltrasio, and J. M. Aguirre, *Tetrahedron Lett.* **36**, 3299 (1995).

AB2 E. C. Ashby and J. R. Boone, *J. Org. Chem.* **41**, 2890 (1976).

AC1 J. W. Apsimon and T. L. Collier, *Tetrahedron* **42**, 5157 (1986).

AC2 P. Andreoli, G. Cainelli, M. Contento, D. Giacomini, G. Marteli, and M. Panunzio, *J. Chem. Soc. Perkin Trans.* I, 945 (1988).

AC3 C. Agami, F. Couty, and C. Lequesne, *Tetrahedron* **51**, 4043 (1995).

AC4 S. K. Armstrong, E. W. Collington, J. G. Knight, A. Naylor, and S. Warren, *J. Chem. Soc. Perkin Trans.* I, 1433 (1993).

AC5 A. F. Abdel-Magid, K. G. Carson, B. D. Harris, C. A. Maryanoff, and R. D. Shah, *J. Org. Chem.* **61**, 3849 (1996).

AD1 E. C. Ashby, F. R. Dobbs, and H. P. Hopkins, *J. Am. Chem. Soc.* **95**, 2823 (1973); **97**, 3158 (1975).

AD2 E. C. Ashby, R. N. de Priest, and T. N. Pham, *Tetrahedron Lett.* **24**, 2825 (1983).

AD3 E. C. Ashby, R. N. de Priest, A. B. Goel, B. Wenderoth, and T. N. Pham, *J. Org. Chem.* **49**, 3545 (1984).

AD4 E. C. Ashby and A. K. Deshpande, *J. Org. Chem.* **59**, 3798 (1994).

AF1 S. G. Alvarez, G. B. Fisher, and B. Singaram, *Tetrahedron Lett.* **36**, 2367 (1995).

AF2 C. Agami, M. Fadlallah, A. Kazakos, and J. Levisalles, *Tetrahedron* **35**, 969 (1979).

AG1 E. C. Ashby, A. B. Goel, and R. N. de Priest, *Tetrahedron Lett.* **22**, 1763 and 3729 (1981).

AG2 E. C. Ashby, A. B. Goel, and R. N. de Priest, *J. Am. Chem. Soc.* **102**, 7779 (1980); E. C. Ashby and A. B. Goel, *J. Org. Chem.* **46**, 3934 (1981).

AH1 A. Arase, M. Hoshi, T. Yamaki, and H. Nakanishi, *J. Chem. Soc. Chem. Comm.,* 855 (1994).

AH2 A. Anantanarayan and M. Hart, *J. Org. Chem.* **56**, 991 (1991).

AK1 C. Agami, A. Kazakos, J. Levisalles, and A. Sevin, *Tetrahedron* **36**, 2977 (1980).

AK2 T. D. Aicher and Y. Kishi, *Tetrahedron Lett.* **28**, 3463 (1987).

AL1 E. C. Ashby and J. J. Lin, *Tetrahedron Lett.,* 4481 (1977); E. C. Ashby, J. J. Lin, and R. Kovar, *J. Org. Chem.* **41**, 1939 (1976); E. C. Ashby, J. J. Lin, and A. B. Goel, *J. Org. Chem.* **43**, 183 (1978).

AM1 A. F. Abdel-Magid, C. Maryanoff, and K. G. Carson, *Tetrahedron Lett.* **31**, 5595 (1990); *Synlett,* 537 (1990).

AM2 A. Abiko and S. Masamune, *Tetrahedron Lett.* **33**, 5517 (1992).

AM3 N. T. Ahn, F. Maurel, and J. M. Lefour, *N. J. Chem.* **19**, 353 (1995).

AM4 A. Abiko, O. Moriya, S. A. Filla, and S. Masamune, *Angew. Chem. Int. Ed. Engl.* **34**, 793 (1995).

AM5 E. Akgün, K. Mahmood, and C. A. Mathis, *J. Chem. Soc. Chem. Comm.,* 761 (1994).

AO1 K. Abe, J. Okumura, T. Tsugoshi, and N. Nakamura, *Synthesis,* 597 (1984).

AS1 G. Adam and D. Seebach, *Synthesis,* 373 (1988).

AS2 A. Krief, D. Surleraux, and M. Frauenrath, *Tetrahedron Lett.* **29**, 6157 (1988).

AW1 E. C. Ashby and C. O. Welder, *Tetrahedron Lett.* **36**, 7171 (1995).

AZ1 S. V. d'Andrea, S. Zhao, S. A. Mizsak, J. P. Freeman, and J. Szmuszkovicz, *Org. Prep. Proc. Int.* **26**, 114 (1994).

B1 J. H. Babler, *Synth. Comm.* **12**, 839 (1982).

B2 P. A. Bartlett, *Tetrahedron* **36**, 2 (1980).

B3 R. C. Bernotas, *Tetrahedron Lett.* **31**, 469 (1990).

B4 S. Bhattacharyya, *J. Org. Chem.* **60**, 4928 (1995).

B5 S. Bhattacharyya, *Tetrahedron Lett.* **35**, 2401 (1994).

B6 S. Bhattacharyya, *Synlett,* 971 (1995).

B7 S. Bhattacharyya, *J. Chem. Soc. Perkin Trans.* I, 1381 (1991).

BA1 J. R. Boone and E. C. Ashby, *Top. Stereochem.* **11**, 53 (1979).

BA2 J. Barluenga, E. Aguilar, J. Joglar, B. Olano, and S. Fustero, *J. Chem. Soc. Chem. Commun.*, 1132 (1989).

BA3 J. Barlengua, E. Aguilar, B. Olano, and S. Fustero, *Synlett,* 463 (1990).

BA4 M. T. Barros, C. M. Alves, A. Gil Santos, L. S. Godinho and C. D. Maycock, *Tetrahedron Lett.* **36**, 2321 (1995).

BB1 R. F. Borch, M. D. Bernstein, and M. D. Durst, *J. Am. Chem. Soc.* **93**, 2897 (1971).

BB2 J. E. Baldwin and K. A. Black, *J. Org. Chem.* **48**, 2778 (1983).

BB3 R. J. Borders and R. A. Bryson, *Chem. Lett.,* 9 (1984).

BB4 M. Bonin, R. Besselièvre, D. S. Grierson, and H. P. Husson, *Tetrahedron Lett.* **24**, 1493 (1983).

BB5 M. G. Brasca, H. B. Broughton, D. Craig, S. V. Ley, A. A. Somovila, and P. L. Toogood, *Tetrahedron Lett.* **29**, 1853 (1988).

BB6 S. Bartel and F. Bohlmann, *Tetrahedron Lett.* **30**, 685 (1989).

BB7 A. S. Bhanu Prasad, J. V. Bhaskar Kanth, and M. Periasamy, *Tetrahedron* **48**, 4623 (1992).

BB8 J. Bach, R. Berenguer, J. Farràs, J. Garcia, J. Meseguer, and J. Vilarrasa, *Tetrahedron Asymmetry* **6**, 2683 (1995); J. Bach, R. Berenguer, J. Garcia, and J. Vilarrasa, *Tetrahedron Lett.* **36**, 3425 (1995).

BB9 J.-M. Brunel and G. Buono, *Synlett,* 177 (1996).

BB10 R. Beugelmans, S. Bourdet, A. Bigot, and J. Zhu, *Tetrahedron Lett.* **35**, 4349 (1994).

BB11 G. Bartoli, M. Bosco, L. Sambri, and E. Marcantoni, *Tetrahedron Lett.* **37**, 7421 (1996).

BB12 M. C. Bernabeu, P. Bonete, F. Caturla, R. Chinchilla, and C. Nájera, *Tetrahedron Asymmetry* **7**, 2475 (1996).

BB13 J. Bach, R. Berenguer, J. Garcia, T. Loscertales, and J. Vilarrasa, *J. Org. Chem.* **61**, 9021 (1996).

BC1 H. C. Brown and Y. M. Choi, *Synthesis,* 439 (1981).

BC2 S. Banfi, S. Colonna, H. Molinari, and S. Julia, *Synth. Comm.* **13**, 901 (1983).

BC3 H. C. Brown, J. S. Cha, B. Nazer, S. C. Kim, S. Krishnamurthy, and C. A. Brown, *J. Org. Chem.* **49**, 885 (1984).

BC4 H. C. Brown, J. S. Cha, B. Nazer, and N. M. Yoon, *J. Am. Chem. Soc.* **106**, 8001 (1984).

BC5 H. C. Brown, J. S. Cha, N. M. Yoon, and B. Nazer, *J. Org. Chem.* **52**, 5400 (1987).

BC6 A. Boudin, G. Cerveau, C. Chuit, R. J. P. Corriu, and C. Reye, *Bull. Chem. Soc. Japan* **61**, 101 (1988).

BC7 K. F. Burri, R. A. Cardone, W. Y. Chen, and P. Rosen, *J. Am. Chem. Soc.* **100**, 7069 (1978).

BC8 D. Bacos, J. P. Célérier, E. Marx, C. Saliou, and G. Lhommet, *Tetrahedron Lett.* **30**, 1081 (1989).

BC9 M. D. Bomann, I. C. Guch, and M. DiMare, *J. Org. Chem.* **60**, 5995 (1995).

BC10 S. Bhattacharyya, A. Chatterjee, and J. S. Williamson, *Synlett,* 1079 (1995).

BC11 A. F. Burchat, J. M. Chong, and N. Nielsen, *J. Org. Chem.* **61**, 7627 (1996).

BD1 J. Brussee, F. Dofferhoff, G. G. Kruse, and A. van der Gen, *Tetrahedron* **46**, 1653 (1990).

BD2 G. Boireau, A. Deberly, and R. Toneva, *Synlett,* 585 (1993).

BD3 A. K. Beck, R. Dahinden, and F. N. M. Kuhnle, *Reductions in Organic Synthesis,* ACS Symposium Series **641**, 53 (1996).

BE1 M. C. Benhamou, G. Etemad-Moghadam, V. Speziale, and A. Lattes, *Synthesis,* 891 (1979).

BF1 G. R. Brown and A. J. Foubister, *J. Chem. Soc. Chem. Commun.,* 455 (1985).

BF2 J. C. Belletire and S. F. Fry, *Synth. Comm.* **18**, 29 (1988).

BF3 C. Bolm and M. Felder, *Tetrahedron Lett.* **34**, 6041 (1993).

BG1 R. Baker, C. L. Gibson, C. J. Swain, and D. J. Tapolczay, *J. Chem. Soc. Perkin Trans.* I, 1509 (1985).

BG2 R. Bloch, E. Guibé-Jampel, and C. Girard, *Tetrahedron Lett.* **26**, 4087 (1985).

BG3 R. Bloch, L. Gilbert, and C. Girard, *Tetrahedron Lett.* **29**, 1021 (1988).

BG4 L. Banfi, G. Guanti, and M. T. Zannetti, *J. Org. Chem.* **60**, 7870 (1995).

BG5 R. Bloch and L. Gilbert, *J. Org. Chem.* **52**, 4603 (1987).

BG6 M. Bellec and J.-C. Guillemin, *Tetrahedron Lett.* **36**, 6883 (1995).

BH1 H. C. Brown and P. Heim, *J. Org. Chem.* **38**, 912 (1973).

BH2 D. H. R. Barton, R. H. Hesse, C. Wilshire, and M. M. Péchet, *J. Chem. Soc. Perkin Trans.* I, 1075 (1977).

BH3 H. C. Brown, J. L. Hubbard, and B. Singaram, *Tetrahedron* **37**, 2359 (1981).

BH4 J. C. Briggs and P. Hodge, *J. Chem. Soc. Chem. Commun.,* 310 (1988).

BH5 C. L. Barney, E. W. Huber, and J. R. Mc Carthy, *Tetrahedron Lett.* **31**, 5547 (1990).

BH6 D. L. Boger, T. Honda, and Q. Dang, *J. Am. Chem. Soc.* **116**, 5619 (1994).

BI1 G. Bram, E. d'Incan, and A. Loupy, *N. J. Chim.* **6**, 573 (1982).

BJ1 H. C. Brown, P. K. Jadhav, and A. K. Mandal, *Tetrahedron* **37**, 3547 (1981).

BJ2 D. M. Bailey and R. E. Johnson, *J. Org. Chem.* **35**, 3574 (1970).

BJ3 P. J. Brown, D. N. Jones, M. A. Khan, and N. A. Meanwell, *Tetrahedron Lett.* **24**, 405 (1983).

BJ4 J. Barluenga, J. Joglar, F. J. Gonzalez, and S. Fustero, *Tetrahedron Lett.* **30**, 2001 (1989).

BJ5 H. C. Brown, N. N. Joshi, C. Pyun, and B. Singaram, *J. Am. Chem. Soc.* **111**, 1754 (1989).

BJ6 G. Beck, H. Jendralla, and K. Kesseler, *Synthesis,* 1014 (1995).

BK1 H. C. Brown and S. Krishnamurthy, *J. Am. Chem. Soc.* **95**, 1669 (1973).

BK2 C. A. Brown, S. Krishnamurthy, and S. C. Kim, *J. Chem. Soc. Chem. Commun.,* 373 (1973).

BK3 H. C. Brown and S. C. Kim, *Synthesis,* 635 (1977).

BK4 H. C. Brown and S. Krishnamurthy, *J. Am. Chem. Soc.* **94**, 7159 (1972).

BK5 H. C. Brown and S. Krishnamurthy, *Tetrahedron* **35**, 567 (1979).

BK6 H. C. Brown, S. C. Kim, and S. Krishnamurthy, *J. Org. Chem.* **45**, 1 (1980).

BK7 (a) H. C. Brown, G. W. Kramer, A. N. Levy, and M. M. Midland, *Organic Syntheses via Boranes* (Wiley-Interscience, New York, 1975); (b) H. C. Brown, *Hydroboration* (Benjamin, New York, 1980).

BK8 J. R. Blanton and S. J. Kruzska, *Microchemical Journal* **47**, 120 (1993).

BK9 B. P. Bandgar, N. S. Kshirsagar, and P. P. Wadgaonkar, *Synth. Comm.* **25**, 941 (1995).

BK10 B. P. Bandgar, N. S. Kshirsagar, and P. P. Wadgaonkar, *Synth. Comm.* **25**, 863 (1995).

BL1 R. Bartnik, S. Lesniak, and A. Laurent, *Tetrahedron Lett.* **22**, 4811 (1981).

BL2 P. Baeckström, L. Li, M. Wickrama-Ratne, and T. Norin, *Synth. Comm.* **20**, 423 (1990).

BL3 F. R. Blase and H. Le, *Tetrahedron Lett.* **36**, 4559 (1995).

BL4 E. Brown, A. Lézé, and J. Touet, *Tetrahedron Asymmetry* **3**, 841 (1992).

BL5 T. Bach and C. Lange, *Tetrahedron Lett.* **37**, 4363 (1996).

BL6 E. Brown, A. Lézé, and J. Touet, *Tetrahedron Asymmetry* **7**, 2029 (1996).

BM1 H. C. Brown, C. P. Mathew, C. Pyun, J. C. Son, and N. M. Yoon, *J. Org. Chem.* **49**, 3091 (1984).

BM2 D. Blondet and C. Morin, *J. Chem. Soc. Perkin Trans.* I, 1085 (1984).

BM3 H. C. Brown and G. A. Molander, *J. Org. Chem.* **51**, 4512 (1986), and references cited.

BM4 R. K. Boeckman and R. Michalak, *J. Am. Chem. Soc.* **96**, 1623 (1974).

BM5 J. M. Brunet, M. Maffei, and G. Buono, *Tetrahedron Asymmetry* **4**, 2255 (1993).

BM6 L. E. Burgess and A. I. Meyers, *J. Am. Chem. Soc.* **113**, 9858 (1991).

BM7 L. E. Burgess and A. I. Meyers, *J. Org. Chem.* **57**, 1656 (1992).

BM8 C. Boga, F. Manescalchi, and D. Savoia, *Tetrahedron* **50**, 4709 (1994).

BM9 B. P. Bandgar, R. K. Modhave, P. P. Wadgaonkar, and A. R. Sande, *J. Chem. Soc. Perkin Trans.* I, 1993 (1996).

BN1 H. C. Brown, S. Narashimhan, and Y. M. Choi, *J. Org. Chem.* **47**, 4702 (1982).

BN2 H. C. Brown, S. Narashimhan, and Y. M. Choi, *Synthesis,* 441 (1981).

BN3 H. C. Brown and S. Narashimhan, *J. Org. Chem.* **49**, 3891 (1984).

BN4 H. C. Brown, S. Narashimhan, and V. Somayaji, *J. Org. Chem.* **48**, 3091 (1983).

BN5 H. C. Brown, B. Nazer, J. S. Cha, and J. A. Sikorski, *J. Org. Chem.* **51**, 5264 (1986).

BN6 S. Bock, H. Noth, and P. Rahm, *Z. Naturforsch.* **43b**, 53 (1988).

BN7 R. Beugelmans, L. Neuville, M. Bois-Choussy, J. Chastanet, and J. Zhu, *Tetrahedron Lett.* **36**, 3129 (1995).

BO1 J. Barluenga, B. Olano, and S. Fustero, *J. Org. Chem.* **50**, 4052 (1985); *J. Chem. Soc. Chem. Comm.,* 410 (1988).

BP1 H. C. Brown, W. S. Park, and B. T. Cho, *J. Org. Chem.* **51**, 3278 (1986).

BP2 H. C. Brown, W. S. Park, B. T. Cho, and P. V. Ramachandran, *J. Org. Chem.* **52**, 5406 (1987).

BP3 A. Bianco, P. Passacantilli, and G. Righi, *Synth. Comm.* **18**, 1765 (1988).

BP4 P. Bravo, E. Piovosi, and G. Resnati, *J. Chem. Soc. Perkin Trans.* I, 1201 (1989).

BP5 J. V. Bhaskar Kanth and M. Periasamy, *J. Org. Chem.* **56**, 5964 (1991).

BP6 J. P. Brunel, O. Pardigon, B. Faure, and G. Buono, *J. Chem. Soc. Chem. Comm.,* 287 (1992).

BR1 M. Bonin, J. R. Romero, D. S. Grierson, and H. P. Husson, *J. Org. Chem.* **49**, 2392 (1984).

BR2 J. Barluenga, J. G. Resa, B. Olano, and S. Fustero, *J. Org. Chem.* **52**, 1425 (1987).

BR3 E. Brown, J. P. Robin, and R. Dhal, *Tetrahedron* **38**, 2569 (1982); **45**, 141 (1989).

BR4 H. C. Brown and P. V. Ramachandran, *Acc. Chem. Res.* **25**, 16 (1992).

BS1 J. H. Babler and S. J. Sarussi, *J. Org. Chem.* **48**, 4416 (1983).

BS2 H. J. Bestmann and R. Schobert, *Angew. Chem. Int. Ed. Engl.* **22**, 780 (1983).

BS3 D. M. Brestensky and J. M. Stryker, *Tetrahedron Lett.* **30**, 5677 (1989).

BS4 J. D. Buynak, J. B. Strickland, T. Hurd, and A. Phan, *J. Chem. Soc. Chem. Commun.,* 89 (1989).

BS5 T. Bottin-Strzalko and J. Seyden-Penne, *Bull. Soc. Chim. Fr. II,* 161 (1984).

BS6 M. C. Barden and J. Schwartz, *J. Org. Chem.* **60**, 5963 (1995).

BS7 C. Bolm, A. Seger, and M. Felder, *Tetrahedron Lett.* **34**, 8079 (1993).

BS8 Y. L. Bennani and K. B. Sharpless, *Tetrahedron Lett.* **34**, 2083 (1993).

BV1 J. Brussee, R. A. T. M. van Benthem, C. G. Kruse, and A. van der Gen, *Tetrahedron Asymmetry* **1**, 163 (1990).

BV2 J. Barluenga, A. L. Viado, E. Aguilar, S. Fustero, and B. Olano, *J. Org. Chem.* **58**, 5972 (1993).

BW1 D. E. Bergbreiter and S. A. Walker, *J. Org. Chem.* **54**, 5138 (1989).

BY1 H. C. Brown and N. M. Yoon, *J. Am. Chem. Soc.* **88**, 1464 (1966).

C1 S. K. Chung, *J. Org. Chem.* **45**, 3513 (1980); S. K. Chung and F. F. Chung, *Tetrahedron Lett.,* 2473 (1979).

C2 S. C. Chen, *Synthesis,* 691 (1974).

C3 M. Cherest, *Tetrahedron* **36**, 1593 (1980).

C4 J. A. Cowan, *Tetrahedron Lett.* **27**, 1205 (1986).

C5 J. S. Cha, *Org. Prep. Proc. Int.* **21,** 453 (1989).

C6 A. S. Cieplak, *J. Am. Chem. Soc.* **103**, 4540 (1981).

CA1 D. L. Comins and A. H. Abdullah, *J. Org. Chem.* **49**, 3392 (1984).

CA2 D. J. Cram and F. A. Abd Elhafez, *J. Am. Chem. Soc.* **74**, 5828 (1952).

CA3 E. J. Corey, M. Azimioara, and S. Sarshar, *Tetrahedron Lett.* **33**, 3429 (1992).

CB1 B. Caro, B. Boyer, G. Lamaty, and G. Jaouen, *Bull. Soc. Chim. Fr. II,* 281 (1983).

CB2 E. J. Corey, R. K. Bakshi, and S. Shibata, *J. Am. Chem. Soc.* **109**, 5551 (1987).

CB3 E. J. Corey, R. K. Bakshi, S. Shibata, C. P. Chen, and V. K. Singh, *J. Am. Chem. Soc.* **109**, 7925 (1987).

CB4 E. J. Corey and R. K. Bakshi, *Tetrahedron Lett.* **31**, 611 (1990).

CB5 J. S. Cha and H. C. Brown, *J. Org. Chem.* **58**, 4727 (1993); *Org. Prep. Proc. Int.* **26**, 459 (1994).

CB6 J. S. Cha and H. C. Brown, *J. Org. Chem.* **58**, 4732 (1993).

CB7 J. S. Cha and H. C. Brown, *J. Org. Chem.* **58**, 3974 (1993).

CB8 J. Cossy, V. Bellosta, and M. C. Müller, *Tetrahedron Lett.* **33**, 5045 (1992).

CC1 P. C. M. Chan and J. M. Chong, *J. Org. Chem.* **53**, 5584 (1988).

CC2 M. Chan, K. Cheng, K. M. Ho, C. T. Ng, T. M. Yam, B. S. L. Wang, and T. Luh, *J. Org. Chem.* **53**, 4466 (1988).

CC3 E. J. Corey, C. P. Chen, and G. A. Reichard, *Tetrahedron Lett.* **30**, 5547 (1989).

CC4 P. C. M. Chan and J. M. Chong, *J. Org. Chem.* **53**, 5584 (1988).

CC5 M. Chini, P. Crotti, L. A. Flippin, and F. Macchia, *J. Org. Chem.* **55**, 4265 (1990).

CC6 B. T. Cho and Y. S. Chun, *Tetrahedron Asymmetry* **3**, 1539 (1992).

CC7 T. Y. Chau, L. J. Chen, and Y. Qin, *Tetrahedron Asymmetry* **6**, 1221 (1995).

CC8 M. Colombini, P. Crotti, V. Di Bussolo, L. Favero, C. Gardelli, F. Macchia, and M. Pineschi, *Tetrahedron* **51**, 8089 (1995).

CC9 B. T. Cho and Y. S. Chun, *J. Chem. Soc. Perkin Trans.* I, 3200 (1990); *Tetrahedron Asymmetry* **3**, 1583 (1992).

CC10 J. S. Cha, S. W. Chang, O. O. Kwon, and J. M. Kim, *Synlett,* 165 (1996).

CC11 J. M. Campelo, R. Chakraborty, and J. M. Marinas, *Synth. Comm.* **26**, 415 (1996).

CD1 M. C. Carreno, E. Dominguez, J. L. Garcia-Ruano, and A. Rubio, *J. Org. Chem.* **52**, 3619 (1987).

CE1 Y. M. Choi, R. W. Emblidge, N. Kucharczyk, and R. D. Sofia, *J. Org. Chem.* **54**, 1194 (1989).

CE2 Y. M. Choi and R. W. Emblidge, *J. Org. Chem.* **54**, 1198 (1989).

CE3 C. Caze, N. El Moualij, P. Hodge, C. J. Lock, and J. Ma, *J. Chem. Soc. Perkin Trans.* I, 345 (1995).

CF1 O. Chiodi, F. Fotiadu, M. Sylvestre, and G. Buono, *Tetrahedron Lett.* **37**, 39 (1996).

CG1 R. J. P. Corriu, C. Guérin, and J. J. E. Moreau, *Top. Stereochem.* **15**, 45 (1984).

CG2 R. J. P. Corriu and C. Guérin, *Adv. Organomet. Chem.* **20**, 265 (1982).

CG3 R. J. P. Corriu and C. Guérin, *J. Organomet. Chem.* **144**, 165 (1978).

CG4 K. M. Chen, K. G. Gunderson, G. E. Hardtmann, K. Prasad, O. Repic, and M. J. Shapiro, *Chem. Lett.,* 1923 (1987).

CG5 M. C. Carreno, J. L. Garcia-Ruano, A. M. Martin, C. Pedregal, J. H. Rodriguez, A. Rubio, J. Sanchez, and G. Solladié, *J. Org. Chem.* **55**, 2120 (1990).

CG6 E. J. Corey and A. V. Gavai, *Tetrahedron Lett.* **29**, 3201 (1988).

CG7 D. Calmes, L. Gorrichon-Guigon, P. Maroni, A. Accary, R. Barret, and J. Huet, *Tetrahedron* **37**, 879 (1981).

CG8 A. B. Charette and A. Giroux, *Tetrahedron Lett.* **37**, 6669 (1996).

CH1 K. M. Chen, G. E. Hardtmann, K. Prasad, O. Repic, and M. J. Shapiro, *Tetrahedron Lett.* **28**, 155 (1987).

CH2 D. L. Comins and J. J. Herrick, *Tetrahedron Lett.* **25**, 1321 (1984).

CH3 S. K. Chung and G. Han, *Synth. Comm.* **12**, 903 (1982).

CH4 R. D. Clark and C. H. Heathcock, *Tetrahedron Lett.,* 2027 (1974).

CH5 J. M. Coxon, K. N. Houk, and R. T. Luibrand, *J. Org. Chem.* **60**, 418 (1995).

CH6 E. J. Corey and C. J. Helal, *Tetrahedron Lett.* **36**, 9153 (1995).

CH7 E. J. Corey and C. J. Helal, *Tetrahedron Lett.* **37**, 4837 (1996).

CH8 E. J. Corey and C. J. Helal, *Tetrahedron Lett.* **37**, 5675 (1996).

CJ1 J. S. Cha, M. K. Jeoung, J. M. Kim, O. O. Kwon, and J. C. Lee, *Org. Prep. Proc. Int.* **26**, 583 (1994).

CJ2 J. M. Chong and J. Johannsen, *Tetrahedron Lett.* **35**, 7197 (1994).

CK1 J. S. Cha and S. S. Kwon, *J. Org. Chem.* **52**, 5487 (1987).

CK2 J. S. Cha, J. E. Kim, S. Y. Oh, and J. D. Kim, *Tetrahedron Lett.* **28**, 4575 (1987).

CK3 J. S. Cha and S. S. Kwon, *J. Org. Chem.* **55**, 1690 (1990).

CK4 D. J. Conn, J. J. Kaminski, D. M. Solomon, and A. T. McPhall, *J. Org. Chem.* **53**, 3265 (1988).

CK5 J. S. Cha, J. M. Kim, M. K. Jeoung, O. O. Kwon, and E. J. Kim, *Org. Prep. Proc. Int.* **27**, 95 (1995).

CK6 E. J. Corey and H. Kigoshi, *Tetrahedron Lett.* **32**, 5025 (1991).

CL1 R. C. Cookson and N. J. Liverton, *J. Chem. Soc. Perkin Trans.* I, 1589 (1985).

CL2 E. J. Corey and J. O. Link, *Tetrahedron Lett.* **30**, 6275 (1989).

CL3 D. L. Comins and D. H. Lamunyon, *Tetrahedron Lett.* **30**, 5053 (1989).

CL4 J. M. Coxon and R. T. Luibrand, *Tetrahedron Lett.* **34**, 7093 (1993).

CL5 J. M. Coxon and R. T. Luibrand, *Tetrahedron Lett.* **34**, 7097 (1993).

CL6 E. J. Corey, J. O. Link, and R. K. Bakshi, *Tetrahedron Lett.* **33**, 7107 (1992).

CL7 E. J. Corey and J. O. Link, *Tetrahedron Lett.* **33**, 3431 (1992).

CL8 E. J. Corey and J. O. Link, *Tetrahedron Lett.* **33**, 4141 (1992).

CL9 E. J. Corey and J. O. Link, *J. Am. Chem. Soc.* **114**, 1906 (1992).

CL10 E. J. Corey and J. O. Link, *J. Org. Chem.* **56**, 442 (1991).

CL11 P. R. Carlier, K. M. Lo, M. M-C. Lo, and I. D. Williams, *J. Org. Chem.* **60**, 7511 (1995).

CM1 J. M. Chong and E. K. Mar, *J. Org. Chem.* **56**, 893 (1991).

CM2 G. Conole, R. J. Mears, H. De Silva, and A. Whiting, *J. Chem. Soc. Perkin Trans.* I, 1825 (1995).

CN1 Z. Cai, B. Nassim, and R. Crabbé, *J. Chem. Soc. Perkin Trans.* I, 1573 (1983).

CN2 H. Chikashita, T. Nikaya, H. Uemura, and K. Itoh, *Bull. Chem. Soc. Japan* **62**, 2121 (1989).

CO1 J. S. Cha, S. Y. Oh, K. W. Lee, M. S. Yoon, and J. C. Lee, *Heterocycles* **27**, 1595 (1988).

CP1 M. P. Cooke and T. M. Parlman, *J. Org. Chem.* **40**, 531 (1975).

CP2 M. Cherest and N. Prudent, *Tetrahedron* **36**, 1599 (1980).

CP3 P. Chautemps and J. L. Pierre, *Tetrahedron* **32**, 549 (1976).

CP4 D. Calvo, M. Port, B. Delpech, and R. Lett, *Tetrahedron Lett.* **37**, 1023 (1996).

CP5 C. Cimarelli, G. Palmieri, and G. Bartoli, *Tetrahedron Asymmetry* **5**, 1455 (1994); G. Bartoli, C. Cimarelli, E. Marcantoni, G. Palmieri, and M. Petrini, *J. Org. Chem.* **59**, 5328 (1994).

CR1 A. R. Chamberlin and S. H. Reich, *J. Am. Chem. Soc.* **107**, 1440 (1975).

CR2 E. J. Corey and K. S. Rao, *Tetrahedron Lett.* **32**, 4623 (1991).

CS1 G. T. Crisp and W. J. Scott, *Synthesis,* 335 (1985).

CS2 C. L. Cavallaro and J. Schwartz, *J. Org. Chem.* **61**, 3863 (1996).

CT1 G. Chauvière, B. Tchoubar, and Z. Welvart, *Bull. Soc. Chim. Fr.,* 1426 (1963).

CT2 C. K. Cheung, L. T. Tseng, M. H. Lin, S. Srivastava, and W. T. Le Noble, *J. Am. Chem. Soc.* **108**, 1598 (1986).

CT3 A. S. Cieplak, B. D. Tait, and C. R. Johnson, *J. Am. Chem. Soc.* **111,** 8447 (1989).

CT4 D. Cai, D. Tschaen, Y. J. Shi, T. R. Verhoeven, R. A. Reamer, and A. W. Douglas, *Tetrahedron Lett.* **34**, 3243 (1993).

CU1 C. Camacho, G. Uribe, and R. Contreras, *Synthesis,* 1027 (1982).

CW1 J. Chen, K. A. Waiman, M. A. Belshe, and M. Di Mare, *J. Org. Chem.* **59**, 523 (1994).

CW2 K. Chibale and S. Warren, *Tetrahedron Lett.* **35**, 3991 (1994).

CW3 J. Clayden and S. Warren, *Angew. Chem. Int. Ed. Engl.* **35**, 241 (1996).

CY1 J. S. Cha, M. S. Yoon, Y. S. Kim, and K. W. Lee, *Tetrahedron Lett.* **29**, 1069 (1988).

CY2 J. Choi and N. M. Yoon, *Synthesis,* 597 (1996).

D1 R. E. Doolittle, *Org. Prep. Proc. Int.* **13**, 179 (1981).

D2 P. Deslongchamps, *Stereoelectronic Effects in Organic Chemistry* (Pergamon Press, Oxford, 1983), pp. 212, 213.

D3 M. DiMare, *J. Org. Chem.* **61**, 8378 (1996).

DA1 W. G. Dauben and J. W. Ashmore, *Tetrahedron Lett.,* 4487 (1978).

DA2 D. Dharanipragada, A. Alarcon, and V. J. Hruby, *Org. Prep. Proc. Int.* **23**, 396 (1991).

DA3 S. E. Denmark and J. Amburgey, *J. Am. Chem. Soc.* **115**, 10386 (1993).

DB1 O. Dirat, T. Berranger, and Y. Langlois, *Synlett,* 935 (1995).

DC1 E. V. Dehmlow and R. Cyrankiewicz, *J. Chem. Res.* (S), 24 (1990).

DC2 J. Das and S. Chandrasekaran, *Synth. Comm.* **20**, 907 (1990).

DD1 S. J. Danishefsky, M. P. De Ninno, and S. Chen, *J. Am. Chem. Soc.* **110**, 3929 (1988).

DD2 M. P. De Ninno, S. J. Danishefsky, and G. Schulte, *J. Am. Chem. Soc.* **110**, 3295 (1988).

DD3 P. Duhamel, L. Duhamel, and J. Gralak, *Tetrahedron Lett.,* 2329 (1972).

DD4 S. G. Davies, G. J.-M. Doisneau, J. C. Prodger, and H. J. Sanganee, *Tetrahedron Lett.* **35**, 2373 (1994).

DF1 P. N. Davey, G. W. J. Fleet, and P. J. C. Harding, *J. Chem. Res.* (S), 336 (1981).

DF2 L. Dubois, J. C. Fiaud, and H. B. Kagan, *Tetrahedron Asymmetry* **6**, 1097 (1995).

DG1 J. Drouin, S. Gauthier, O. Patricola, P. Lantéri, and R. Longeray, *Synlett,* 791 (1993).

DG2 T. J. Donohoe, R. Garg, and C. A. Stevenson, *Tetrahedron Asymmetry* **7**, 317 (1996).

DJ1 S. E. Denmark and T. K. Jones, *J. Org. Chem.* **47**, 4595 (1982).

DK1 A. R. Daniewski, J. Kiegel, E. Piotrowska, T. Warchol, and W. Wojciechowska, *Annalen,* 593 (1988); A. R. Daniewski, J. Kiegel, *Synth. Comm.* **18**, 115 (1988).

DK2 P. Delair, A. M. Kanazawa, M. B. M. de Avezedo, and A. E. Greene, *Tetrahedron Asymmetry* **7**, 2707 (1996).

DK3 P. Dosa, I. Kronish, J. McCallum, J. Schwartz, and M. C. Barden, *J. Org. Chem.* **61**, 4886 (1996).

DK4 U. P. Dhokte, S. V. Kulkarni, and H. C. Brown, *J. Org. Chem.* **61**, 5140 (1996).

DL1 L. Dai, B. Lou, Y. Zhang, and G. Guo, *Tetrahedron Lett.* **27**, 4343 (1986).

DL2 C. Denis, B. Laignel, D. Plusquellec, J. Y. Le Marouille, and A. Botrel, *Tetrahedron Lett.* **37**, 53 (1996).

DL3 E. Didier, B. Loubinoux, G. M. Ramos Tombo, and G. Rihs, *Tetrahedron* **47**, 4941 (1991).

DM1 J. I. Dickstein and S. I. Miller, *The Chemistry of the Carbon–Carbon Triple Bond, Part 2,* S. Patai, Ed. (John Wiley and Sons, Chichester, 1977), p. 853.

DN1 S. E. Denmark, N. Nakajima, O. J.-C. Nicaise, A.-M. Faucher, and J. P. Edwards, *J. Org. Chem.* **60**, 4884 (1995).

DN2 E. V. Dehmlow, T. Niemann, and A. Kraft, *Synth. Comm.* **26**, 1467 (1996).

DR1 V. H. Dahanukar and S. D. Rychnovsky, *J. Org. Chem.* **61**, 8317 (1996).

DR2 P. Deprez, J. Royer, and H.-P. Husson, *Tetrahedron* **49**, 3781 (1993).

DS1 G. Durrant and J. K. Sutherland, *J. Chem. Soc. Perkin Trans.* I, 2582 (1972).

DS2 N. De Kimpe, G. Staneova, R. Verhe, and N. Schamp, *Synthesis,* 587 (1988).

DS3 S. E. Denmark, J. A. Sternberg, and R. Lueoend, *J. Org. Chem.* **53**, 1251 (1988).

DS4 W. A. Donaldson and L. Shang, *Tetrahedron Lett.* **36**, 1575 (1995).

DS5 L. Deloux and M. Srebnik, *Chem. Rev.* **93**, 763 (1993).

DS6 S. E. Denmark, M. E. Schnute, L. R. Marcin, and A. Thorarensen, *J. Org. Chem.* **60**, 3205 (1995).

DS7 B. Di Simone, D. Savoia, E. Tagliavini, and A. Umani-Ronchi, *Tetrahedron Asymmetry* **6**, 301 (1995).

DT1 A. W. Douglas, D. M. Tschaen, R. A. Reamer, and Y.-J. Shi, *Tetrahedron Asymmetry* **7**, 1303 (1996).

DT2 S. E. Denmark, A. Thorarensen, and D. S. Middleton, *J. Am. Chem. Soc.* **118**, 8266 (1996).

DW1 A. R. Daniewski and W. Wojciechowska, *J. Org. Chem.* **47**, 2993 (1982).

E1 E. L. Eliel, in *Asymmetric Synthesis, Vol. 2,* J. D. Morisson, Ed. (Academic Press, New York, 1983), p. 125.

E2 E. L. Eliel, *Record Chem. Prog.* **22**, 129 (1961).

EB1 E. L. Eliel, V. G. Badding, and M. N. Rerick, *J. Am. Chem. Soc.* **84**, 2371 (1961).

EB2 D. A. Evans and J. Bartroli, *Tetrahedron Lett.* **23**, 807 (1982).

EB3 D. Entwistle, P. Boehm, R. A. W. Johnstone, and R. P. Telford, *J. Chem. Soc. Perkin Trans.* I, 27 (1980).

EB4 G. Etemad-Moghadam, M. C. Benhamou, V. Speziale, A. Lattes, and A. Bielawska, *N. J. Chim.* **4**, 727 (1980).

EB5 D. A. Evans, S. L. Bender, and J. Morris, *J. Am. Chem. Soc.* **110**, 2506 (1988).

EC1 D. A. Evans and K. T. Chapman, *Tetrahedron Lett.* **27**, 5939 (1986).

EC2 D. A. Evans, K. T. Chapman, and E. M. Carreira, *J. Am. Chem. Soc.* **110**, 3560 (1988).

ED1 E. L. Eliel and D. W. Del Monte, *J. Am. Chem. Soc.* **80**, 1744 (1958).

ED2 D. A. Evans, M. J. Dart, and J. L. Duffy, *Tetrahedron Lett.* **35**, 8541 (1994).

EE1 D. A. Evans, M. D. Ennis, and D. J. Mathre, *J. Am. Chem. Soc.* **104**, 1737 (1982).

EF1 E. L. Eliel, S. V. Frye, E. R. Hortelano, X. Chen, and X. Bai, *Pure Appl. Chem.* **63**, 1591 (1991).

EF2 D. A. Evans and G. C. Fu, *J. Org. Chem.* **55**, 5678 (1990).

EG1 A. J. Elliott and H. Guzik, *Tetrahedron Lett.* **23**, 1983 (1982).

EG2 D. A. Evans, J. A. Gauchet-Prunet, E. M. Carreira, and A. B. Charette, *J. Org. Chem.* **56**, 741 (1991).

EG3 D. A. Evans, J. R. Gage, and J. L. Leighton, *J. Am. Chem. Soc.* **114**, 9434 (1992).

EH1 J. Elliott, D. Hall, and S. Warren, *Tetrahedron Lett.* **30**, 601 (1989).

EH2 E. L. Eliel and X. C. He, *J. Org. Chem.* **55**, 2114 (1990); X. C. He and E. L. Eliel, *Tetrahedron* **43**, 4973 (1987).

EH3 D. A. Evans and A. H. Hoveyda, *J. Org. Chem.* **55**, 5190 (1990).

EK1 D. Enders, M. Knopp, J. Runsink, and G. Raabe, *Annalen,* 1095 (1996).

ER1 D. A. Evans, A. M. Ratz, B. E. Huff, and G. S. Sheppard, *J. Am. Chem. Soc.* **117**, 3448 (1995).

ES1 O. Eisenstein, H. B. Schlegel, and M. M. Kayser, *J. Org. Chem.* **47**, 2886 (1982).

ES2 D. A. Evans, E. B. Sjogren, J. Bartroli, and R. L. Dow, *Tetrahedron Lett.* **27**, 4957 (1986).

EW1 J. Elliott and S. Warren, *Tetrahedron Lett.* **27**, 645 (1986).

F1 I. Fleming, in *Comprehensive Organic Chemistry, Vol. 3,* D. Barton and W. Ollis, Eds., 1979, p. 541.

FB1 J. C. Fuller, C. M. Belisle, C. T. Goralski, and B. Singaram, *Tetrahedron Lett.* **35**, 5389 (1994).

FB2 J. M. J. Fréchet, E. Bald, and P. Lecavalier, *J. Org. Chem.* **51**, 3462 (1986).

FB3 C. A. M. Fraga and E. J. Barreiro, *Synth. Comm.* **25**, 1133 (1995).

FC1 J. A. Fehrentz and B. Castro, *Synthesis,* 676 (1983).

FC2 A. Fadel, J. L. Canet, and J. Salaun, *Tetrahedron Lett.* **30**, 6687 (1989).

FC3 J. A. Fehrentz, J. C. Califano, M. Amblard, A. Loffet, and J. Martinez, *Tetrahedron Lett.* **35**, 369 (1994).

FD1 O. Froelich, P. Desos, M. Bonin, J.-C. Quirion, and H.-P. Husson, *J. Org. Chem.* **61**, 6700 (1996).

FF1 G. B. Fisher, J. C. Fuller, J. Harrison, C. T. Goralski, and B. Singaram, *Tetrahedron Lett.* **34**, 1091 (1993).

FF2 G. B. Fisher, J. C. Fuller, J. Harrison, S. G. Alvarez, E. R. Burkhardt, C. T. Goralski, and B. Singaram, *J. Org. Chem.* **59**, 6378 (1994).

FG1 L. A. Flippin, D. W. Gallagher, and K. Jalali-Araghi, *J. Org. Chem.* **54**, 1430 (1989).

FH1 G. W. J. Fleet and P. J. C. Harding, *Tetrahedron Lett.* **22**, 675 (1981).

FH2 G. W. J. Fleet, P. J. C. Harding, and M. J. Whitcombe, *Tetrahedron Lett.* **21**, 4031 (1980).

FH3 J. A. Fehrentz, A. Heitz, and B. Castro, *Int. J. Peptide Protein Res.* **26**, 236 (1985).

FH4 F. A. Davis, M. S. Haque, and R. M. Przeslawski, *J. Org. Chem.* **54**, 2021 (1989).

FH5 G. B. Fisher, J. Harrison, J. C. Fuller, C. T. Goralski, and B. Singaram, *Tetrahedron Lett.* **33**, 4533 (1992).

FK1 J. M. Finan and Y. Kishi, *Tetrahedron Lett.* **23**, 2719 (1982).

FK2 R. Frenette, M. Kakushima, R. Zamboni, R. N. Young, and T. R. Verhoeven, *J. Org. Chem.* **52**, 304 (1987).

FK3 T. Fujisawa, E. Kojima, T. Itoh, and T. Sato, *Tetrahedron Lett.* **26**, 6089 (1985).

FK4 G. Frenking, K. F. Kohler, and M. T. Reetz, *Tetrahedron* **47**, 8991 and 9005 (1992).

FM1 A. H. Fray and A. I. Meyers, *J. Org. Chem.* **61**, 3362 (1996).

FO1 H. Fujii, K. Oshima, and K. Utimoto, *Tetrahedron Lett.* **32**, 6147 (1991).

FR1 S. P. Forsey, D. Rajapaksa, N. J. Taylor, and R. Rodrigo, *J. Org. Chem.* **54**, 4280 (1989).

FR2 R. Fargues-Sakellariou, M. Rivière, and A. Lattes, *N. J. Chim.* **9**, 95 (1985).

FS1 J. C. Fuller, E. L. Stangeland, T. C. Jackson, and B. Singaram, *Tetrahedron Lett.* **35**, 1515 (1994).

FS2 J. C. Fuller, E. L. Stangeland, C. T. Goralski, and B. Singaram, *Tetrahedron Lett.* **34**, 257 (1993).

FS3 C. Franot, G. B. Stone, P. Engeli, C. Spöndlin, and E. Waldvogel, *Tetrahedron Asymmetry* **6**, 2755 (1995).

G1 B. Ganem, *J. Org. Chem.* **40**, 2846 (1975); J. M. Fortunato and B. Ganem, *J. Org. Chem.* **41**, 2194 (1976).

G2 B. Giese, *Angew. Chem. Int. Ed. Engl.* **24**, 553 (1985); *Radicals in Organic Synthesis: Formation of Carbon–Carbon Bonds* (Pergamon Press, Oxford, 1986).

G3 T. W. Greene, *Protective Groups in Organic Synthesis* (Wiley, New York, 1981), p. 297.

G4 J. Gorzynski Smith, *Synthesis,* 629 (1984).

G5 N. Greeves, in *Comprehensive Organic Chemistry,* B. M. Trost and I. Fleming, Eds. (Pergamon Press, London, 1991), Vol. 8, Chapter 1.1.

G6 T. Gadja, *Tetrahedron Asymmetry* **5**, 1965 (1994).

GB1 G. Guanti, L. Banfi, E. Narisano, and C. Scolastico, *Tetrahedron Lett.* **25**, 4693 (1984).

GB2 R. B. Gammil, L. T. Bell, and S. A. Nash, *J. Org. Chem.* **49**, 3039 (1984).

GB3 G. Guanti, L. Banfi, E. Narisano, and C. Scolastico, *Tetrahedron* **44**, 3671 (1988).

GB4 G. Guanti, L. Banfi, A. Guaragna, and E. Narisano, *J. Chem. Soc. Perkin Trans.* I, 2369 (1988).

GB5 H. W. Gibson and F. C. Bailey, *J. Chem. Soc. Chem. Comm.,* 815 (1977).

GB6 O. W. Gooding and R. P. Bansai, *Synth. Comm.* **25**, 1155 (1995).

GB7 G. Guanti, L. Banfi, R. Riva, and M. T. Zannetti, *Tetrahedron Lett.* **34**, 5483 (1993).

GB8 G. Guanti, L. Banfi, and R. Riva, *Tetrahedron* **51**, 10343 (1995).

GB9 E. H. Gould and E. Babad, *J. Org. Chem.* **37**, 2208 (1972).

GC1 A. K. Ghosh and Y. Chen, *Tetrahedron Lett.* **36**, 6811 (1995).

GC2 A. G. Giumani, G. Chiavari, M. M. Musiani, and P. Rossi, *Synthesis,* 743 (1980).

GD1 G. Giffels, C. Dreisbach, U. Kragl, M. Weigerding, H. Waldmann, and C. Wandrey, *Angew. Chem. Int. Ed. Engl.* **34**, 2005 (1995).

GE1 W. C. Guida, E. E. Entreken, and A. R. Guida, *J. Org. Chem.* **49**, 3024 (1984).

GF1 G. W. Gribble and D. C. Ferguson, *J. Chem. Soc. Chem. Comm.,* 535 (1975).

GH1 E. R. Grandbois, S. I. Howard, and J. D. Morrison, in *Asymmetric Synthesis,* Vol. 2, J. Morrison, Ed. (Academic Press, New York, 1983), p. 71.

GJ1 G. W. Gribble, J. L. Johnson, and M. G. Saulnier, *Heterocycles* **16**, 2109 (1981).

GJ2 G. W. Gribble, J. M. Jasinski, J. T. Pellicone, and J. A. Panetta, *Synthesis,* 766 (1978).

GK1 G. W. Gribble, W. J. Kelly, and S. E. Emery, *Synthesis,* 763 (1978).

GK2 T. Goto, M. Konno, M. Saito, and T. Sato, *Bull. Chem. Soc. Japan* **62**, 1205 (1989).

GL1 A. L. Gemal and J. L. Luche, *J. Org. Chem.* **44**, 4187 (1979); J. L. Luche and A. L. Gemal, *J. Am. Chem. Soc.* **101**, 5848 (1979).

GL2 A. L. Gemal and J. L. Luche, *J. Am. Chem. Soc.* **103**, 5454 (1981), and references cited.

GL3 A. K. Ganguly, Y. T. Liu, and O. Sarre, *J. Chem. Soc. Chem. Comm.,* 1166 (1983).

GL4 V. Gevorgyan and E. Lukevics, *J. Chem. Soc. Chem. Comm.,* 1234 (1985).

GL5 M. Gonzales-Sierra, M. Laborde, and E. A. Ruveda, *Synth. Comm.* **17**, 431 (1987).

GL6 J. T. Gupton and W. J. Layman, *J. Org. Chem.* **52**, 3683 (1987).

GL7 A. K. Ghosh and W. Liu, *J. Org. Chem.* **60**, 6198 (1995).

GL8 J. L. Garcia Ruano, A. Lorente, and J. H. Rodriguez, *Tetrahedron Lett.* **33**, 5637 (1992).

GM1 G. Gondos, L. G. McGirr, C. R. Jablonski, W. Snedden, and J. C. Orr, *J. Org. Chem.* **53**, 3057 (1988).

GM2 G. J. Mc Garvey, J. A. Mathys, K. J. Wilson, K. R. Overly, P. T. Buonora, and P. Grant-Spoors, *J. Org. Chem.* **60**, 7778 (1995).

GM3 G. J. Mc Garvey, J. M. Williams, R. N. Hiner, Y. Matsubara, and T. Oh, *J. Am. Chem. Soc.* **108**, 4943 (1986).

GM4 T. G. Gant and A. I. Meyers, *Tetrahedron* **50**, 2297 (1994).

GN1 G. W. Gribble and C. F. Nutaitis, *Org. Prep. Proc. Int.* **17**, 317 (1985).

GN2 G. W. Gribble, C. F. Nutaitis, and R. M. Leese, *Heterocycles* **22**, 379 (1984).

GN3 G. W. Gribble and C. F. Nutaitis, *Synthesis,* 709 (1987).

GO1 G. Gondos and J. C. Orr, *J. Chem. Soc. Chem. Comm.,* 1239 (1982).

GO2 B. Ganem and J. O. Osby, *Chem. Rev.* **86**, 763 (1986).

GP1 J. L. Garcia-Ruano, C. Pedregal, and J. H. Rodriguez, *Tetrahedron* **43**, 4407 (1987).

GP2 N. Galeotti, J. Poncet, L. Chiche, and P. Jouin, *J. Org. Chem.* **58**, 5370 (1993).

GR1 L. Guerrier, J. Royer, D. Grierson, and H. P. Husson, *J. Am. Chem. Soc.* **105**, 7754 (1983).

GR2 D. S. Grierson, J. Royer, L. Guerrier, and H. P. Husson, *J. Org. Chem.* **51**, 4475 (1986).

GS1 R. B. Gammil, D. M. Sobieray, and P. M. Gold, *J. Org. Chem.* **46**, 3555 (1981).

GS2 A. Giannis and K. Sandhoff, *Angew. Chem. Int. Ed. Engl.* **28**, 218 (1989).

GS3 Y. Gao and K. B. Sharpless, *J. Am. Chem. Soc.* **110**, 7538 (1988).

GS4 Y. Gao and K. B. Sharpless, *J. Org. Chem.* **53**, 4081 (1988).

GT1 H. L. Goering and C. C. Tseng, *J. Org. Chem.* **46**, 5250 (1981).

GW1 S. H. Graham and A. J. S. Williams, *Tetrahedron* **21**, 3263 (1965).

GW2 N. Greeves and S. Warren, *Tetrahedron Lett.* **27**, 259 (1986).

GW3 N. M. Goudgaon, P. P. Wadgaonkar, and G. W. Kabalka, *Synth. Comm.* **19**, 805 (1989).

GW4 G. J. McGarvey, J. M. Williams, R. N. Hiner, Y. Matsubara, and T. Oh, *J. Am. Chem. Soc.* **108**, 4943 (1986).

GZ1 B. W. Gung, Z. Zhu, and D. A. Mareska, *J. Org. Chem.* **58**, 1367 (1993).

H1 T. L. Ho, *Synth. Comm.* **12**, 339 (1982).

H2 A. Hajos, *Complex Hydrides* (Elsevier, Amsterdam, 1979).

H3 M. Hudlicky, *Reductions in Organic Chemistry* (Ellis Horwood Ltd., Chichester, 1984).

H4 C. H. Heathcock, in *Comprehensive Organic Chemistry,* B. M. Trost and I. Fleming, Eds. (Pergamon Press, London, 1991), Vol. 2, Chapter 1.6.

H5 R. W. Hoffmann, in *Stereocontrolled Organic Synthesis,* B. M. Trost, Ed. (Blackwell Science, Oxford, 1994), p. 259.

HA1 J. Harrison, S. G. Alvarez, G. Godjoian, and B. Singaram, *J. Org. Chem.* **59,** 7193 (1994).

HA2 R. O. Hutchins, J. Adams, and M. C. Rutledge, *J. Org. Chem.* **60**, 7396 (1995).

HA3 R. O. Hutchins, A. Abdel-Magid, Y. P. Stercho, and A. Wambsgans, *J. Org. Chem.* **52**, 704 (1987).

HB1 F. A. Hicks, S. C. Berk, and S. L. Buchwald, *J. Org. Chem.* **61**, 2713 (1996).

HC1 R. O. Hutchins and F. Cistone, *Org. Prep. Proc. Int.* **13**, 225 (1981).

HC2 D. Hernanz, F. Camps, A. Guerrero, and A. Delgado, *Tetrahedron Asymmetry* **6**, 2291 (1995).

HD1 T. Hamada, K. Daikai, R. Irie, and T. Katsuki, *Tetrahedron Asymmetry* **6**, 2441 (1995).

HE1 T. D. Hubert, D. P. Eyman, and D. F. Wiemer, *J. Org. Chem.* **49**, 2279 (1984).

HF1 J. Harrison, J. C. Fuller, C. T. Goralski, and B. Singaram, *Tetrahedron Lett.* **35**, 5201 (1994).

HG1 M. Hedayatullah and A. Guy, *Synthesis,* 357 (1978).

HG2 Y. Hong, Y. Gao, X. Nie, and C. M. Zepp, *Tetrahedron Lett.* **35**, 5551 and 6631 (1994).

HH1 P. F. Hudrlik and A. M. Hudrlik, in *The Chemistry of the Carbon–Carbon Triple Bond,* S. Patai, Ed. (Wiley, Chichester, 1978), pp. 199–219.

HH2 R. O. Hutchins, D. Hoke, J. Heogh, and D. Koharski, *J. Org. Chem.* **36**, 1568 (1971).

HH3 R. O. Hutchins and M. G. K. Hutchins, in *Reduction of Triple Bonded Groups,* S. Patai, Ed. (Wiley, Chichester, 1983), p. 571.

HH4 T. Hanamoto and T. Hiyama, *Tetrahedron Lett.* **29**, 6467 (1988).

HH5 D. J. Hart, W. P. Hong, and L. Y. Hsu, *J. Org. Chem.* **52**, 4665 (1987).

HI1 J. Hiratake, M. Inagaki, Y. Yamamoto, and J. Oda, *J. Chem. Soc. Perkin Trans.* I, 1053 (1987).

HJ1 B. J. Hussey, R. A. W. Johnstone, P. Boehm, and I. D. Entwistle, *Tetrahedron* **38**, 3769 (1982).

HJ2 L. M. Harwood and M. Julia, *Synthesis,* 456 (1980).

HJ3 D. A. Horne and A. Jordan, *Tetrahedron Lett.,* 1357 (1978).

HJ4 B. D. Harris and M. M. Joullié, *Tetrahedron* **44**, 3489 (1988).

HJ5 G. Hutton, T. Jolliff, H. Mitchell, and S. Warren, *Tetrahedron Lett.* **36**, 7905 (1995).

HK1 R. O. Hutchins, D. Kandasamy, C. A. Maryanoff, D. Masilamani, and B. E. Maryanoff, *J. Org. Chem.* **42**, 82 (1977).

HK2 R. O. Hutchins and D. Kandasamy, *J. Org. Chem.* **40**, 2530 (1975).

HL1 R. O. Hutchins and K. Learn, *J. Org. Chem.* **47**, 4380 (1982).

HL2 K. M. Ho, C. H. Lam, and T. Y. Luh, *J. Org. Chem.* **54**, 4474 (1989).

HL3 D. H. Hua, N. Lagneau, H. Wang, and J. Chen, *Tetrahedron Asymmetry* **6**, 349 (1995).

HL4 D. J. Hart and V. Leroy, *Tetrahedron* **51**, 5757 (1995).

HL5 D. H. Hua, N. M. Lagneau, Y. Chen, P. M. Robben, G. Clapham, and P. D. Robinson, *J. Org. Chem.* **61**, 4508 (1996).

HM1 R. O. Hutchins and M. Markowitz, *J. Org. Chem.* **46**, 3574 (1981).

HM2 R. O. Hutchins and M. Markowitz, *Tetrahedron Lett.* **21**, 813 (1980).

HM3 H. Haubenstock and T. Mester, *J. Org. Chem.* **48**, 945 (1983).

HM4 J. Huet, Y. Maroni-Barnaud, N. T. Anh, and J. Seyden-Penne, *Tetrahedron Lett.,* 159 (1976).

HM5 C. J. Helal, P. A. Magriotis, and E. J. Corey, *J. Am. Chem. Soc.* **118**, 10938 (1996).

HM6 D. R. J. Hose, M. F. Mahon, K. C. Molloy, T. Raynham, and M. Wills, *J. Chem. Soc. Perkin Trans.* I, 691 (1996).

HN1 R. O. Hutchins and N. R. Natale, *Org. Prep. Proc. Int.* **11**, 201 (1979).

HN2 T. Hirabe, M. Nojima, and S. Kusabayashi, *J. Org. Chem.* **49**, 4084 (1984).

HN3 R. O. Hutchins, N. R. Natale, and I. M. Taffer, *J. Chem. Soc. Chem. Comm.,* 1088 (1978).

HN4 M. Harre, K. Nickisch, and J. Westermann, *Tetrahedron Lett.* **34**, 3123 (1993).

HP1 H. Handel and J. L. Pierre, *Tetrahedron* **31**, 997 (1975); J. L. Pierre, H. Handel, and R. Perraud, *Tetrahedron* **31**, 2795 (1975).

HP2 Y. He, X. Pan, H. Zhao, and S. Wang, *Synth. Comm.* **19**, 3051 (1989).

HP3 H. N. Houk, M. N. Paddon-Row, N. G. Rondan, Y. D. Wu, F. K. Brown, D. C. Spellmeyer, J. T. Matz, Y. Li, and R. J. Loncharich, *Science* **231**, 1108 (1986).

HP4 D. Horton, J. P. Roski, and P. Norris, *J. Org. Chem.* **61**, 3783 (1996).

HR1 R. O. Hutchins, D. Rotstein, N. Natale, J. Fanelli, and D. Dimmel, *J. Org. Chem.* **41**, 3328 (1976).

HR2 R. O. Hutchins and M. C. Rutledge, *Tetrahedron Lett.* **28**, 5619 (1987).

HS1 R. O. Hutchins and W. Y. Su, *Tetrahedron Lett.* **25**, 695 (1984); R. O. Hutchins, W. Y. Su, R. Sivakumar, F. Cistone, and Y. P. Stercho, *J. Org. Chem.* **48**, 3412 (1983).

HS2 D. M. Hrubowchak and F. X. Smith, *Tetrahedron Lett.* **24**, 4951 (1983).

HS3 D. H. Hua, G. Sinai-Zingde, and S. Venkataraman, *J. Am. Chem. Soc.* **107**, 4089 (1985).

HS4 Y. Hamada, M. Shibata, T. Sugiura, S. Kato, and T. Shioiri, *J. Org. Chem.* **52**, 1252 (1987).

HS5 A. Heine and D. Stalke, *Angew. Chem. Int. Ed. Engl.* **31**, 854 (1992).

HS6 J. B. Hendrickson, M. Singer, and M. S. Hussoin, *J. Org. Chem.* **58**, 6913 (1993).

HS7 J. E. Hong, W. S. Shin, W. B. Jang, and D. Y. Oh, *J. Org. Chem.* **61**, 2199 (1996).

HT1 R. O. Hutchins, I. M. Taffer, and W. Burgoyne, *J. Org. Chem.* **46**, 5214 (1981).

HT2 D. J. Hart and Y. Tsai, *J. Org. Chem.* **47**, 4403 (1982).

HT3 C. Hoffman, R. S. Tanke, and M. J. Miller, *J. Org. Chem.* **54**, 3750 (1989).

HW1 K. N. Houk and Y. Wu, in *Stereochemistry of Organic and Bioorganic Transformations,* Vol. 17, W. Bartmann, K. B. Sharpless, Eds. (Verlag Chemie, Weinheim, 1987), p. 247.

HW2 J. Hatem and B. Waegell, *Tetrahedron* **46**, 2789 (1990).

HW3 K. J. Hodgetts, C. J. Wallis, and T. W. Wallace, *Synlett,* 1235 (1995).

HY1 M. Hayashi, T. Yoshiga, K. Nakatani, K. Ono, and N. Oguni, *Tetrahedron* **50**, 2821 (1994).

HZ1 Y. He, H. Zhao, X. Pan, and S. Wang, *Synth. Comm.* **19**, 3047 (1989).

I1 J. Ipaktschi, *Chem. Ber.* **117**, 856 (1984).

I2 J. Ipaktschi, *Chem. Ber.* **117**, 3320 (1984).

IA1 K. Iseki, D. Asada, M. Takahashi, T. Nagai, and Y. Kobayashi, *Tetrahedron Lett.* **36**, 3711 (1995).

IK1 Y. Ito, T. Katsuki, and M. Yamaguchi, *Tetrahedron Lett.* **26**, 4643 (1985).

IK2 K. Ishizumi, K. Koga, and S. Yamada, *Chem. Pharm. Bull.* **16**, 492 (1968).

IK3 T. Ishihara, M. Kuroboshi, K. Yamaguchi, and Y. Okada, *J. Org. Chem.* **55**, 3107 (1990).

IL1 E. d'Incan, A. Loupy, A. Restelli, J. Seyden-Penne, and P. Viout, *Tetrahedron* **38**, 1755 (1982).

IL2 E. d'Incan and A. Loupy, *Tetrahedron* **37**, 1171 (1981); E. d'Incan, A. Loupy, and A. Maia, *Tetrahedron Lett.* **22**, 941 (1981).

IM1 K. Ishihara, A. Mori, I. Arai, and H. Yamamoto, *Tetrahedron Lett.* **27**, 983 (1986).

IM2 K. Ishihara, A. Mori, and H. Yamamoto, *Tetrahedron Lett.* **28**, 6613 (1987).

IM3 K. Ishihara, A. Mori, and H. Yamamoto, *Tetrahedron* **46**, 4595 (1990).

IN1 S. Itsuno, M. Nakano, K. Miyazaki, H. Masuda, K. Ito, A. Hirao, and S. Nakahama, *J. Chem. Soc. Perkin Trans.* I, 2039 (1985).

IS1 S. Itsuno, Y. Sakurai, and K. Ito, *Synthesis,* 995 (1988).

IS2 S. Itsuno, Y. Sakurai, K. Shimizu, and K. Ito, *J. Chem. Soc. Perkin Trans.* I, 1548 (1989).

IS3 G. Iwasaki, M. Sano, M. Sodeoka, K. Yoshida, and M. Shibasaki, *J. Org. Chem.* **53**, 4864 (1988).

IT1 T. Iida, T. Tamura, T. Matsumoto, and F. C. Chang, *Synthesis,* 957 (1984).

IT2 T. Imai, T. Tamura, A. Yamamuro, T. Sato, T. A. Wollmann, R. M. Kennedy, and S. Masamune, *J. Am. Chem. Soc.* **108**, 7402 (1986).

IT3 M. Ihara, M. Toyota, K. Fukumoto, and T. Kametani, *J. Chem. Soc. Perkin Trans.* I, 2151 (1986).

IW1 S. Itsuno, T. Wakasugi, K. Ito, A. Hirao, and S. Nakahama, *Bull. Chem. Soc. Japan* **58**, 1669 (1985).

IY1 H. Iida, N. Yamazaki, and C. Kibayashi, *J. Org. Chem.* **51**, 3769 (1986).

IY2 H. Iida, N. Yamazaki, and C. Kibayashi, *J. Chem. Soc. Chem. Comm.,* 746 (1987).

IY3 T. Ishihara, K. Yamaguchi, M. Kuroboshi, and K. Utimoto, *Tetrahedron Lett.* **35**, 5263 (1994).

J1 S. Jarosz, *Carbohydr. Res.* **183**, 201 (1988).

JB1 V. Jager and V. Buss, *Annalen,* 101 (1980).

JB2 T. Janecki, R. Bodalski, M. Wieczorek, and G. Bujacz, *Tetrahedron* **51**, 1721 (1995).

JD1 S. Jegham and B. C. Das, *Tetrahedron Lett.* **30**, 2801 (1989).

JD2 T. K. Jones and S. E. Denmark, *Org. Synth. Coll. Vol. VII,* 524 (1990).

JF1 J. Jacques, C. Fouquey, and R. Viterbo, *Tetrahedron Lett.,* 4617 (1971).

JF2 S. Jegham, J. L. Fourrey, and B. C. Das, *Tetrahedron Lett.* **30**, 1959 (1989).

JG1 V. Jager, H. Grund, V. Buss, W. Schwab, I. Muller, R. Schohe, R. Franz, and R. Ehrler, *Bull. Soc. Chim. Belg.* **92**, 1039 (1983).

JJ1 W. R. Jackson, H. A. Jacobs, B. R. Matthews, G. S. Jayatilake, and K. G. Watson, *Tetrahedron Lett.* **31**, 1447 (1990).

JJ2 C. A. Jones, I. G. Jones, M. North, and C. R. Pool, *Tetrahedron Lett.* **36**, 7885 (1995).

JK1 J. C. Jochims, Y. M. Kobayashi, and E. Skrzelewski, *Tetrahedron Lett.,* 571 (1974).

JL1 D. K. Jones, D. C. Liotta, I. Shinkai, and D. J. Mathre, *J. Org. Chem.* **58**, 799 (1993).

JM1 G. Jaouen and A. Meyer, *J. Am. Chem. Soc.* **97**, 4667 (1975).

JM2 V. Jager, I. Muller, and E. F. Paulus, *Tetrahedron Lett.* **26**, 2997 (1985).

JM3 T. K. Jones, J. J. Mohan, L. C. Xavier, T. J. Blacklock, D. J. Mathre, P. Sohar, E. T. T. Jones, R. A. Reamer, F. E. Roberts, and E. J. J. Grabowski, *J. Org. Chem.* **56**, 763 (1991).

JS1 R. Johansson and B. Samuelsson, *J. Chem. Soc. Chem. Comm.,* 201 (1984).

JS2 V. Jager, W. Schwab, and V. Buss, *Angew. Chem. Int. Ed. Engl.* **20**, 601 (1981).

JS3 C. R. Johnson and C. J. Stark, *J. Org. Chem.* **47**, 1196 (1982).

JS4 I. Jabre, M. Saquet, and A. Thuillier, *J. Chem. Res. (S),* 106 (1990).

JT1 R. A. W. Johnstone and R. P. Telford, *J. Chem. Soc. Chem. Comm.,* 354 (1978).

JU1 M. E. Jung, Y. Usui, and C. T. Vu, *Tetrahedron Lett.* **28**, 5977 (1987).

JW1 S. A. Jaeger, M. D. Ward, and C. A. Martin, *Tetrahedron* **40**, 2691 (1984).

K1 S. Krishnamurthy, *J. Org. Chem.* **45**, 2550 (1980).

K2 Y. Kishi, *Pure Appl. Chem.* **53**, 1163 (1981); H. Nagaoka, W. Rutsch, G. Schmid, H. Ho, M. R. Johnson, and Y. Kishi, *J. Am. Chem. Soc.* **102**, 7962 (1980).

K3 Y. Kikugawa, *J. Chem. Res. (S),* 212 (1977); 184 (1978).

K4 S. Krishnamurthy, *J. Org. Chem.* **46**, 4628 (1981).

K5 G. Kotokos, *Synthesis,* 299 (1990).

KA1 S. Kim and K. H. Ahn, *J. Org. Chem.* **49**, 1717 (1984).

KA2 V. Kumar, A. Amann, G. Ourisson, and B. Luu, *Synth. Comm.* **17**, 1279 (1987).

KA3 S. Krishnamurthy and D. Aimino, *J. Org. Chem.* **54**, 4458 (1989).

KA4 M. Kaselj, J. L. Adcock, H. Luo, H. Zhang, H. Li, and W. J. Le Noble, *J. Am. Chem. Soc.* **117**, 7088 (1995).

KA5 N. M. Khan, V. Arumugan, and S. Balasubramanian, *Tetrahedron Lett.* **37**, 4819 (1996).

KB1 S. Krishnamurthy and H. C. Brown, *J. Org. Chem.* **41**, 3064 (1976).

KB2 S. Krishnamurthy and H. C. Brown, *J. Org. Chem.* **40**, 1834 (1975); **42**, 1197 (1977).

KB3 S. Krishnamurthy and H. C. Brown, *J. Org. Chem.* **47**, 276 (1982).

KB4 G. W. Kabalka and J. D. Baker, *J. Org. Chem.* **40**, 1834 (1975); G. W. Kabalka and S. T. Summers, *J. Org. Chem.* **46**, 1217 (1981).

KB5 S. Krishnamurthy and H. C. Brown, *J. Org. Chem.* **48**, 3085 (1983).

KB6 M. M. Kayser, L. Breau, S. Eliev, P. Morand, and H. S. Ip, *Can J. Chem.* **64**, 104 (1986).

KB7 G. W. Kabalka, J. D. Baker, and G. W. Neal, *J. Org. Chem.* **42**, 512 (1977).

KB8 S. Krishnamurthy and H. C. Brown, *J. Am. Chem. Soc.* **98**, 3383 (1976).

KE1 K.-Y. Ko and E. L. Eliel, *J. Org. Chem.* **51**, 5353 (1986).

KF1 Y. Kumar and L. Florvall, *Synth. Comm.* **13**, 489 (1983).

KF2 F. A. Kraus and K. Frazier, *J. Org. Chem.* **45**, 4262 (1980).

KF3 Y. Kawanami, I. Fujita, Y. Taniguchi, T. Katsuki, and M. Yamaguchi, *Chem. Lett.,* 2021 (1987).

KF4 K. Y. Ko, W. J. Frazee, and E. L. Eliel, *Tetrahedron* **40**, 1333 (1984).

KF5 Y. Kawanami, I. Fujita, S. Asahara, T. Katsuki, and M. Yamaguchi, *Bull. Chem. Soc. Jap.* **62**, 3598 (1989).

KG1 D. M. Ketcha and G. W. Gribble, *J. Org. Chem.* **50**, 5451 (1985).

KG2 N. Klempier, P. Geymayer, P. Stadler, K. Faber, and H. Griengl, *Tetrahedron Asymmetry* **1**, 111 (1990).

KH1 S. Kim, C. Y. Hong, and C. Yang, *Angew. Chem. Int. Ed. Engl.* **22**, 562 (1983).

KH2 Y. Kamitori, M. Hojo, R. Masuda, T. Inoue, and T. Izumi, *Tetrahedron Lett.* **24**, 2575 (1983); *Synthesis,* 387 (1983).

KH3 Y. Kamitori, M. Hojo, R. Masuda, T. Inoue, and T. Izumi, *Tetrahedron Lett.* **23**, 4585 (1982).

KH4 F. L. Koerwitz, G. B. Hammond, and D. F. Wiemer, *J. Org. Chem.* **54**, 738 (1989).

KH5 S. Kobayashi, I. Hachiya, and M. Yasuda, *Tetrahedron Lett.* **37**, 5569 (1996).

KI1 Y. Kirugawa, S. Ikegami, and S. I. Yamada, *Chem. Pharm. Bull.* **17**, 98 (1969).

KJ1 L. R. Krepski, K. M. Jensen, S. M. Heilman, and J. K. Rasmussen, *Synthesis,* 301 (1986).

KJ2 M. J. Kang, J. S. Jang, and S. G. Lee, *Tetrahedron Lett.* **36**, 8829 (1995).

KK1 S. Kim, Y. J. Kim, and K. H. Anh, *Tetrahedron Lett.* **22**, 3369 (1983).

KK2 Y. Kikugawa, M. Kuramoto, I. Saito, and S. Yamada, *Chem. Pharm. Bull.* **21**, 1927 (1973).

KK3 H. Kosugi, H. Konta, and H. Uda, *J. Chem. Soc. Chem. Comm.,* 211 (1985).

KK4 S. Kiyooka, H. Kuroda, and Y. Shimasaki, *Tetrahedron Lett.* **27**, 3009 (1986).

KK5 S. Kim, Y. J. Kim, C. H. Oh, and K. H. Ahn, *Bull. Korean Chem. Soc.* **5**, 202 (1984).

KK6 S. Kim and J. S. Ko, *Synth. Comm.* **15**, 603 (1985).

KK7 T. Kitazume, T. Kobayashi, T. Yamamoto, and T. Yamazaki, *J. Org. Chem.* **52**, 3218 (1987).

KK8 Y. Kikugawa and M. Kawase, *Synth. Comm.* **9**, 49 (1979).

KK9 S. H. Kim, J. H. Kim, and N. M. Yoon, *Bull. Korean Chem. Soc.* **10**, 117 (1989).

KK10 H. Kosugi, M. Kitaoka, K. Tagami, A. Takahashi, and H. Uda, *J. Org. Chem.* **52**, 1078 (1987).

KK11 M. Kawase and Y. Kikugawa, *J. Chem. Soc. Perkin Trans.* I, 643 (1979).

KK12 A. M. P. Koskinen and P. M. Koskinen, *Tetrahedron Lett.* **34**, 6765 (1993).

KL1 D. M. Ketcha, B. A. Lieurance, D. F. J. Homan, and G. W. Gribble, *J. Org. Chem.* **54**, 4350 (1989).

KL2 D. M. Ketcha and B. A. Lieurance, *Tetrahedron Lett.* **30**, 6833 (1989).

KL3 Y. M. Kobayashi, J. Lambrecht, J. C. Jochims, and U. Burkert, *Chem. Ber.* **111**, 3442 (1978).

KL4 J. Kang, J. W. Lee, J. I. Kim, and C. Pyun, *Tetrahedron Lett.* **36**, 4265 (1995).

KL5 B. C. Kim and W. K. Lee, *Tetrahedron* **52**, 12117 (1996).

KM1 M. M. Kayser and P. Morand, *Can. J. Chem.* **58**, 2484 (1980); **56**, 1524 (1978).

KM2 S. Kim, Y. C. Moon, and K. H. Ahn, *J. Org. Chem.* **47**, 3311 (1982).

KM3 N. Katagiri, M. Muto, M. Nomura, T. Higashikawa, and C. Kaneko, *Chem. Pharm. Bull.* **39**, 1112 (1991).

KN1 Y. S. Kulkarni, M. Niwa, E. Ron, and B. B. Snider, *J. Org. Chem.* **52**, 1568 (1987).

KN2 K. Kondo, A. Negishi, and D. Tunemoto, *Angew. Chem. Int. Ed. Engl.* **13**, 407 (1974).

KN3 T. Katoh, Y. Nagata, Y. Kobayashi, K. Arai, J. Minami, and S. Terashima, *Tetrahedron Lett.* **34**, 5743 (1993).

KN4 G. Kokotos and C. Noula, *J. Org. Chem.* **61**, 6994 (1996).

KO1 S. Kim, C. H. Oh, J. S. Ko, K. H. Ahn, and Y. J. Kim, *J. Org. Chem.* **50**, 1927 (1985).

KP1 F. G. Kathawala, B. Prager, K. Prasad, O. Repic, M. J. Shapiro, R. S. Stabler, and L. Widler, *Helv. Chim. Acta* **69**, 803 (1986).

KP2 K. E. Kim, S. B. Park, and N. M. Yoon, *Synth. Comm.* **18**, 89 (1988).

KP3 Y. H. Kim, D. H. Park, I. S. Byun, I. K. Yoon, and C. S. Park, *J. Org. Chem.* **58**, 4511 (1993).

KR1 E. Keinan and Z. Roth, *J. Org. Chem.* **48**, 1769 (1983).

KS1 N. G. Kundu, S. Sikdar, R. P. Herzberg, S. A. Schmitz, and S. G. Khatri, *J. Chem. Soc. Perkin Trans.* I, 1295 (1985).

KS2 P. Kocienski and S. D. A. Street, *Synth. Comm.* **14**, 1087 (1984).

KS3 T. Katsuki and K. B. Sharpless, *J. Am. Chem. Soc.* **102**, 5974 (1980).

KS4 A. Kubo, N. Saito, N. Kawakami, Y. Matsuyama, and T. Miwa, *Synthesis,* 824 (1987).

KS5 P. F. Keusenkothen and M. B. Smith, *Tetrahedron Lett.* **30**, 3369 (1989).

KS6 S. S. Kulp and R. Szarko, *J. Org. Chem.* **53**, 5573 (1988).

KS7 D. M. Krein, P. J. Sullivan, and K. Turnbull, *Tetrahedron Lett.* **37**, 7213 (1996).

KS8 S. W. Kaldor, M. G. Siegel, J. E. Fritz, B. A. Dressman, and P. J. Hahn, *Tetrahedron Lett.* **37**, 7193 (1996).

KT1 T. Kametani, M. Tsubuki, Y. Tatsuzaki, and T. Honda, *J. Chem. Soc. Perkin Trans.* I, 639 (1990).

KT2 S. Kano, Y. Tanaka, E. Sugino, and S. Hibino, *Synthesis,* 695 (1980).

KU1 H. Kotsuki, Y. Ushio, N. Yoshimura, and M. Ochi, *J. Org. Chem.* **52**, 2594 (1987).

KU2 H. Kotsuki, Y. Ushio, I. Kadota, and M. Ochi, *J. Org. Chem.* **54**, 5153 (1989).

KU3 H. Kotsuki, Y. Ushio, I. Kadota, and M. Ochi, *Chem. Lett.,* 927 (1988).

KU4 H. Kotsuki, Y. Ushio, N. Yoshimura, and M. Ochi, *Bull. Chem. Soc. Japan* **61**, 2684 (1988).

KW1 D. M. Kalvin and R. W. Woodard, *Tetrahedron* **40**, 3387 (1984).

KW2 G. W. Kabalka, P. P. Wadgaonkar, and N. Chatla, *Synth. Comm.* **20**, 293 (1990).

KW3 T. H. Keller and L. Weiler, *J. Am. Chem. Soc.* **112**, 450 (1990).

KW4 B. M. Kim, S. F. Williams, and S. Masamune, in *Comprehensive Organic Chemistry,* B. M. Trost and I. Fleming, Eds. (Pergamon Press, London, 1991), Vol. 2, Chapter 1.7.

KW5 T. H. Keller and L. Weiler, *Tetrahedron Lett.* **31**, 6307 (1990).

KY1 T. Koisumi, N. Yamamoto, and E. Yoshii, *Chem. Pharm. Bull.* **21**, 312 (1973).

KY2 H. Kotsuki, N. Yoshimura, I. Kadota, Y. Ushii, and M. Ochi, *Synthesis,* 401 (1990).

KY3 M. Koreeda and Z. You, *J. Org. Chem.* **54**, 5195 (1989).

L1 C. F. Lane, *Synthesis,* 135 (1975).

L2 C. F. Lane, *Chem. Rev.* **76**, 773 (1976).

L3 A. Lattes, personal communication.

LA1 N. Langlois and R. Andriamialisoa, *Tetrahedron Lett.* **29**, 3259 (1988).

LB1 G. Lewin, M. Bert, J. C. Dauguet, C. Schaeffer, J. C. Guinamant, and J. P. Volland, *Tetrahedron Lett.* **30**, 7049 (1989).

LC1 M. Lautens, P. Chiu, and J. T. Colucci, *Angew. Chem. Int. Ed. Engl.* **32**, 281 (1993).

LD1 C. K. Lau, C. Dufresne, P. C. Belanger, S. Pietre, and J. Scheigetz, *J. Org. Chem.* **51**, 3038 (1986).

LD2 C. W. Lindsley and M. Di Mare, *Tetrahedron Lett.* **35**, 5141 (1994).

LD3 N. Langlois and N. Dahuron, *Tetrahedron Lett.* **37**, 3993 (1996).

LG1 M. Le Corre, E. Gheerbrant, and H. Le Deit, *J. Chem. Soc. Chem. Comm.,* 313 (1989).

LG2 E. Lukevics, V. N. Gevorgyan, and Y. S. Goldberg, *Tetrahedron Lett.* **25**, 1415 (1984).

LH1 N. H. Lin, Y. He, and H. Kopecka, *Tetrahedron Lett.* **36**, 2563 (1995).

LK1 R. S. Lenox and J. A. Katzenellenbogen, *J. Am. Chem. Soc.* **95**, 957 (1973).

LK2 J. P. Leeds and H. A. Kirst, *Synth. Comm.* **18**, 777 (1988).

LL1 P. T. Lansbury and R. E. Mc Leay, *J. Am. Chem. Soc.* **87**, 831 (1965).

LL2 M. Larchevêque and J. Lalande, *Bull. Soc. Chim. Fr.,* 116 (1987); *J. Chem. Soc. Chem. Comm.,* 83 (1985).

LL3 J. M. Lefour and A. Loupy, *Tetrahedron* **34**, 2597 (1978).

LL4 H. J. Liu and W. Luo, *Synth. Comm.* **19**, 387 (1989).

LL5 H. Li and W. J. Le Noble, *Rec. Trav. Chim. Pays-Bas* **111**, 199 (1992).

LM1 P. T. Lansbury and C. A. Mojica, *Tetrahedron Lett.* **27**, 3967 (1986).

LM2 Y. Lu, Y. W. C. Miet, N. Kunesch, and J. Poisson, *Tetrahedron Asymmetry* **4**, 893 (1993).

LM3 N. Lewis, A. Mc. Killop, R. J. K. Taylor, and R. J. Watson, *Synth. Comm.* **25**, 561 (1995).

LO1 M. Lögers, L. E. Overman, and G. S. Welmaker, *J. Am. Chem. Soc.* **117**, 9139 (1995).

LP1 M. Lanier and R. Pastor, *Tetrahedron Lett.* **36**, 2491 (1995).

LP2 I. Ling, B. Podányi, T. Hámori, and S. Sólyom, *J. Chem. Soc. Perkin Trans.* I, 1423 (1995).

LR1 Y. Landais, J. P. Robin, and A. Lebrun, *Tetrahedron* **47**, 3787 (1991).

LR2 N. Langlois and A. Rojas, *Tetrahedron Lett.* **34**, 2477 (1993).

LR3 A. J. Locke and C. J. Richards, *Tetrahedron Lett.* **37**, 7861 (1996).

LS1 A. Loupy and J. Seyden-Penne, *Tetrahedron* **36**, 1937 (1980).

LS2 A. Loupy, J. Seyden-Penne, and B. Tchoubar, *Tetrahedron Lett.,* 1677 (1976).

LS3 A. Liguori, G. Sindona, and N. Ucella, *Tetrahedron* **39**, 683 (1983).

LS4 Y. Liu and J. Schwartz, *J. Org. Chem.* **59**, 940 (1994).

LS5 Y. Liu and J. Schwartz, *Tetrahedron* **51**, 4471 (1995).

LS6 L. P. Linney, C. R. Self, and I. H. Williams, *Tetrahedron Asymmetry* **5**, 813 (1994); *J. Chem. Soc. Chem. Comm.,* 1651 (1994).

LS7 M. Laxman and M. M. Sharma, *Synth. Comm.* **20**, 111 (1990).

LT1 C. K. Lau, S. Tardif, C. Dufresne, and J. Scheigetz, *J. Org. Chem.* **54**, 491 (1989).

LT2 I. D. Linney, H. Tye, M. Wills, and R. J. Butlin, *Tetrahedron Lett.* **35**, 1785 (1994).

LU1 B. H. Lipshutz, C. S. Ung, and S. Sengupta, *Synlett,* 64 (1990).

LV1 P. T. Lansbury and J. P. Vacca, *Tetrahedron Lett.* **23**, 2623 (1982).

LY1 J. T. Lin, T. Yamazaki, and T. Kitazume, *J. Org. Chem.* **52**, 3211 (1987).

LZ1 R. Lin and Y. Zhang, *Synth. Comm.* **17**, 1403 (1987).

M1 J. Malek, *Org. React.* **34**, 1 (1985).

M2 M. M. Midland, in *Asymmetric Synthesis, Vol. 2,* J. Morrison, Ed. (Academic Press, New York, 1983), p. 45.

M3 J. Malek, *Org. React.* **36**, 249 (1988).

M4 M. Madesclaire, *Tetrahedron* **44**, 6537 (1988).

M5 L. N. Mander, in *Stereochemistry of Organic Compounds,* E. L. Eliel and S. W. Wilen, Eds. (Wiley, New York, 1995), pp. 875–87.

M6 P. Mohr, *Tetrahedron Lett.* **32**, 2219 (1991).

M7 A. E. Moorman, *Synth. Comm.* **23**, 789 (1993).

M8 J. McGill, *Synthesis,* 1089 (1993).

MA1 T. Mikami, H. Asano, and O. Mitsonobu, *Chem. Lett.,* 2033 (1987).

MA2 K. Maruoka, Y. Araki, and H. Yamamoto, *J. Am. Chem. Soc.* **110**, 2650 (1988).

MB1 S. Masamune, G. S. Bates, and P. E. Georghiou, *J. Am. Chem. Soc.* **96**, 3686 (1974).

MB2 W. S. Mahoney, D. M. Brestensky, and J. M. Stryker, *J. Am. Chem. Soc.* **110**, 291 (1988).

MB3 G. A. Molander, K. L. Bobbitt, and C. K. Murray, *J. Am. Chem. Soc.* **114**, 2759 (1992).

MB4 G. A. Molander and K. L. Bobbitt, *J. Am. Chem. Soc.* **115**, 7517 (1993).

MB5 A. I. Meyers, M. Bos, and D. A. Dickman, *Org. Synth.* **67**, 60 (1988); A. I. Meyers and T. R. Elworthy, *J. Org. Chem.* **57**, 4732 (1992).

MC1 J. Malek and M. Cerny, *Synthesis,* 217 (1972).

MC2 J. A. Marshall and R. D. Carroll, *J. Org. Chem.* **30**, 2748 (1965).

MC3 S. A. Miller and A. R. Chamberlin, *J. Org. Chem.* **54**, 2502 (1989).

MC4 S. Mabic and N. Castagnoli, *J. Org. Chem.* **61**, 309 (1996).

ME1 P. M. Maitlis, P. Espinet, and M. J. H. Russell, in *Comprehensive Organometallic Chemistry,* G. Wilkinson, Ed. (Pergamon Press, Oxford, 1982), p. 487.

MF1 A. Mori, J. Fujiwara, K. Maruoka, and H. Yamamoto, *Tetrahedron Lett.* **24**, 4581 (1983).

MG1 S. B. Mandal, V. S. Giri, and S. C. Pakrashi, *Synthesis,* 1128 (1987).

MG2 S. B. Mandal, V. S. Giri, M. S. Sabeena, and S. C. Pakrashi, *J. Org. Chem.* **53**, 4236 (1988).

MH1 A. I. Meyers, R. J. Himmelsbach, and M. Reuman, *J. Org. Chem.* **48**, 4053 (1983).

MH2 M. L. Mancini and J. F. Honek, *Tetrahedron Lett.* **24**, 4295 (1983).

MH3 R. Moriwake, S. Hamano, D. Miki, S. Saito, and S. Torii, *Chem. Lett.,* 815 (1986).

MH4 G. J. McGarvey, R. N. Hiner, Y. Matsubara, and T. Oh, *Tetrahedron Lett.* **24**, 2733 (1983).

MH5 A. G. Myers, M. Hammond, and Y. Wu, *Tetrahedron Lett.* **37**, 3083 (1996).

MI1 A. Mori, K. Ishihara, I. Arai, and H. Yamamoto, *Tetrahedron* **43**, 755 (1987).

MI2 K. Matsuki, H. Inoue, and M. Takeda, *Tetrahedron Lett.* **34**, 1167 (1993).

MJ1 D. J. Mathre, T. K. Jones, L. C. Xavier, T. J. Blacklock, R. A. Reamer, J. J. Mohan, E. T. T. Jones, K. Hoogsteen, M. W. Baum, and E. J. J. Grabowski, *J. Org. Chem.* **56**, 751 (1991).

MK1 M. M. Midland and Y. C. Kwon, *J. Am. Chem. Soc.* **105**, 3725 (1983).

MK2 G. J. McGarvey and M. Kimura, *J. Org. Chem.* **47**, 5422 (1982).

MK3 A. B. Mikkilineni, P. Kumar, and E. Abushanab, *J. Org. Chem.* **53**, 6005 (1988).

MK4 Y. Mori, M. Kuhara, A. Takeuchi, and M. Suzuki, *Tetrahedron Lett.* **29**, 5419 (1988).

MK5 G. Mehta, F. A. Khan, B. Ganguly, and J. Chandrasekhar, *J. Chem. Soc. Chem. Comm.*, 1711 (1992).

MK6 P. Munier, A. Krusinski, D. Picq, and D. Anker, *Tetrahedron* **51**, 1229 (1995).

ML1 F. Montanari, D. Landini, and F. Rolla, *Top. Curr. Chem.* **101**, 149 (1982).

ML2 C. Meier and W. H. G. Laux, *Tetrahedron Asymmetry* **6**, 1089 (1995).

ML3 J. M. McGill, E. S. LaBell, and M. A. Williams, *Tetrahedron Lett.* **37**, 3977 (1996).

ML4 C. Meier and W. H. G. Laux, *Tetrahedron* **52**, 589 (1996).

MM1 B. E. Maryanoff, D. F. Mc Comsey, and S. O. Nortey, *J. Org. Chem.* **46**, 355 (1981).

MM2 P. Ma, V. S. Martin, S. Masamune, K. B. Sharpless, and S. M. Viti, *J. Org. Chem.* **47**, 1378 (1982).

MM3 M. Muraki and T. Mukaiyama, *Chem. Lett.*, 1447 (1974).

MM4 A. J. McAlees, R. McCrindle, and D. W. Sneddon, *J. Chem. Soc. Perkin Trans.* I, 2037 (1977).

MM5 K. Maruoka, T. Miyazaki, M. Ando, Y. Matsumura, S. Sakane, K. Hattori, and H. Yamamoto, *J. Am. Chem. Soc.* **105**, 2831 (1983).

MM6 T. Mehler and J. Martens, *Tetrahedron Asymmetry* **4**, 1983 (1993).

MM7 T. Mehler and J. Martens, *Tetrahedron Asymmetry* **4**, 2299 (1993).

MM8 M. J. McKennon and A. I. Meyers, *J. Org. Chem.* **58**, 3568 (1993).

MM9 M. J. Munchhof and A. I. Meyers, *J. Org. Chem.* **60**, 7084 (1995).

MN1 F. Manescalchi, A. R. Nardi, and D. Savoia, *Tetrahedron Lett.* **35**, 2775 (1994).

MN2 L. E. Martinez, W. A. Nugent, and E. Jacobsen, *J. Org. Chem.* **61**, 7963 (1996).

MO1 S. Morita, K. Otsubo, J. Matsubara, T. Ohtani, and M. Uchida, *Tetrahedron Asymmetry* **6**, 245 (1995).

MP1 R. J. Mattson, K. M. Pham, D. J. Leuck, and K. A. Cowen, *J. Org. Chem.* **55**, 2552 (1990).

MP2 E. M. Marlett and W. S. Park, *J. Org. Chem.* **55**, 2968 (1990).

MP3 L. Manzoni, T. Pilati, G. Poli, and C. Scolastico, *J. Chem. Soc. Chem. Comm.*, 1027 (1992).

MP4 M. Misun and A. Pfaltz, *Helv. Chim. Acta* **79**, 961 (1996).

MQ1 L. Micouin, J.-C. Quirion, and H.-P. Husson, *Tetrahedron Lett.* **37**, 849 (1996).

MR1 G. Maier, C. Roth, and R. K. Schmitt, *Chem. Ber.* **118**, 704 (1985).

MR2 J. L. Marco, J. Royer, and H.-P. Husson, *Synth. Comm.* **17**, 669 (1987).

MR3 M. A. Makhlouf and B. Rickborn, *J. Org. Chem.* **46**, 4810 (1981).

MR4 J. Mauger and A. Robert, *Tetrahedron* **44**, 2493 (1988).

MS1 G. Maier, R. K. Schmitt, and U. Seipp, *Chem. Ber.* **118**, 722 (1985).

MS2 J. C. Melendez and M. M. Sollhuber, *Heterocycles* **29**, 313 (1989).

MS3 G. Maier, M. Schneider, and T. Sayrac, *Chem. Ber.* **111**, 3412 (1978).

MS4 Y. Mori and M. Suzuki, *Tetrahedron Lett.* **30**, 4383 (1989).

MS5 J. A. Marshall, J. M. Salovich, and B. G. Shearer, *J. Org. Chem.* **55**, 2398 (1990).

MS6 W. S. Mahoney and J. M. Stryker, *J. Am. Chem. Soc.* **111**, 8818 (1989).

MS7 G. Maier, U. Seipp, H. O. Kalinowski, and M. Henrich, *Chem. Ber.* **117**, 1427 (1994).

MS8 A. I. Meyers, R. F. Spohn, and R. J. Lindeman, *J. Org. Chem.* **50**, 3633 (1985).

MS9 M. Masui and T. Shioiri, *Tetrahedron* **30**, 8363 (1995); *Synlett,* 49 (1996).

MS10 J. M. Mattalia, A. Samat, and M. Chanon, *J. Chem. Soc. Perkin Trans.* I, 1769 (1991); J. M. Mattalia, Y. Berchadsky, E. Péralez, J. C. Négrel, P. Tordo, and M. Chanon, *Tetrahedron Lett.* **37**, 4717 (1996).

MT1 P. J. Maurer, H. Takahata, and H. Rapoport, *J. Am. Chem. Soc.* **106**, 1095 (1984).

MT2 T. Mukaiyama, S. Tanaka, and M. Asami, *Chem. Lett.,* 433 (1982).

MT3 Y. Mori, A. Takeuchi, H. Kageyama, and M. Suzuki, *Tetrahedron Lett.* **29**, 5423 (1988).

MT4 J. A. Marshall, J. D. Trometer, and D. G. Cleary, *Tetrahedron* **45**, 391 (1989).

MT5 Y. Matsuda, S. Tanimoto, T. Okamoto, and S. M. Ali, *J. Chem. Soc. Perkin Trans.* I, 1279 (1989).

MT6 D. J. Mathre, A. S. Thompson, A. W. Douglas, K. Hoogsteen, J. D. Carroll, E. G. Corley, and E. J. J. Grabowski, *J. Org. Chem.* **58**, 2880 (1993).

MT7 K. Miyashita, T. Toyoda, H. Miyabe, and T. Inanishi, *Synlett,* 1229 (1995).

MV1 M. S. Mourad, R. S. Varma, and G. W. Kabalka, *Synthesis,* 654 (1985).

MV2 M. S. Mourad, R. S. Varma, and G. W. Kabalka, *Synth. Comm.* **14**, 1099 (1984).

MV3 J. R. Mohrig, P. J. Vreede, S. C. Schultz, and C. A. Fierke, *J. Org. Chem.* **46**, 4655 (1981).

MV4 M. S. Mourad, R. S. Varma, and G. W. Kabalka, *J. Org. Chem.* **50**, 133 (1985).

MW1 D. Mukherjee, Y. D. Wu, F. R. Fronczek, and K. N. Houk, *J. Am. Chem. Soc.* **110**, 3328 (1988).

MW2 J. A. Marshall and X. Wang, *J. Org. Chem.* **56**, 3211 (1991).

MY1 V. P. Miller, D. Yang, T. M. Weigel, O. Han, and H. Liu, *J. Org. Chem.* **54**, 4175 (1989).

MY2 A. G. Myers, B. H. Yang, H. Chen, and J. L. Gleason, *J. Am. Chem. Soc.* **116**, 9361 (1994).

MY3 T. Mukaiyama, H. Yamashita, and M. Asami, *Chem. Lett.,* 385 (1983).

MY4 A. G. Myers, B. H. Yang, and D. J. Kopecky, *Tetrahedron Lett.* **37**, 3623 (1996).

MZ1 G. Majetich, Y. Zhang, and K. Wheless, *Tetrahedron Lett.* **35**, 8727 (1994).

N1 S. Narasimhan, *Heterocycles* **18**, 131 (1982).

N2 Nguyen Trong Anh, *Top. Curr. Chem.* **88**, 146 (1980); *Orbitales Frontières* (InterEditions/CNRS Editions, Paris, 1995).

N3 K. M. Nicholas, *Acc. Chem. Res.* **20**, 207 (1987).

N4 V. Nevalainen, *Tetrahedron Asymmetry* **2**, 1133 (1991) and quoted references.

N5 M. Nogradi, *Stereoselective Synthesis,* 2nd Edition, VCH, Weinheim, Chapter 3 (1995).

NB1 J. M. Nuzillard, A. Boumendjel, and G. Massiot, *Tetrahedron Lett.* **30**, 3779 (1989).

NB2 C. F. Nutaitis and J. E. Bernardo, *Synth. Comm.* **20**, 487 (1990).

NB3 C. F. Nutaitis and J. E. Bernardo, *J. Org. Chem.* **54**, 5629 (1989).

NF1 T. Nakata, M. Fukui, and T. Oishi, *Tetrahedron Lett.* **29**, 2219 (1988).

NG1 C. F. Nutaitis and G. W. Gribble, *Tetrahedron Lett.* **24**, 4287 (1983).

NG2 C. F. Nutaitis and G. W. Gribble, *Org. Prep. Proc. Int.* **17**, 11 (1985).

NH1 M. Narisada, I. Horibe, F. Watanabe, and K. Takeda, *J. Org. Chem.* **54**, 5308 (1989).

NK1 Y. Nagao, K. Kawabata, K. Seno, and E. Fujita, *J. Chem. Soc. Perkin Trans.* I, 2470 (1980); Y. Nagao, W. M. Dai, M. Ochiai, S. Tsukagoshi, and E. Fujita, *J. Am. Chem. Soc.* **110**, 289 (1988); S. Sano, Y. Kobayashi, T. Kondo, M. Takebayashi, S. Maruyama, T. Fujita, and Y. Nagao, *Tetrahedron Lett.* **36**, 2097 (1995).

NK2 P. Nebois, N. Kann, and A. E. Greene, *J. Org. Chem.* **60**, 7690 (1995).

NM1 G. R. Newkome, V. K. Majestic, and J. D. Sauer, *Org. Prep. Proc. Int.* **12**, 345 (1980); *Tetrahedron Lett.* **22**, 3039 (1981).

NM2 K. Nonoshita, K. Maruoka, and H. Yamamoto, *Bull. Chem. Soc. Japan* **61**, 2241 (1988).

NM3 S. Narasimhan, S. Madhavan, and K. G. Prasad, *J. Org. Chem.* **60**, 5314 (1995).

NM4 S. Narasimhan, S. Madhavan, and K. G. Prasad, *Synth. Comm.* **26**, 703 (1996).

NN1 M. Nishizawa and R. Noyori, in *Comprehensive Organic Chemistry,* B. M. Trost and I. Fleming, Eds. (Pergamon Press, London, 1991), Vol. 8, Chapter 1.7.

NO1 E. G. Neeland, J. P. Ounsworth, R. J. Sims, and L. Weiler, *J. Org. Chem.* **59**, 7383 (1994).

NP1 K. Narasaka and F. C. Pai, *Tetrahedron* **40**, 2233 (1984).

NS1 A. Nag, A. Sarkar, S. K. Sarkar, and S. K. Paut, *Synth. Comm.* **17**, 1007 (1987).

NS2 V. Nair and A. K. Sinhababu, *J. Org. Chem.* **45**, 1893 (1980).

NS3 Y. Nakagawa and R. V. Stevens, *J. Org. Chem.* **53**, 1871 (1988).

NS4 J. A. Nikles and C. N. Sukenik, *Tetrahedron Lett.* **23**, 4211 (1982).

NS5 T. Nagamitsu, T. Sunazuka, R. Obata, H. Tomoda, H. Tanaka, Y. Harigaya, S. Omura, and A. B. Smith, *J. Org. Chem.* **60**, 8126 (1995).

NT1 T. Nakata, T. Tanaka, and T. Oishi, *Tetrahedron Lett.* **22**, 4723 (1981).

NT2 H. Nöth, S. Thomas, and M. Schmidt, *Chem. Ber.* **129**, 451 (1996).

NU1 K. Narasaka and Y. Ukaji, *Chem. Lett.,* 147 (1984).

NU2 V. Nevalainen, R. Uggla, and M. R. Sundberg, *Tetrahedron Asymmetry* **6**, 1431 (1995).

NW1 S. Nahm and S. M. Weinreb, *Tetrahedron Lett.* **22**, 3815 (1981).

NW2 A. Nelson and S. Warren, *Tetrahedron Lett.* **37**, 1501 (1996).

NY1 K. Narasaka, S. Yamazaki, and Y. Ukaji, *Chem. Lett.,* 2065 (1984).

NY2 T. Nagata, K. Yorozu, T. Yamada, and T. Mukaiyama, *Angew. Chem. Int. Ed. Engl.* **34**, 2145 (1995).

OB1 L. E. Overman and R. M. Burk, *Tetrahedron Lett.* **25**, 5737 (1984).

OB2 W. Oppolzer, J. Blagg, I. Rodriguez, and E. Walter, *J. Am. Chem. Soc.* **112**, 2767 (1990).

OB3 W. Oppolzer, C. G. Bochet, and E. Merifield, *Tetrahedron Lett.* **35**, 7015 (1994).

OC1 R. Ostwald, P. Y. Chavant, H. Stadtmüller, and P. Knochel, *J. Org. Chem.* **59**, 4143 (1994).

OF1 K. Ogura, M. Fujita, T. Inaba, K. Takahashi, and H. Iida, *Tetrahedron Lett.* **24**, 503 (1983).

OH1 A. Ono and H. Hayakawa, *Chem. Lett.,* 853 (1987).

OI1 K. Otsuka, K. Ito, and T. Katsuki, *Synlett,* 429 (1995).

OK1 R. Ozegowski, A. Kunath, and H. Schick, *Annalen,* 1443 (1996).

OM1 L. E. Overman and R. S. McCready, *Tetrahedron Lett.* **23**, 2355 (1982).

OM2 G. Ohloff, B. Maurer, B. Winter, and W. Griersch, *Helv. Chim. Acta* **66**, 192 (1983).

OM3 W. Oppolzer, R. J. Mills, and M. Réglier, *Tetrahedron Lett.* **27**, 183 (1986).

ON1 T. Oishi and T. Nakata, *Acc. Chem. Res.* **17**, 338 (1984).

OP1 M. E. Osborn, J. F. Pegues, and L. A. Paquette, *J. Org. Chem.* **45**, 167 (1980).

OP2 W. Oppolzer, G. Poli, C. Starkemann, and G. Bernardinelli, *Tetrahedron Lett.* **29**, 3559 (1988).

OS1 A. Ookawa and K. Soai, *J. Chem. Soc. Perkin Trans.* I, 1465 (1987).

OS2 A. Ono, N. Suzuki, and J. Kamimura, *Synthesis,* 736 (1987).

OS3 T. Olsson, K. Stern, and S. Sundell, *J. Org. Chem.* **53**, 2468 (1988).

OS4 S. Ono, S. Shuto, and A. Matsuda, *Tetrahedron Lett.* **37**, 221 (1996); S. Shuto, S. Ono, Y. Hase, N. Kamiyama, H. Takada, K. Yamasihita, and A. Matsuda, *J. Org. Chem.* **61**, 915 (1996).

OT1 T. Okada, S. Tomita, and M. Oda, *Bull. Chem. Soc. Japan* **62**, 459 and 2342 (1989).

OT2 K. Okada, S. Tomita, and M. Oda, *Tetrahedron Lett.* **27**, 2645 (1986).

OW1 G. A. Olah, A. Wu, and O. Farooq, *J. Org. Chem.* **53**, 5143 (1988).

OW2 G. A. Olah, A. Wu, and O. Farooq, *J. Org. Chem.* **54**, 1452 (1989).

OW3 P. O'Brien and S. Warren, *Tetrahedron Lett.* **37**, 3051 (1996).

P1 A. R. Pinder, *Synthesis,* 425 (1980).

P2 I. Paterson, in *Comprehensive Organic Chemistry,* B. M. Trost and I. Fleming, Eds. (Pergamon Press, London, 1991), Vol. 2, Chapter 1.8.

P3 A. Pfaltz, *Acc. Chem. Res.* **26**, 339 (1993).

PA1 F. Pecquet and J. d'Angelo, *Tetrahedron Lett.* **23**, 2777 (1982).

PB1 M. Petrini, R. Ballini, and G. Rosini, *Synthesis,* 713 (1987).

PB2 K. Pal, M. L. Behnke, and L. Tong, *Tetrahedron Lett.* **34**, 6205 (1993).

PB3 R. P. Polniaszek and S. E. Belmont, *J. Org. Chem.* **55**, 4688 (1990).

PC1 S. V. Park, S. K. Chung, and M. Newcomb, *J. Org. Chem.* **52**, 3275 (1987).

PC2 L. A. Paquette, K. D. Combrink, S. W. Elmore, and R. D. Rogers, *J. Am. Chem. Soc.* **113**, 1335 (1991).

PC3 K. Prasad, K. M. Chen, O. Repic, and G. E. Hardtmann, *Tetrahedron Asymmetry* **1**, 307 (1990).

PC4 I. Paterson and J. A. Channon, *Tetrahedron Lett.* **33**, 797 (1992).

PD1 T. D. Penning, S. W. Djuric, R. A. Haack, V. J. Kalish, J. M. Miyashiro, B. W. Rowell, and S. S. Yu, *Synth. Comm.* **20**, 307 (1990).

PD2 M. Periasamy, A. Devasagayaraj, N. Satyanarayana, and C. Narayana, *Synth. Comm.* **19**, 565 (1989).

PE1 C. Pedregal, J. Esquerra, A. Escribano, M. C. Carreño, and J. L. Garcia-Ruano, *Tetrahedron Lett.* **35**, 2053 (1994).

PF1 L. A. Paquette, D. Friedrich, E. Pinard, J. P. Williams, D. St. Laurent, and B. A. Roden, *J. Am. Chem. Soc.* **115**, 4377 (1993).

PH1 T. B. Patrick, S. Hosseini, and S. Bains, *Tetrahedron Lett.* **31**, 179 (1990).

PJ1 J. Powell, N. James, and S. J. Smith, *Synthesis,* 338 (1986).

PJ2 K. R. K. Prasad and N. N. Joshi, *J. Org. Chem.* **61**, 3888 (1996).

PK1 R. P. Polniaszek and C. R. Kaufman, *J. Am. Chem. Soc.* **111**, 4859 (1989).

PL1 J. S. Prasad and L. S. Liebeskind, *Tetrahedron Lett.* **28**, 1857 (1987).

PL2 K. A. Parker and M. W. Ledeboer, *J. Org. Chem.* **61**, 3214 (1996).

PM1 V. Peper and J. Martens, *Tetrahedron Lett.* **37**, 8351 (1996).

PM2 V. Peper and J. Martens, *Chem. Ber.* **129**, 691 (1996).

PN1 E. Péralez, J. C. Négrel, and M. Chanon, *Tetrahedron Lett.* **36**, 6457 (1995).

PP1 J. C. Philips and C. Perianayagam, *Tetrahedron Lett.,* 3263 (1975).

PP2 R. Polt, M. A. Peterson, and L. De Young, *J. Org. Chem.* **57**, 5469 (1992).

PP3 A. Pasquarello, G. Poli, and C. Scolastico, *Synlett,* 93 (1992).

PP4 L. Pégorier, Y. Petit, and M. Larchevêque, *J. Chem. Soc. Chem. Comm.,* 633 (1994).

PP5 I. Paterson and M. V. Perkins, *Tetrahedron Lett.* **33**, 801 (1992).

PR1 M. N. Paddon-Row, N. C. Rondan, and K. N. Houk, *J. Am. Chem. Soc.* **104**, 7162 (1982).

PR2 A. Pelter, R. M. Rosser, and S. Mills, *J. Chem. Soc. Perkin Trans.* I, 717 (1984).

PR3 J. P. Poyser, F. de Reinach-Hirzbach, and G. Ourisson, *J. Chem. Soc. Perkin Trans.* I, 378 (1974).

PR4 I. Paterson and D. J. Rawson, *Tetrahedron Lett.* **30**, 7463 (1989).

PR5 L. A. Paquette, R. J. Ross, and J. P. Springer, *J. Am. Chem. Soc.* **110**, 6192 (1988).

PR6 J. C. Plaquevent and A. Ravard, *J. Organomet. Chem.* **361**, C51 (1989).

PR7 R. A. Pilli, S. Russowsky, and L. C. Dias, *J. Chem. Soc. Perkin Trans.* I, 1213 (1990).

PR8 S. V. Pansare and R. Gnana Ravi, *Tetrahedron Lett.* **36**, 5959 (1995).

PS1 A. Pelter and K. Smith, in *Comprehensive Organic Chemistry, Vol. 3,* D. Barton and W. D. Ollis, Eds. (Pergamon Press, Oxford, 1979), p. 695.

PS2 D. de Peretti, T. Strzalko-Bottin, and J. Seyden-Penne, *Bull. Soc. Chim. Fr.,* 2925 (1974).

PV1 H. V. Patel, K. A. Vyas, S. P. Pandey, and P. S. Fernandes, *Org. Prep. Proc. Int.* **27**, 81 (1995).

PW1 I. Paterson and D. J. Wallace, *Tetrahedron Lett.* **35**, 9087 (1994).

PW2 M. N. Paddon-Row, Y.-D. Wu, and K. N. Houk, *J. Am. Chem. Soc.* **114**, 10638 (1992).

QB1 G. J. Quallich, J. F. Blake, and T. M. Woodall, *J. Am. Chem. Soc.* **116**, 8516 (1994).

QK1 G. J. Quallich, K. N. Keavey, and T. M. Woodall, *Tetrahedron Lett.* **36**, 4729 (1995).

QW1 G. J. Quallich and T. M. Woodall, *Tetrahedron Lett.* **34**, 785 (1993).

QW2 G. J. Quallich and T. M. Woodall, *Tetrahedron Lett.* **34**, 4145 (1993).

QW3 G. J. Quallich and T. M. Woodall, *Synlett,* 929 (1993).

R1 F. Rolla, *J. Org. Chem.* **47**, 4327 (1982).

R2 M. Ramaiah, *Tetrahedron* **43**, 3576 (1987).

R3 B. C. Ranu, *Synlett,* 885 (1993).

R4 M. T. Reetz, *Angew. Chem. Int. Ed. Engl.* **30**, 1531 (1991).

R5 D. P. Rotella, *Tetrahedron Lett.* **36**, 5453 (1995).

RB1 U. Rosentreter, L. Born, and J. Kurz, *J. Org. Chem.* **51**, 1165 (1986).

RB2 R. A. Rhodes and D. W. Boykin, *Synth. Comm.* **18**, 681 (1988).

RB3 K. S. Ravikumar, S. Baskaran, and S. Chandrasekaran, *J. Org. Chem.* **58**, 5981 (1993).

RC1 C. S. Rao, R. T. Chakrasali, H. Ila, and H. Junjappa, *Tetrahedron* **46**, 2195 (1990).

RC2 K. S. Ravikumar and S. Chandrasekaran, *J. Org. Chem.* **61**, 826 (1996).

RC3 K. S. Ravikumar and S. Chandrasekaran, *Tetrahedron* **52**, 9137 (1996).

RD1 M. T. Reetz, M. W. Drewes, B. R. Matthews, and K. Lennick, *J. Chem. Soc. Chem. Comm.,* 1474 (1989).

RE1 S. Ram and R. E. Ehrenkaufer, *Tetrahedron Lett.* **26**, 5367 (1985).

RE2 K. L. Rinehart, A. F. Ellis, C. J. Michejda, and P. A. Kittle, *J. Am. Chem. Soc.* **82**, 4112 (1960).

RG1 D. J. Raber and W. C. Guida, *J. Org. Chem.* **41**, 690 (1976); D. J. Raber, W. C. Guida, and D. C. Shoenberger, *Tetrahedron Lett.* **22**, 5107 (1981).

RG2 A. V. Rama Rao, M. K. Gurjar, P. A. Sharma, and V. Kaiwar, *Tetrahedron Lett.* **31**, 2341 (1990).

RG3 D. J. Ramôn, G. Guillena, and D. Seebach, *Helv. Chim. Acta* **79**, 875 (1996).

RH1 G. Rucker, H. Horster, and W. Gajewski, *Synth. Comm.* **10**, 623 (1980).

RJ1 G. A. Russell, W. Jiang, S. S. Hu, and R. K. Khanna, *J. Org. Chem.* **51**, 5498 (1986).

RJ2 J. M. Ramanjulu and M. M. Joullié, *Synth. Comm.* **26**, 1379 (1996).

RK1 K. S. Reddy, O. H. Ko, D. Ho, P. E. Persons, and J. M. Cassady, *Tetrahedron Lett.* **28**, 3075 (1987).

RK2 M. T. Reetz, F. Kayser, and K. Harms, *Tetrahedron Lett.* **35**, 8769 (1994).

RK3 A. Ryglowski and P. Kafarski, *Tetrahedron* **52**, 10685 (1996).

RL1 M. Rodriguez, M. Llinares, S. Doulut, A. Heitz, and J. Martinez, *Tetrahedron Lett.* **32**, 923 (1991).

RL2 V. Roubaud, F. Le Moigne, A. Mercier, and P. Tordo, *Synth. Comm.* **26**, 1507 (1996).

RM1 R. Rawson and A. I. Meyers, *J. Chem. Soc. Chem. Comm.,* 494 (1992).

RM2 I. Reiner, J. Martens, S. Schwarz, and H. Henkel, *Tetrahedron Asymmetry* **7**, 1763 (1996).

RO1 S. Ramaswamy and A. C. Oehlschlager, *Can. J. Chem.* **67**, 794 (1989).

RP1 A. S. Rao, S. K. Paknikar, and J. G. Kirtane, *Tetrahedron* **39**, 2323 (1983).

RP2 P. D. Ren, S. F. Pan, T. W. Dong, and S. H. Wu, *Synth. Comm.* **26**, 763 (1996).

RP3 P. D. Ren, S. F. Pan, T. W. Dong, and S. H. Wu, *Synth. Comm.* **25**, 3395 and 3799 (1995).

RP4 P.D. Ren, S.F. Pan, T.W. Dong, and S.H. Wu, *Synth. Comm.* **26**, 763 (1996).

RR1 R. C. Roemmele and H. Rapoport, *J. Org. Chem.* **54**, 1886 (1989).

RR2 R. Romagnoli, E. C. Ross, H. Hiemstra, M. J. Moolenaar, W. N. Speckamp, B. Kaptein, and H. E. Shoemaker, *Tetrahedron Lett.* **35**, 1087 (1994).

RS1 B. C. Ranu, A. Sarkar, and R. Chakraborty, *J. Org. Chem.* **59**, 4114 (1994).

RS2 M. J. Robins, V. Samano, and M. D. Johnson, *J. Org. Chem.* **55**, 410 (1990).

RW1 I. Reiners, J. Wilken, and J. Martens, *Tetrahedron Asymmetry* **6**, 3063 (1995).

S1 T. Sheradsky, in *The Chemistry of the Azido Group,* S. Patai, Ed. (Wiley, Chichester, 1971), Chapter 6.

S2 J. Seyden-Penne, in *Reductions in Organic Chemistry,* ACS Symposium Series **641**, 70 (1996).

S3 J. Seyden-Penne, *Chiral Auxiliaries and Ligands in Asymmetric Synthesis* (Wiley, New York, 1995).

S4 V. K. Singh, *Synthesis,* 605 (1991).

S5 G. B. Stone, *Tetrahedron Asymmetry* **5**, 465 (1994).

S6 G. Solladié, *Pure Appl. Chem.* **60**, 1699 (1988).

S7 M. B. Sassaman, *Tetrahedron* **52**, 10835 (1996).

SA1 C. Schmidt, K. L. Adams, and U. Fechner, *Can. J. Chem.* **52**, 1732 (1974).

SA2 G. Solladié and A. Almario, *Tetrahedron Lett.* **35**, 1937 (1994); **33**, 2477 (1992).

SA3 H. Suzuki, S. Aoyagi, and C. Kibayashi, *Tetrahedron Lett.* **36**, 6709 (1995).

SA4 T. B. Sim, J. H. Ahn, and N. M. Yoon, *Synthesis,* 324 (1996).

SB1 M. Schlosser and Z. Brich, *Helv. Chim. Acta* **61**, 1903 (1978).

SB2 A. K. Sinhababu and R. T. Borchardt, *Tetrahedron Lett.* **24**, 227 (1983).

SB3 A. M. Salunkhe and H. C. Brown, *Tetrahedron Lett.* **36**, 7987 (1995).

SB4 A. Solladié-Cavallo and M. Bencheqroun, *Tetrahedron Asymmetry* **2**, 1165 (1991).

SB5 A. K. Szardenings, T. S. Burkoth, G. C. Look, and D. A. Campbell, *J. Org. Chem.* **61**, 6720 (1996).

SC1 Y. J. Shi, D. Cai, U. H. Dolling, A. W. Douglas, D. M. Tschaen, and T. R. Verhoeven, *Tetrahedron Lett.* **35**, 6409 (1994).

SC2 C. R. Sarko, S. E. Collibee, A. L. Knorr, and M. DiMare, *J. Org. Chem.* **61**, 868 (1996).

SD1 G. Solladié, G. Demailly, and C. Greck, *Tetrahedron Lett.* **26**, 435 (1985).

SD2 D. C. Sarkar, A. R. Das, and B. C. Ranu, *J. Org. Chem.* **55**, 5799 (1990).

SE1 P. F. Schuda, C. B. Ebner, and T. M. Morgan, *Tetrahedron Lett.* **27**, 2567 (1986).

SF1 E. Santaniello, P. Ferraboschi, and P. Sozzani, *J. Org. Chem.* **46**, 4584 (1981).

SF2 E. Santaniello, A. Fiecchi, A. Manzocchi, and P. Ferraboschi, *J. Org. Chem.* **48**, 3074 (1983).

SG1 G. Solladié, C. Greck, G. Demailly, and A. Solladié-Cavallo, *Tetrahedron Lett.* **23**, 5047 (1982).

SG2 C. R. Sarko, I. C. Guch, and M. DiMare, *J. Org. Chem.* **59**, 705 (1994).

SG3 J.-P. Surivet, J. Goré, and J.-M. Vatèle, *Tetrahedron Lett.* **37**, 371 (1996).

SH1 S. Saito, T. Hasegawa, M. Inaba, R. Nishida, T. Fujii, S. Nomizu, and T. Moriwake, *Chem. Lett.,* 1389 (1984).

SH2 W. N. Speckamp and H. Hiemstra, *Tetrahedron* **41**, 4367 (1985).

SH3 D. Seebach, I. M. P. Huber, and M. A. Syfrig, *Helv. Chim. Acta* **70**, 1357 (1987).

SH4 J. Seyden-Penne, A. Habert-Somny, and A.-M. Cohen, *Bull. Soc. Chim. Fr.,* 700 (1965).

SH5 W. Sucrow, R. Heider, and N. Joraschek, *Chem. Ber.* **121**, 1039 (1988).

SH6 M. Shimazaki, H. Hara, and K. Suzuki, *Tetrahedron Lett.* **30**, 5447 (1989).

SH7 K. J. Shea, R. G. Higby, and J. W. Gilman, *Tetrahedron Lett.* **31**, 1221 (1990).

SH8 R. J. Sundberg, G. Hamilton, and C. Trindle, *J. Org. Chem.* **51**, 3672 (1986).

SI1 K Soai, T. Isoda, H. Hasegawa, and M. Ishizaki, *Chem. Lett.*, 1897 (1986).

SI2 H. Sakuraba, N. Inomata, and Y. Tanaka, *J. Org. Chem.* **54**, 3482 (1989).

SJ1 D. D. Sternbach and W. C. L. Jamison, *Tetrahedron Lett.* **22**, 3331 (1981).

SJ2 M. Solomon, W. C. L. Jamison, M. Mc Cormick, D. Liotta, D. A. Cherry, J. E. Mills, R. D. Shan, J. D. Rodgers, and C. A. Maryanoff, *J. Am. Chem. Soc.* **110**, 3702 (1988).

SJ3 A. R. Sande, M. H. Jagadale, R. B. Mane, and M. M. Salunkhe, *Tetrahedron Lett.* **25**, 3501 (1984).

SK1 K. Suzuki, E. Katayama, and G. Tsuchihashi, *Tetrahedron Lett.* **25**, 2479 (1984).

SK2 M. Shibuya, M. Kuretani, and S. Kubota, *Tetrahedron* **38**, 2659 (1982).

SK3 N. Suzuki, Y. Kaneko, T. Tsukanaka, T. Nomoto, Y. Ayaguchi, and Y. Izawa, *Tetrahedron* **41**, 2387 (1985).

SK4 B. K. Shull and M. Koreeda, *J. Org. Chem.* **55**, 99 (1990).

SK5 L. Schwink and P. Knochel, *Tetrahedron Lett.* **37**, 25 (1996).

SK6 S. Saito, A. Kuroda, K. Tanaka, and R. Kimura, *Synlett,* 231 (1996).

SL1 G. Solladié and O. Lohse, *J. Org. Chem.* **58**, 4555 (1993).

SL2 J. Szymoniak, H. Lefranc, and C. Moïse, *J. Org. Chem.* **61**, 3926 (1996).

SM1 M. Shimagaki, T. Maeda, Y. Matsuzaki, I. Hori, T. Nakata, and T. Oishi, *Tetrahedron Lett.* **25**, 4775 (1984).

SM2 A. K. Saksena and P. Mangiaracina, *Tetrahedron Lett.* **24**, 273 (1983).

SM3 K. Suzuki, M. Miyazawa, M. Shimazaki, and G. Tsuchihashi, *Tetrahedron Lett.* **27**, 6237 (1986).

SM4 S. Sasatani, T. Miyazaki, K. Maruoka, and H. Yamamoto, *Tetrahedron Lett.* **24**, 4711 (1983).

SM5 S. Saito, F. Matsuda, and S. Terashima, *Tetrahedron Lett.* **29**, 6301 (1988).

SM6 H. G. Schmalz, A. Majdalani, T. Geller, J. Hollander, and J. W. Bats, *Tetrahedron Lett.* **36**, 4777 (1995).

SM7 A. Sudo, M. Matsumoto, Y. Hashimoto, and K. Saigo, *Tetrahedron Asymmetry* **6**, 1853 (1995).

SM8 I. Shimizu, Y. Matsumoto, K. Shoji, T. Ono, A. Satake, and A. Yamamoto, *Tetrahedron Lett.* **37**, 7115 (1996).

SN1 M. Sawamura, Y. Nakayama, T. Kato, and Y. Ito, *J. Org. Chem.* **60**, 1727 (1995).

SO1 K. Soai, H. Oyamada, M. Takase, and A. Ookawa, *Bull. Chem. Soc. Japan* **57**, 1948 (1984); K. Soai, H. Oyamada, and A. Ookawa, *Synth. Comm.* **12**, 463 (1982).

SO2 K. Soai and H. Oyamada, *Synthesis,* 605 (1984).

SO3 K. Soai and A. Ookawa, *J. Org. Chem.* **51**, 4000 (1986).

SO4 H. Sai, T. Ogiku, T. Nishitani, H. Hiramatsu, H. Horikawa, and T. Iwasaki, *Synthesis,* 582 (1995).

SP1 T. N. Sorrell and P. S. Pearlman, *Tetrahedron* **21**, 3963 (1980); *J. Org. Chem.* **45**, 3449 (1980).

SP2 S. Searles, K. A. Pollart, and E. F. Lutz, *J. Am. Chem. Soc.* **79**, 948 (1957).

SR1 A. Sarkar, B. R. Rao, and M. M. Konar, *Synth. Comm.* **19**, 2313 (1989).

SR2 A. Srikrishna, T. Jagadeeswar, and R. Viswajanani, *Tetrahedron* **52**, 1631 (1996).

SS1 M. F. Semmelhack, R. D. Stauffer, and A. Yamashita, *J. Org. Chem.* **42**, 3180 (1977).

SS2 K Suzuki, M. Shimazaki, and G. Tsuchihashi, *Tetrahedron Lett.* **27**, 6233 (1986).

SS3 R. G. Salomon, N. D. Sachinvala, S. R. Raychaudhuri, and D. B. Miller, *J. Am. Chem. Soc.* **106**, 2211 (1984).

SS4 M. Shimagaki, M. Shiokawa, K. Sugai, T. Teranaka, T. Nakata, and T. Oishi, *Tetrahedron Lett.* **29**, 659 (1988).

SS5 A. Srikrihna, J. A. Sattigeri, R. Viswajanani, and C. V. Yelamaggad, *J. Org. Chem.* **60**, 2260 (1995).

SS6 A. Solladié-Cavallo, J. Suffert, A. Adib, and G. Solladié, *Tetrahedron Lett.* **31**, 6649 (1990).

SS7 A. Solladié-Cavallo, M. C. Simon-Wermeister, and D. Farkhani, *Helv. Chim. Acta* **74**, 390 (1991).

SS8 J. Seyden-Penne and C. Schaal, *Bull. Soc. Chim. Fr.,* 3653 (1969).

SS9 M. Shi, Y. Satoh, T. Makihara, and Y. Masaki, *Tetrahedron Asymmetry* **6**, 2109 (1995).

ST1 R. J. Sims, S. A. Tischler, and L. Weiler, *Tetrahedron Lett.* **24**, 253 (1983).

SV1 A. Srikrihna, R. Viswajanani, J. A. Sattigeri, and C. V. Yelamaggad, *Tetrahedron Lett.* **36**, 2347 (1995).

SV2 A. Srikrishna and R. Viswajanani, *Tetrahedron* **51**, 3339 (1995).

SW1 J. Suwiński and P. Wagner, *Tetrahedron* **52**, 9541 (1996).

SW2 G. Stork, F. West, H. Y. Lee, R. C. A. Isaacs, and S. Manabe, *J. Am. Chem. Soc.* **118**, 10660 (1996).

SY1 K. Soai, S. Yokoyama, and Y. Mochida, *Synthesis,* 647 (1987).

SY2 K. Soai, T. Yamanoi, H. Mikima, and H. Oyamada, *J. Chem. Soc. Chem. Comm.,* 138 (1985).

SY3 K. Soai, S. Yokoyama, and A. Ookawa, *Synthesis,* 48 (1987).

SY4 H. Suginome and S. Yamada, *J. Org. Chem.* **50**, 2489 (1985).

SY5 T. Sakai, F. Yan, and K. Uneyama, *Synlett,* 753 (1995).

SY6 S. Saito and H. Yamamoto, *J. Org. Chem.* **61**, 2928 (1996).

SY7 T. Sakai, F. Yan, F. Kashino, and K. Uneyama, *Tetrahedron* **52**, 233 (1996).

T1 E. Toromanoff, *Tetrahedron* **36**, 2809 (1980).

T2 M. Tramontini, *Synthesis,* 605 (1982).

TA1 S. Taikano, M. Akiyama, S. Sato, and K. Ogasawara, *Chem. Lett.,* 1593 (1983).

TA2 J. J. Tufariello and S. Asrof Ali, *Tetrahedron Lett.,* 4647 (1978).

TA3 D. M. Tschaen, L. Abramson, D. Cai, R. Desmond, U. H. Dolling, L. Frey, S. Karady, Y. J. Shi, and T. R. Verhoeven, *J. Org. Chem.* **60**, 4324 (1995).

TA4 S. D. Toto, B. W. Arbuckle, P. K. Bharadwaj, J. T. Doi, and W. K. Muster, *Phosphorus and Sulfur* **56**, 27 (1991).

TB1 R. D. Tillyer, C. Boudreau, D. Tschaen, U. H. Dolling, and P. J. Reider, *Tetrahedron Lett.* **36**, 4337 (1995).

TC1 S. P. Tanis, Y. H. Chuang, and D. B. Head, *J. Org. Chem.* **53**, 4929 (1988).

TD1 D. F. Taber, P. B. Deker, and M. D. Gaul, *J. Am. Chem. Soc.* **109**, 7488 (1987).

TF1 D. M. Tal, G. D. Frisch, and W. H. Elliott, *Tetrahedron* **40**, 851 (1984).

TF2 M. Taniguchi, H. Fujii, K. Oshima, and K. Utimoto, *Tetrahedron* **51**, 679 (1995).

TG1 C. M. Thompson, D. L. C. Green, and R. Kubas, *J. Org. Chem.* **53**, 5389 (1988).

TH1 T. Tsuda, T. Hayashi, H. Satomi, T. Kawamoto, and T. Saegusa, *J. Org. Chem.* **51**, 537 (1986).

TH2 B. M. Trost and R. I. Higuchi, *J. Am. Chem. Soc.* **118**, 10094 (1996).

TK1 J. F. Templeton, V. P. S. Kumar, R. S. Kim, and F. S. Labella, *J. Chem. Soc. Perkin Trans.* I, 1361 (1987).

TK2 F. Toda, K. Kiyochige, and M. Yagi, *Angew. Chem. Int. Ed. Engl.* **28**, 320 (1989).

TK3 H. Takayanagi, Y. Kitano, and Y. Morinaka, *J. Org. Chem.* **59**, 2700 (1994).

TL1 B. Tamami and M. M. Lakouraj, *Synth. Comm.* **25**, 3089 (1995).

TM1 T. Takahashi, M. Miyazawa, and T. Tsuji, *Tetrahedron Lett.* **26**, 5139 (1985).

TM2 S. H. J. Thompson, M. F. Mahon, K. C. Molloy, M. S. Hadley, and T. Gallagher, *J. Chem. Soc. Perkin Trans.* I, 379 (1995).

TM3 J. Tsuji and T. Mandai, *Synthesis,* 1 (1996).

TN1 M. Takasu, Y. Naruse, and H. Yamamoto, *Tetrahedron Lett.* **29**, 1947 (1988).

TO1 T. Takahashi, A. Ootake, H. Yamada, and J. Tsuji, *Tetrahedron Lett.* **26**, 69 (1985).

TO2 T. Teresawa and T. Okada, *J. Chem. Soc. Perkin Trans.* I, 1252 (1978).

TP1 S. M. H. Tabaei and C. U. Pittman, *Tetrahedron Lett.* **34**, 3263 (1993).

TR1 E. W. Thomas, R. H. Rynbrandt, D. C. Zimmerman, L. T. Bell, C. R. Muchmore, and E. W. Yankee, *J. Org. Chem.* **54**, 4535 (1989).

TR2 S. Tsay, J. A. Robl, and J. R. Hwu, *J. Chem. Soc. Perkin Trans.* I, 757 (1990).

TR3 B. Trost, G. T. Rivers, and J. M. Gold, *J. Org. Chem.* **45**, 1835 (1980).

TS1 T. Tsuda, H. Satomi, T. Hayashi, and T. Saegusa, *J. Org. Chem.* **52**, 439 (1987).

TS2 P. C. Thieme, H. Sauter, and G. Reissenweber, *Chem. Ber.* **121**, 1059 (1988).

TT1 A. Toshimitsu, K. Terao, and S. Uemura, *J. Org. Chem.* **52**, 2018 (1987).

TT2 T. Tokoroyama, M. Tsukamoto, T. Asada, and H. Ho, *Tetrahedron Lett.* **28**, 6645 (1987).

TU1 L. F. Tietze and J. Utecht, *Synthesis,* 957 (1993).

TV1 P. J. Tirel, M. Vaultier, and R. Carrié, *Tetrahedron Lett.* **30**, 1947 (1989).

TY1 H. Toi, Y. Yamamoto, A. Sonoda, and S. I. Murahashi, *Tetrahedron* **37**, 2261 (1981).

TY2 T. Tsuda, T. Yoshida, T. Kawamoto, and T. Saegusa, *J. Org. Chem.* **52**, 1624 (1987).

TY3 N. Toyooka, Y. Yoshida, and T. Momose, *Tetrahedron Lett.* **36**, 3715 (1995).

UA1 H. Urabe, Y. Aoyama, and F. Sato, *J. Org. Chem.* **57**, 5056 (1992).

UC1 G. V. Ullas, C. K. Chu, M. K. Ahn, and Y. Kosugi, *J. Org. Chem.* **53**, 2413 (1988).

UI1 N. Umino, T. Iwakuma, and N. Itoh, *Tetrahedron Lett.,* 2875 (1976).

V1 S. M. Viti, *Tetrahedron Lett.* **23**, 4541 (1982).

VB1 E. F. J. de Vries, J. Brussee, and A. van der Gen, *J. Org. Chem.* **59**, 7133 (1994).

VB2 E. F. J. de Vries, J. Brussee, C. G. Kruse, and A. van der Gen, *Tetrahedron Asymmetry* **5**, 377 (1994).

VG1 G. Verardo, A. G. Giumani, P. Strazzolini, and M. Poiana, *Synthesis,* 121 (1993).

VK1 R. S. Varma and G. W. Kabalka, *Synth. Comm.* **15**, 985 (1985).

VK2 R. S. Varma and G. W. Kabalka, *Synth. Comm.* **14**, 1093 (1984).

VK3 R. S. Varma and G. W. Kabalka, *Synth. Comm.* **15**, 151 (1985).

VK4 H. Vestmijze, H. Klein, and P. Vermeer, *Synthesis,* 430 (1979).

VM1 A. Viger, A. Marquet, D. H. R. Barton, W. B. Motherwell, and S. Z. Zard, *J. Chem. Soc. Perkin Trans.* I, 1937 (1982).

VN1 H. van Derwel, N. M. M. Nibbering, and M. M. Kayser, *Can. J. Chem.* **66**, 2587 (1988).

VO1 M. C. Venuti and O. Ort, *Synthesis,* 985 (1988).

W1 E. Winterfeldt, *Synthesis,* 617 (1975).

W2 D. C. Wigfield, *Tetrahedron* **35**, 449 (1976).

W3 E. R. H. Walker, *Chem. Soc. Rev.* **5**, 23 (1976).

W4 R. C. Wade, *J. Mol. Catal.* **18**, 273 (1983).

W5 J. Wardell, *Comprehensive Organometallic Chemistry, Vol. 2,* G. Wilkinson, Ed. (Pergamon Press, Oxford, 1982), p. 863.

W6 C. I. F. Watt, *Adv. in Phys. Org. Chem.* **24**, 58 (1988).

W7 E. Winterfeldt, in *Stereocontrolled Organic Synthesis,* B. M. Trost, Ed. (Blackwell Science, Oxford, 1994), p. 275.

W8 A. Wells, *Synth. Comm.* **26**, 1143 (1996).

WB1 P. M. Wovkulich, A. D. Batcho, and M. R. Uskokovic, *Helv. Chim. Acta* **67**, 612 (1984).

WB2 R. M. Williams, E. J. Brunner, and M. R. Sabol, *Synthesis,* 963 (1988).

WD1 J. P. Wiebe, C. Deline, K. D. Buckingham, V. Dave, and J. B. Stothers, *Steroids* **45**, 39 (1985).

WD2 J. G. H. Willems, F. J. Dommerholt, J. B. Hammink, A. M. Vaarhorst, L. Thijs, and B. Zwanenburg, *Tetrahedron Lett.* **36**, 603 (1995).

WF1 E. Winterfeld and R. Freund, *Annalen,* 1262 (1986).

WF2 J. Wright, L. Frambes, and P. Reeves, *J. Organomet. Chem.* **476**, 215 (1994).

WG1 J. E. Wrobel and B. Ganem, *Tetrahedron Lett.* **22**, 3447 (1981).

WG2 R. M. Williams, T. Glinka, and E. Kwast, *J. Am. Chem. Soc.* **110**, 5927 (1988).

WH1 Y. Wu and K. N. Houk, *J. Am. Chem. Soc.* **109**, 906 and 908 (1987).

WH2 Y. Wu, K. N. Houk, and B. M. Trost, *J. Am. Chem. Soc.* **109**, 5560 (1987).

WH3 Y. D. Wu, K. N. Houk, J. Florez, and B. M. Trost, *J. Org. Chem.* **56**, 3656 (1991).

WH4 Y. D. Wu, K. N. Houk, and M. N. Paddon-Row, *Angew. Chem. Int. Ed. Engl.* **31**, 1019 (1992).

WH5 K. T. Wanner and G. Höfner, *Tetrahedron* **47**, 1895 (1991).

WI1 T. Watkamatsu, H. Inaki, A. Ogawa, M. Watanabe, and Y. Ban, *Heterocycles* **14**, 1441 (1980).

WK1 P. Wipf, Y. Kim, and D. M. Goldstein, *J. Am. Chem. Soc.* **117**, 11106 (1995).

WL1 M. Weissenberg and J. Levisalles, *Tetrahedron* **51**, 5711 (1995).

WM1 O. Wheeler and J. L. Mateos, *Can. J. Chem.* **36**, 1049 (1958).

WM2 S. Wallbaum and J. Martens, *Tetrahedron Asymmetry* **3**, 1475 (1992).

WM3 L. J. Westrum and A. I. Meyers, *Tetrahedron Lett.* **35**, 973 (1994).

WN1 C. D. Weis and G. R. Newkome, *Synthesis,* 1053 (1995).

WO1 D. R. Williams, M. H. Osterhout, and J. P. Reddy, *Tetrahedron Lett.* **34**, 3271 (1993).

WP1 R. A. Wade and D. T. Price, *Tetrahedron Lett.* **30**, 1185 (1989).

WP2 S. S. Wong and M. N. Paddon-Row, *J. Chem. Soc. Chem. Comm.,* 456 (1990).

WR1 D. E. Ward and C. K. Rhee, *Can. J. Chem.* **67**, 1206 (1989).

WR2 D. E. Ward and C. K. Rhee, *Synth. Comm.* **18**, 1927 (1988).

WR3 H. H. Wasserman and V. Rusiecki, *Tetrahedron Lett.* **29**, 4977 (1988).

WR4 D. E. Ward, C. K. Rhee, and W. M. Zoghaib, *Tetrahedron Lett.* **29**, 517 (1988).

WR5 S. A. Weissman and P. V. Ramachandran, *Tetrahedron Lett.* **37**, 3791 (1996).

WS1 K. E. Wiegers and S. G. Smith, *J. Am. Chem. Soc.* **99**, 1480 (1978); *J. Org. Chem.* **43**, 1126 (1978).

WS2 S. S. Wang and C. N. Sukenick, *J. Org. Chem.* **50**, 653 (1985).

WT1 Y. D. Wu, J. A. Tucker, and K. N. Houk, *J. Am. Chem. Soc.* **113**, 5018 (1991).

WV1 H. Wyss, U. Vogeli, and R. Scheffold, *Helv. Chim. Acta* **64**, 775 (1981).

WW1 H. Weidmann, N. Wolf, and W. Timpe, *Carbohydr. Res.* **24**, 184 (1972).

WY1 G. J. Wells, T. H. Yan, and L. A. Paquette, *J. Org. Chem.* **49**, 3604 (1984).

XS1 Y. B. Xiang, K. Snow, and M. Belley, *J. Org. Chem.* **58**, 993 (1993).

YA1 S. Yamada and H. Akimoto, *Tetrahedron Lett.,* 3105 (1969).

YA2 N. M. Yoon, J. H. Ahn, D. K. An, and Y. S. Shon, *J. Org. Chem.* **58**, 1941 (1993).

YC1 N. M. Yoon and J. S. Cha, *J. Korean Chem. Soc.* **22**, 259 (1978).

YC2 N. M. Yoon and J. Choi, *Synlett,* 135 (1993).

YC3 N. M. Yoon, J. Choi, and Y. S. Show, *Synth. Comm.* **23**, 3047 (1993).

YG1 N. M. Yoon and Y. S. G. Young, *J. Org. Chem.* **50**, 2443 (1985).

YH1 H. Yamataka and T. Hanafusa, *J. Am. Chem. Soc.* **108**, 6643 (1986).

YH2 N. M. Yoon, Y. S. Hwang, and H. S. Yang, *Bull. Korean Chem. Soc.* **10**, 120 (1989).

YH3 L.-X. Yang and K. G. Hofer, *Tetrahedron Lett.* **37**, 6081 (1996).

YI1 K. Yamada, N. Itoh, and T. Iwakuma, *J. Chem. Soc. Chem. Comm.,* 1089 (1978).

YK1 S. Yamaguchi, K. Kabuto, and F. Yasuhara, *Chem. Lett.,* 461 (1981).

YK2 N. M. Yoon, S. K. Kim, and Y. S. G. Young, *Bull. Korean Chem. Soc.* **7**, 323 (1986).

YK3 N. Yamazaki and C. Kibayashi, *J. Am. Chem. Soc.* **111**, 1396 (1989).

YK4 N. Yamazaki and C. Kibayashi, *Tetrahedron Lett.* **29**, 5767 (1988).

YK5 N. M. Yoon, E. G. Kim, H. S. Son, and J. Choi, *Synth. Comm.* **23**, 1595 (1993).

YK6 Z. You and M. Koreeda, *Tetrahedron Lett.* **34**, 2745 (1993).

YL1 N. M. Yoon, H. J. Lee, J. Kong, and J. S. Chung, *J. Korean Chem. Soc.* **19**, 468 (1975).

YL2 N. M. Yoon and W. S. Lee, *Bull. Korean Chem. Soc.* **7**, 296 (1986).

YL3 B. Yadagiri and J. W. Lown, *Synth. Comm.* **20**, 175 (1990).

YL4 I. K. Youn, S. W. Lee, and C. S. Pak, *Tetrahedron Lett.* **29**, 4453 (1988).

YL5 S. Yoo and S. Lee, *Synlett,* 419 (1990).

YM1 S.-I. Yamada, Y. Mori, K. Morimatsu, Y. Ishizu, Y. Ozaki, R. Yoshioka, T. Nakatani, and H. Seko, *J. Org. Chem.* **61**, 8586 (1996).

YN1 H. Yoda, T. Nakajima, and K. Takabe, *Tetrahedron Lett.* **37**, 5531 (1996).

YO1 N. M. Yoon, I. H. Oh, K. I. Choi, and H. J. Lee, *Heterocycles* **22**, 39 (1984).

YO2 M. Yamashita and I. Ojima, *J. Am. Chem. Soc.* **105**, 6339 (1983).

YO3 M. Yatagai and T. Ohnuki, *J. Chem. Soc. Perkin Trans.* I, 1826 (1990).

YP1 N. M. Yoon, C. S. Pak, H. C. Brown, S. Krishnamurthy, and T. P. Stocky, *J. Org. Chem.* **38**, 2786 (1973).

YP2 N. M. Yoon, H. M. Park, B. T. Cho, and I. H. Oh, *Bull. Korean Chem. Soc.* **4**, 287 (1983).

YP3 N. M. Yoon, K. B. Park, and Y. S. G. Young, *Tetrahedron Lett.* **24**, 5367 (1983).

YR1 C. Yue, J. Royer, and H.-P. Husson, *J. Org. Chem.* **55**, 1140 (1990).

YS1 N. M. Yoon and T. B. Sim, *Bull. Korean Chem. Soc.* **14**, 749 (1993).

YS2 J. Yu and J. B. Spencer, *J. Org. Chem.* **61**, 3234 (1996).

YU1 K. Yamamoto, K. Ueno, and K. Naemura, *J. Chem. Soc. Perkin Trans.* I, 2607 (1991).

YY1 N. M. Yoon, H. S. Yang, and Y. S. Hwang, *Bull. Korean Chem. Soc.* **10**, 205 (1989).

Z1 P. Zlatoidsky, *Tetrahedron Lett.* **36**, 7281 (1995).

ZA1 R. Zimmer, T. Arnold, K. Homann, and H. U. Reissig, *Synthesis,* 1050 (1994).

ZC1 H. C. Zhang, M. J. Costanzo, and B. E. Maryanoff, *Tetrahedron Lett.* **35**, 4891 (1994).

ZH1 Q. C. Zhu, R. O. Hutchins, and M. G. K. Hutchins, *Org. Prep. Proc. Int.* **26**, 193 (1994).

ZH2 H.-C. Zhang, B. D. Harris, C. A. Maryanoff, and B. E. Maryanoff, *Tetrahedron Lett.* **37**, 7897 (1996).

ZY1 F.-Y. Zhang, C.-W. Yip, and A. S. C. Chan, *Tetrahedron Asymmetry* **7**, 2463 (1996).

Index